Weather Radar Polarimetry

Weather Radar Polarimetry

Guifu Zhang

CRC Press
Taylor & Francis Group
Boca Raton London New York

CRC Press is an imprint of the
Taylor & Francis Group, an **informa** business

CRC Press
Taylor & Francis Group
6000 Broken Sound Parkway NW, Suite 300
Boca Raton, FL 33487-2742

First issued in paperback 2019

© 2017 by Taylor & Francis Group, LLC
CRC Press is an imprint of Taylor & Francis Group, an Informa business

No claim to original U.S. Government works

ISBN-13: 978-1-4398-6958-1 (hbk)
ISBN-13: 978-0-367-86653-2 (pbk)

Library of Congress Cataloging-in-Publication Data

Names: Zhang, Guifu, 1962-
Title: Weather radar polarimetry.
Description: Boca Raton : Taylor & Francis, 2016. | "A CRC title." | Includes
bibliographical references and index.
Identifiers: LCCN 2016005615 | ISBN 9781439869581 (alk. paper)
Subjects: LCSH: Radar meteorology. | Polarimetry. | Radar--Interference. |
Electromagnetic waves--Scattering. | Remote sensing.
Classification: LCC QC973.5 .Z43 2016 | DDC 551.63/53--dc23
LC record available at http://lccn.loc.gov/2016005615

Visit the Taylor & Francis Web site at
http://www.taylorandfrancis.com

and the CRC Press Web site at
http://www.crcpress.com

Contents

Foreword

Although weather radars have been in operation for more than half a century and Doppler weather radars for a few decades, polarimetric weather radars, which retain Doppler capability, have only recently achieved operational status. Nevertheless, polarimetry offers more than an incremental improvement in the understanding of cloud physics, in quantitative measurements of precipitation, and in the discrimination of scatterer types. In fact polarimetric weather radars are likely to have a broader impact on improving the interpretation and quantitative measurement of weather phenomena than that achieved with Doppler radars. Thus users of polarimetric radar need to understand the basics of polarimetry so they can better interpret the data provided by this instrument. There are several texts describing the theory and application of weather radars including Doppler radars, but relatively few that focus on polarimetry. At the University of Oklahoma, Dr. Guifu Zhang created a popular course containing the theory and application of polarimetry to interpret polarimetric radar echoes from various types of precipitation as well as those from biological scatterers. *Weather Radar Polarimetry* is rooted in this course, offered to students in the School of Meteorology as well as those in the School of Electrical and Computer Engineering. Thus this book has benefited from the interactions of Dr. Zhang with students from two disciplines, as well as from his collaboration with researchers at the National Center for Atmospheric Research and the National Oceanic and Atmospheric Administration's National Severe Storms Laboratory. In *Weather Radar Polarimetry* Dr. Zhang takes a unique approach to teaching weather echo processing, polarimetric theory, and the application of theory to the interpretation of polarimetric weather radar observations. Homework assignments and examples with real data offer hands-on experience and demystify this difficult subject. Those who put the time into working out the details will be well prepared to enter this field either as developers of polarimetric radars or interpreters of polarimetric data.

Richard J. Doviak
Dusan S. Zrnić
National Severe Storms Laboratory
National Oceanic and Atmospheric Administration

Preface

Radar has proven to be an indispensable tool for weather studies, as has been well documented. Radar reflectivity and Doppler measurements have demonstrated their value in weather observation, quantification, and forecasting. Now, we have another set of measurements we can use to better study weather: polarimetric radar data (PRD). After decades of research and development, weather radar polarimetry has now matured to the point that the national NEXRAD (WSR-88D) network has been upgraded with dual-polarization capability. Furthermore, other national weather radar networks have radars capable of producing multiparameter PRD. My understanding of weather radar polarimetry is as follows:

> There is polarimetry in radar innovation,
> which makes weather observations more accurate.
> It brought excitement with hydrometeor classification,
> and improvement in quantitative precipitation estimation.
> Multiparameter measurements contain rich information,
> leading to a deeper understanding of cloud physics.
> It helps model initialization and parameterization,
> with great potential for the future of weather prediction!

Although the technology of radar polarimetry has matured and PRD are available nationally and worldwide, radar polarimetry is still in its initial stages for operational usage. There is a lot of room for research and development, especially in using PRD for weather forecasts. It is important to know the principles of radar polarimetry and of PRD estimation and improvement, as well as information content, and error characterization. There is a growing need for a textbook that meteorology students, scholars, and scientists can use to obtain this knowledge. Based on the weather radar polarimetry classes taught by the author at the University of Oklahoma, this book was written to provide readers with the fundamentals and tools to effectively and optimally use the available PRD. Please find supporting data and tools for the homework exercises at http://weather.ou.edu/~guzhang/page/book.html.

MATLAB® is a registered trademark of The MathWorks, Inc. For product information, please contact:

The MathWorks, Inc.
3 Apple Hill Drive
Natick, MA 01760-2098 USA
Tel: 508 647 7000
Fax: 508-647-7001
E-mail: info@mathworks.com
Web: www.mathworks.com

Acknowledgments

I express my appreciation to the pioneers in developing weather radar polarimetry, to my colleagues who further my understanding, to my students who motivated me to write this book, to my friends for their encouragement, and to my family for their support. In particular, I wish to thank Drs. Richard J. Doviak, Dusan S. Zrnić, Alexander Ryzhkov, Jidong Gao, and Terry Schuur at the National Severe Storm Laboratory; Drs. J. Vivekanandan, Edward Brandes, and Juanzhen Sun at the National Center for Atmospheric Research; Profs. Ming Xue, Howard B. Bluestein, Yan (Rockee) Zhang, and Drs. Shaya Karimkashi and Boon Leng Cheong at the University of Oklahoma (OU); and Prof. V. Bringi at Colorado State University.

I thank the following graduate students (whom I advised or co-advised), who have completed or are pursuing their PhD in radar polarimetry at OU: Youngsun Jung (2008), Qing Cao (2009), Yinguang Li (2013), Lei Lei (2014), Petar Bukovcic, Vivek Mahale, and Bryan Putnam. Some of their dissertation works are included in this book. My friend Dr. Yasser Al-Rashid at Lockheed Martin Corporation connected me with my publisher. My visiting student, Hao Huang from Nanjing University, helped with the bibliography and figures.

Financial support from National Science Foundation grants and the National Oceanic and Atmospheric Administration are greatly appreciated. I would also like to express my appreciation for support from OU.

Guifu Zhang
University of Oklahoma

About the Author

Guifu Zhang, PhD, is a professor in the School of Meteorology at the University of Oklahoma, Norman, Oklahoma, where he has formulated theories on weather radar interferometry and phased array radar polarimetry. Currently, he is working on topics such as the optimal use of polarimetric radar data (PRD) in quantitative precipitation estimation and quantitative precipitation forecast and the research and development of polarimetric phased array radars for weather measurements and multimission capability.

Dr. Zhang earned his BS in physics from Anhui University, Hefei, China, in 1982, his MS in radio physics from Wuhan University, Wuhan, China, in 1985, and his PhD in electrical engineering from the University of Washington, Seattle, Washington, in 1998. From 1985 to 1993, Dr. Zhang was an assistant professor and associate professor in the Space Physics Department at Wuhan University. In 1989, he worked as a visiting scholar with Dr. Tomohiro Oguchi at the Communication Research Laboratory in Japan. From 1993 to 1998, Dr. Zhang studied and worked with professors Leung Tsang, Yasuo Kuga, and Akira Ishimaru in the Department of Electrical Engineering at the University of Washington, where he was first a visiting scientist and later a PhD student. He was a scientist with the National Center for Atmospheric Research (NCAR) between 1998 and 2005. In 2005, he joined the School of Meteorology at the University of Oklahoma, where he is now a professor.

Dr. Zhang's dissertation work focused on the modeling and calculation of wave scattering from targets buried under rough surfaces, and he explored the detection of targets in the presence of clutter using angular correlation functions. He also studied wave scattering from fractal trees. At NCAR and the University of Oklahoma, he developed algorithms for retrieving raindrop size distributions. He led the development of the spectrum-time estimation and processing algorithm to improve the quality of weather radar data and of the PRD simulators that link weather physics state variables to radar variables. His additional research interests span topics that include wave propagation and scattering in random and complex media, remote sensing theory and technology for geophysical applications, algorithms for retrieving physical states and processes, cloud and precipitation microphysics and model parameterization, target detection and classification, clutter identification and filtering, radar signal processing, and optimal estimation.

Dr. Zhang has three US patents and filed more than 10 intellectual property disclosures for his research work. The formulations, methods, and theories in his publications (over 100 refereed papers) are in use across the US and worldwide. He is an active member of the weather and radar communities, including the American Meteorological Society and Institute of Electrical and Electronics Engineers. In addition to his research and service at the University of Oklahoma, he has developed teaching material for courses such as METR/ECE6613, Weather Radar Polarimetry; METR3223, Physical Meteorology II: Cloud Physics, Atmospheric Electricity and Optics; METR5233, Cloud and Precipitation Physics; and METR6803/ECE6973, Wave Interactions with Geophysical Media. This textbook draws extensively upon Dr. Zhang's research and teaching experience.

1 Introduction

Radar is an efficient remote sensing system that measures targeted media with four-dimensional information and fine resolution. It plays a critical role in weather observation, detection of hazards, classification and quantification of precipitation, and forecasting. Polarimetric radar provides multiparameter measurements with unprecedented quality and information. In this introduction, the historical development of the weather radar polarimetry is simply summarized, and the motivation and organization of the book are provided.

1.1 HISTORICAL DEVELOPMENT

The history of weather radar development and application is well documented in the first 18 chapters of *Radar in Meteorology*, edited by Atlas (1990). Here, we briefly summarize the early development and background information that will allow us to progress to weather radar polarimetry.

Radar is an acronym for "*r*adio *d*etecting *a*nd *r*anging of objects," a remote sensing system that transmits and receives electromagnetic waves to detect and locate targets. Although its history can be traced earlier to the use of radio waves to obtain target information, it was during World War II that radar began to receive a great deal of attention; its technology advanced rapidly in response to the need to detect aircraft. According to Nobel laureate I. I. Rabi, it was radar, not the atomic bomb, that contributed most to the Allies' victory in the war.

Prior to the 1930s, the highest radio frequencies used were less than 400 MHz, due to limitations in microwave power generation. The invention of the cavity magnetron made microwave frequency (10 cm and 3 cm wavelength) radars possible and substantially improved detection accuracy and reliability. Once microwave radars were in operation, the marriage between weather and radar occurred naturally. It was immediately observed that rain echoes showed up on the radar displays as clutter, and this clutter sometimes obscured aircraft. One of the early challenges was to remove weather, ground, and sea clutter from radar displays. However, weather measurement in addition to aircraft detection came to be of interest.

At that time, most radar systems were developed by the Radiation Laboratory at the Massachusetts Institute of Technology (MIT). Due to the urgent need for knowledge of radar operation in weather conditions, the Weather Radar Research Project was started in the Department of Meteorology at MIT in 1946. Two radars (10 cm SCT-615 and 3 cm AN/TSP-10) were used to make cloud observations, and their measurements were compared with the direct in situ measurements of hydrometeor particles by instrumented aircraft. A year later, the first radar meteorology conference was held at MIT. D. Atlas was one of ninety attendees. Along with L. Battan, Atlas attended the Harvard–MIT radar school. He then worked at AFCRL (Air Force Cambridge Research Laboratories), where a premier group of scientists and engineers conducted systematic weather radar research in the 1950s. He summarized the early progress in his monograph, *Advances in Radar Meteorology*, in 1964. L. J. Battan led the Thunderstorm Project at the University of Chicago and completed his PhD work, titled "Observations on Formation of Precipitation in Convective Clouds," in 1953. Extending his PhD work, in 1959 he wrote *Radar Meteorology*, the first textbook in the field, with a revision published in 1973. Battan realized that frozen hydrometeors are not spherical, which can cause differences in wave scattering for different polarizations. His book highlighted this fact, with polarization vectors drawn on the cover page.

To address problems of air traffic in the presence of convective weather and to provide timely detection and warning of severe storms, the US Weather Bureau established a national network of 10-cm wavelength radars (WSR-57s). One of the WSR-57s was installed in Norman, Oklahoma, to support research efforts to understand thunderstorms and their effect on the safety of flight. When the National Severe Storms Laboratory (NSSL) was officially formed in Norman in 1962, one of the main objectives was to develop radar technology for severe storm observation. NSSL developed the first 10-cm Doppler weather radar for research to support the National Weather Service and later for polarimetric Doppler radar research. Doppler radar technology was used in the WSR-88D national radar network, which was upgraded with dual-polarization capability by 2013.

The early interaction between NSSL meteorologists and engineers and University of Oklahoma (OU) professors (G. Walker of the EE department and Y. Sasaki of the School of Meteorology) was instrumental in developing the OU–NSSL relationship, which continues to this day. This interaction led to a course in radar meteorology and the textbook *Doppler Radar and Weather Observations* by Doviak and Zrnić, first published in 1984, with a second edition discussing polarimetry published in 1993. (A paperback Dover edition is now available.) In 1990, an undergraduate textbook was published, *Radar for Meteorologists*, by Rinehart (1990, 1991, 1997, 2004). While the Doppler technique was a breakthrough for weather observations, radar polarimetry was the next advancement: polarimetric radar data (PRD) with Doppler fields significantly advanced the microphysical and dynamical quantification of clouds and precipitation.

Weather radar polarimetry is defined as the obtaining of weather information through radar measurements with polarization diversity. The major and minor lengths of hydrometeors of various shapes are typically horizontally and vertically oriented, respectively; these different lengths yield differences in wave scattering for different polarization. The polarization of electromagnetic waves is the orientation/direction of the transverse electric field vibrations, which was first explained by Young and Fresnel in the early 1800s. Although polarization had been documented in the Maxwell equations in 1861, weather radar polarimetry did not receive serious consideration until the 1970s. Many works represented and calculated wave polarization and propagation, such as using the Poincaré sphere (Poincaré 1892) and the Stokes parameters (Stokes 1852a, 1852b). The theoretical foundations for wave scattering include the Rayleigh approximation (Rayleigh 1871), Mie theory (Mie 1908), and Gans theory (Gans 1912). The book *Light Scattering by Small Particles* by Van De Hulst (1957) thoroughly documented these theoretical foundations. Ryde (1941) made the first calculation of hydrometeor scattering and used it to explain the attenuation of radar echoes. Newell and Geotis (1955) summarized earlier studies of meteorological measurements with different polarizations. These studies, which introduced weather radar polarimetry to the radar meteorology community, were mainly focused on the cancellation ratio of power return between circular and linear polarizations and the ratio between the cross-polar and co-polar returns, called the *linear depolarization ratio* (LDR). Oguchi (1960, 1964, 1983) and Waterman (1965, 1969) successfully made numerical calculations of wave scattering by nonspherical raindrops, which served as the basis for the interpretation of polarimetric data. Doviak and Sirmans (1973) noted the additional information that polarization diversity can provide. McCormick and Hendry (1974) studied the polarization properties of wave propagation in precipitation over a communication link, as well as the power and correlation properties of rain from a polarization diversity radar using circular polarization (McCormick and Hendry 1976). Seliga and Bringi (1976) proposed to improve rain estimation by measuring differential reflectivity (Z_{DR}) between horizontal and vertical polarizations, which received wide attention from the community. The first measurement of differential reflectivity in rain in the "slow-switched" mode was successfully made with the CHILL radar during the summer of 1977 and later reported by Seliga et al. (1979). The first "fast-switched" Z_{DR} data were obtained in 1978 with the RAL S-band radar in the United Kingdom by Hall et al. (1980). Seliga and Bringi (1978) also proposed that the differential propagation phase could be used to estimate the rain rate, in combination with Z_{DR}.

Largely due to the fast-switched Z_{DR} measurements (Hall et al. 1980), during the 1980s, the weather community in the United States conducted extensive research and implementation of radar polarimetry. The CHILL radar was modified for fast-switched Z_{DR} measurements (Seliga and Bringi 1982). Statistical analysis of measurement error for differential reflectivity was performed by Bringi et al. (1983). At the same time, the differential reflectivity measurement technique was implemented on the CP-2 radar operated by the National Center for Atmospheric Research (NCAR), and differential reflectivity was successfully measured and used for hail detection (Bringi et al. 1984).

Mueller (1984) provided a general formulation for calculating differential propagation phase from a fast-switched polarimetric radar. Meanwhile, NSSL upgraded their Cimarron radar with dual-polarization capabilities and demonstrated on time series data how to obtain Doppler and polarimetric variables on radars with switchable polarization (Sachidananda and Zrnić 1985). In addition NSSL scientists and engineers proposed measurements of differential phase (ϕ_{DP}) and co-polar cross-correlation coefficient (ρ_{hv}) (Sachidananda and Zrnić 1986, 1989). An additional advance was made when multiparameter radar measurements including LDR were made and used for graupel melting and hail detection studies (Bringi et al. 1986a, 1986b). These early research activities were summarized by Bringi and Hendry (1990) and Seliga et al. (1990).

In the 1990s, research into weather radar polarimetry was extended to the demonstration of the properties of multiparameter polarimetric weather radars. Bringi et al. (1990) examined propagation effects on radar reflectivity measurement. Zahrai and Zrnić (1993) demonstrated, for the first time, real-time measurement of full polarimetric radar parameters, including the differential phase, co-polar correlation coefficient, and co-polar and cross-polar correlation coefficients (ρ_h and ρ_v). At that time, the specific differential phase (K_{DP}), which was the range derivative of ϕ_{DP}, received a great deal of attention because it is approximately linearly related to rainfall rate and immune to partial beam blockage and power calibration error (Balakrishnan and Zrnić 1990; Ryzhkov and Zrnić 1996). A number of polarimetric radar rain estimators were developed based on simulations and studies with simulated or observed raindrop size distributions (DSDs) (Ryzhkov and Zrnić 1995; Sachidananda and Zrnić 1987; Seliga et al. 1986). Improved quantitative precipitation estimation (QPE) was achieved with real data (Brandes et al. 2002; Giangrande and Ryzhkov 2008; Ryzhkov et al. 2005a, 2005b).

The applications of PRD in cloud and precipitation microphysics studies were further motivated by the success the fuzzy logic technique had in classifying radar echoes (Straka et al. 2000; Vivekanandan et al. 1999; Zrnić et al. 2001). A number of field projects that involved S-band polarimetric radar were conducted, including CASES-91 in Kansas, PRECIP98 in Florida (Brandes et al. 2002), and STEPS-00 in Colorado. The data and results from these projects and those from the CSU–CHILL radar and NSSL's Cimarron and KOUN (NSSL's R&D WSR-88D upgraded to have polarimetric capability) radars further justified the value of multiparameter PRD in weather service and the need for a dual-polarization upgrade of the national network of NEXRAD radars (Doviak et al. 2000). The book *Polarimetric Doppler Weather Radar: Principles and Application* was published in 2001, authored by Bringi and Chandrasekar.

To improve model microphysics parameterization and weather forecast, DSDs were retrieved from PRD (Brandes et al. 2004a; Bringi et al. 2002; Cao et al. 2008, 2010; Zhang et al. 2001). A more rigorous way to retrieve microphysics states and to improve quantitative precipitation forecast (QPF) using PRD is to combine them with a numerical weather prediction (NWP) model through data assimilation (Jung et al. 2008a, 2008b). Jung et al. (2008a, 2008b) showed that the addition of PRD (Z_{DR} or K_{DP}) has a positive impact in determining state variables. The polarimetric radar signatures of severe storms were realistically generated from NWP model simulation (Jung et al. 2010a, 2010b; Xue et al. 2010). Radar polarimetry's greatest potential lies in its ability to accurately retrieve model states and microphysical parameters, which can be used to improve weather prediction. This book is meant to lay a foundation for the efficient and optimal use of PRD for weather quantification and forecasting.

1.2   OBJECTIVES AND ORGANIZATION OF THE BOOK

This book provides the fundamentals and principles of weather radar polarimetry by exploring wave scattering and propagation in cloud and precipitation conditions. The purpose and objectives of this book are illustrated in Figure 1.1. The relations between physical state parameters and radar observables are established so that readers understand the polarimetric radar signatures of typical weather conditions. Advanced signal processing and retrieval algorithms are briefly discussed to help readers better interpret PRD. Data analysis tools and methods are provided for readers to get hands-on experience in dealing with real data.

This book starts with the characterization and representation of hydrometeors in Chapter 2. The physical, statistical, and electromagnetic properties of raindrops,

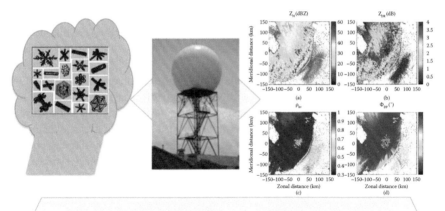

FIGURE 1.1 Sketch of the purpose and objectives of the book.

snowflakes, hailstones, and cloud particles are described. The modeling and representation of particle size distribution, shape, and density/composition are presented. Calculations of dielectric constants are included using the Debye model for water and ice and the Maxwell-Garnett and the Polder–van Santern mixing formulas for air–ice–water mixtures.

Chapter 3 provides the fundamentals and calculations of electromagnetic wave scattering from a single particle, which can be a raindrop, snowflake, hailstone, or cloud particle. The scattering representations are included for both spherical and non-spherical particles. The Rayleigh scattering approximation and Mie theory are introduced. In addition, the T-matrix method and its results for hydrometeor scattering calculations are provided.

Chapter 4 describes wave scattering and propagation in clouds and precipitation composed of randomly distributed particles. Coherent and incoherent scattering, as well as their applications, are explained. The wave statistics are illustrated in terms of mean, covariance, and probability density function and linked to polarimetric radar variables. The propagation effects of attenuation, differential attenuation, propagation phase, and differential phase are also included.

Chapter 5 presents signal processing methods for obtaining and improving polarimetric radar measurements, as well as error quantification. Advanced signal processing for improving data quality is introduced, including multilag processing, dual-polarization, and/or dual-scan clutter detection and mitigation.

Chapter 6 documents conventional uses of weather radar polarimetry and shows how cloud and precipitation microphysics can be studied qualitatively and quantitatively. Furthermore, the polarization signatures of typical weather events are presented.

The fuzzy logic method for hydrometeor classification is described. QPE and constrained methods for DSD retrieval, as well as attenuation correction, are discussed.

Chapter 7 introduces advanced methods to optimally utilize PRD for QPE and QPF. It describes simultaneous attenuation correction and DSD retrieval, statistical retrieval of rain DSDs, variational analysis, and the challenges and potential of assimilating PRD into NWP models to improve weather forecast.

Chapter 8 discusses phased array weather radar polarimetry that is under development. This includes the theoretical formulation of the phased array polarimetry and the issues, challenges, and possible solutions in designing and developing polarimetric phased array radars for fast date update and accurate weather measurements in future.

2 Characterization of Hydrometeors

The characteristics of polarimetric weather radar measurements depend on the physical (e.g., size, shape, and orientation), statistical (e.g., the distribution of sizes and orientations), and electromagnetic (EM) properties (e.g., dielectric constant and conductivity that cause wave scattering) of water particles (raindrops, snowflakes, hailstones, cloud droplets, and ice crystals), as well as other objects (aerosols, insects, birds, etc.) that might be in the radar resolution volume through the main lobe or sidelobes. To correctly relate polarimetric radar echo characteristics to the hydrometeors within the resolution volume, it is necessary to understand cloud/precipitation physics and polarimetric radar measurements. In this chapter, we provide the physical, statistical, and electrical descriptions of raindrops, snowflakes, hailstones, and cloud particles.

2.1 PHYSICAL AND STATISTICAL PROPERTIES

There are a variety of hydrometeors, which are either suspended water or ice particles or water–ice mixtures, whether in clouds or as precipitation. A cloud is a collection of water (liquid, ice, or both) particles condensed from water vapor. The microphysical states of clouds and precipitation and the processes that lead to them are shown in Figure 2.1. Cloud particles form from cloud/ice nuclei and grow through condensation/deposition of water vapor/liquid water. The further growth of cloud particles through collision–coalescence, aggregation, or accretion may yield precipitation in the form of rain, snow, or hail/graupel that reaches the ground. The polarimetric properties of the radar echoes are determined by the physical properties (e.g., size, composition, density, shape, and orientation) of hydrometeors along the propagation path and within the resolution volume.

The physical and statistical properties of hydrometeors and cloud particles are determined from in situ observations made with instrumented aircraft carrying particle probes or by ground-based disdrometers. The commonly used probes are the Particle Measuring Systems series, including the forward scattering spectrometer probe for droplet diameters of 2–47 μm and 2D optical array probes (2D-OAP) for clouds (2D-C: 25–800 μm) and precipitation (2D-P: 200–6400 μm). The newly developed cloud particle imager (CPI: 10–300 μm) by Spec Inc. shows potential for providing high-quality images. Figure 2.2 shows a picture of CPI (upper left) and some sample measurements from it.

Precipitation particles falling near the ground are usually observed with disdrometers (e.g., the impact disdrometer [Joss and Walvogel 1969], the PARSIVEL optical disdrometer [Löffler-Mang 2000], and the two-dimensional video disdrometer

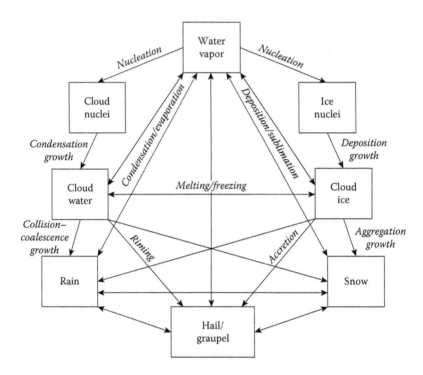

FIGURE 2.1   Sketch of cloud and precipitation microphysical states and processes.

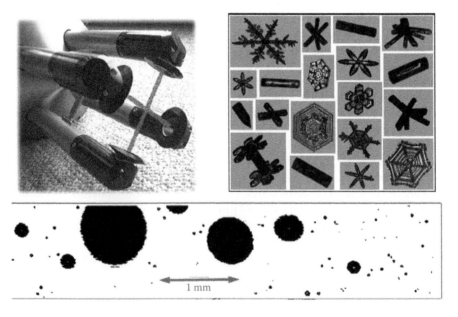

FIGURE 2.2   A picture of CPI (upper left) and some sample measurements from it. (Picture and images are provided by Dr. Paul Lawson at SPECINC, http://www.specinc.com.)

TABLE 2.1

Typical Characteristics of Cloud and Precipitation Particles

	Cloud Droplets	Ice Crystals	Raindrops	Snowflakes	Hailstones/ Graupels
Size, mm	0.001–0.1	0.1–1.0	0.1–8.0	1.0–30	1.0–100
Density, g/cm³	1.0	0.92	1.0	~0.1	0.5–0.92
Shape	Sphere	Varies	Spheroidal	Irregular	Irregular
Orientation	N/A	Varies	~Horizontal	Random $\sigma^a = 20°$	Random $\sigma = 60°$ $(1 - 0.8 f_w)$
Concentration, #/m³	10^6–10^9	10^4–10^6	10–10^4	10–10^3	1–100
Velocity, m/s	<0.1	<0.1	0.2–10	~1.0	1.0–100

[a] σ is the standard deviation of the distribution of particle orientations; f_w is the water fraction.

or 2DVD [Schönhuber 1997; Kruger and Krajewski 2002]). Of these, the 2DVD disdrometer provides the most accurate measurements. 2DVD data from OU and NCAR are used in this book.

Typical microphysical characteristics based on in situ measurements of water particles are summarized in Table 2.1. As shown in the table, the microphysical properties vary greatly from one form to another. Typical sizes range from a few micrometers for cloud droplets to several centimeters for hailstones. The number concentration, the total number of particles in a cubic meter of the atmosphere, varies from ~10^9 for cloud droplets to a few for hailstones. Other physical parameters of hydrometeors, such as density, shape, and orientation, also vary from one species to another. Most of these variations also exist within a species. For example, cloud droplet sizes range over two orders, whereas raindrop number concentrations vary from a few big drops for leading convection to over 10,000 drops in convective storm centers. It is therefore important to know and accurately represent the statistical distributions of these physical parameters for each species in order to correctly interpret radar measurements, as shown in the following subsections.

2.1.1 RAIN

Rain is the most common type of precipitation and was also the first to be identified by radar (Bent et al. 1943). A quantitative measure of rain intensity at the ground is the rainfall rate, which is the depth of rainwater that accumulates on the ground in a unit of time. Rain intensity is described as *light* (<2.5 mm/hr), *moderate* (2.5–7.6 mm/hr), and *heavy* (>7.6 mm/hr). Rainfall rate estimation is still one of the primary applications for weather radar, and it is the area in which radar polarimetry has shown the greatest impact.

Rain is characterized by its type and intensity. Based on the spatial structure (size and shape of the rain region), rain is usually classified as *convective* or *stratiform*. *Convective rain* refers to that produced from local convection with a small horizontal scale of tens of kilometers. Cumulus clouds form as moist air is lifted up and reach super-saturation, and the water vapor condenses to become cloud droplets. When the cloud droplets grow in size, they start to fall and grow bigger through collision–coalescence. Collision–coalescence/accretion and collision breakup are the main physical processes that are responsible for rain microphysical properties (e.g., raindrop size and number density). Stratiform rain is widespread and has relatively uniform rain rates in the horizontal direction, a result of the melting of stratus ice clouds that are formed from slowly rising warm air. One important signature of stratiform rain is the strong echo at the melting level, called *bright band*, in radar reflectivity images.

2.1.1.1 Drop Size Distribution

The rainfall rate, however, is not a complete description of rain microphysics and does not provide sufficient information to assist radar data analysis. Ten small raindrops could produce the same rainfall rate as one big drop and vice versa, but their radar responses are very different. The fundamental description of rain is the drop size distribution (DSD). DSD is defined as the number of drops in a unit volume for each unit diameter bin, which is a function of drop diameter D, expressed as $N(D)$ (# m^{-3} mm^{-1}). DSD, along with shape, orientation, and terminal velocity, provides a more complete description of rain and supplies the information needed to calculate wave scattering and polarimetric radar variables.

DSDs are determined from disdrometer observations. An example of a DSD measurement made by the OU 2DVD is shown in Figure 2.3, including a picture (upper left) of its outside units (i.e., the sensing unit), a sample display (upper right), a histogram (lower left), and a DSD (lower right) calculated from the number of drops for each bin number. The 2DVD consists of two horizontally pointing line-scan cameras whose beams are separated in the vertical by approximately 7 mm. The distance divided by the time difference is used to estimate the drop terminal velocity. Measurements are made within a common area of about 10×10 cm^2. The recorded information for each raindrop includes orthogonal silhouette images and estimates of equivalent volume (equivolume) diameter, oblateness, and fall velocity, v. Observed drops were partitioned into 41 size bins of $\Delta D = 0.2$ mm width and having central diameters of 0.1 to 8.1 mm; the DSD was then estimated by adding all the drops in each size bin and normalizing by their corresponding volume $Av_l\Delta t$ and the bin size ΔD, written as follows:

$$N(D) = \sum_{l=1}^{L} \frac{1}{Av_l \Delta t \Delta D}. \tag{2.1}$$

Figure 2.4 shows a series of DSDs collected on May 13, 2005, for a squall line followed by stratiform rain passing over the NCAR 2DVD deployed at the Kessler Farm Field Laboratory in Washington, Oklahoma. Rain microphysical properties and their time evolution are characterized in terms of the number and size of drops. Strong convective rain contains both large and small drops and has a broad DSD with the decaying stage dominated by small drops. Stratiform rain contains relatively larger drops but has a low number concentration for a given rain rate.

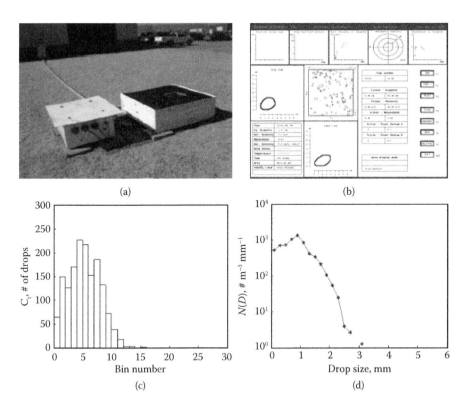

FIGURE 2.3 An example of drop size distribution measurement with a two-dimensional video disdrometer (2DVD): (a) a picture of the OU 2DVD; (b) the display of 2DVD processing; (c) a histogram of observed drops during 1 min; and (d) the derived DSD. In (b) there are two images of the same drop as it passes each of the two imaging planes.

The DSDs are also shown in a 3D plot in Figure 2.4b (top row), along with distributions of mass (row 2) and polarimetric variables (reflectivity Z_H and differential reflectivity Z_{DR}) (rows 3 and 4). Mass distribution $m(D) = \frac{\pi}{6}\rho_w D^3 N(D)$ is defined as the water mass in a unit volume per unit drop size bin. Although the total number concentration (top row of Figure 2.4b) is dominated by small drops, mass distributions peak at medium drop diameters (along with specific attenuation and differential phase), and reflectivity factor and differential reflectivity peak at large drop diameters. This indicates that radar measurements are insensitive to small particles when large particles are present. The polarimetric parameters are defined and explained in Chapter 4. Once the DSD is known, integral physics parameters such as total number concentration N_t, rain water content W, and rainfall rate R can be calculated. They are expressed by the following:

$$N_t = \int_0^{D_{max}} N(D)dD \quad \left[\# \, m^{-3} \right] \tag{2.2}$$

$$W = \frac{\pi}{6}\rho_w \int_0^{D_{max}} D^3 N(D)\,dD = \frac{\pi}{6} \times 10^{-3} \int_0^{D_{max}} D^3 N(D)\,dD \quad \left[\text{g m}^{-3}\right] \quad (2.3)$$

$$R = \frac{\pi}{6} \int_0^{D_{max}} D^3 \left[\text{mm}^3\right] v(D)\left[\text{m/s}\right] N\left[\#\,\text{m}^{-3}\text{mm}^{-1}\right] dD \quad \left[\text{mm}\right]$$

$$= 6\pi \times 10^{-4} \int_0^{D_{max}} D^3 v(D) N(D)\,dD \quad \left[\text{mm hr}^{-1}\right]. \quad (2.4)$$

The nth moment of the DSD, the integral of the product of D's nth power multiplied by $N(D)$, is expressed by the following:

$$M_n = \int_0^{D_{max}} D^n N(D)\,dD. \quad (2.5)$$

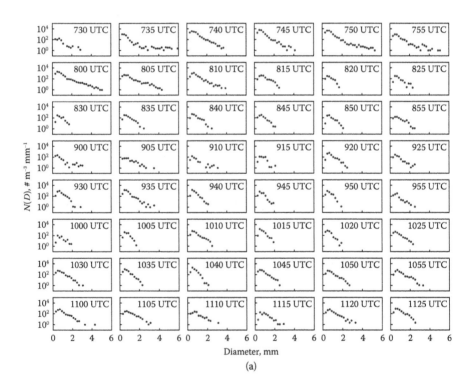

FIGURE 2.4 An example of a time series of raindrop size distributions for a squall line followed by stratiform precipitation, on May 13, 2005, observed by the National Center for Atmospheric Research 2DVD deployed at the Kessler Farm Field Laboratory. (a) 2D plots.
(Continued)

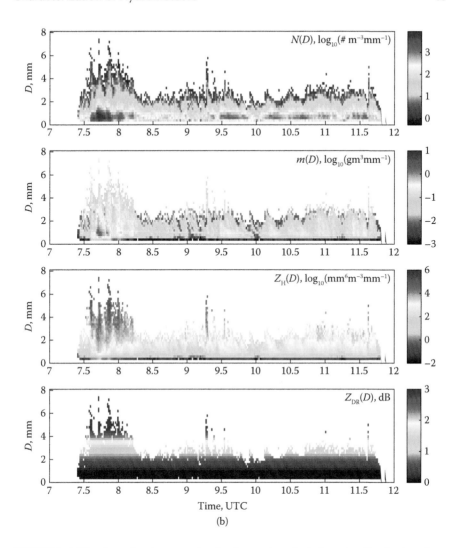

FIGURE 2.4 (Continued) An example of a time series of raindrop size distributions for a squall line followed by stratiform precipitation, on May 13, 2005, observed by the National Center for Atmospheric Research 2DVD deployed at the Kessler Farm Field Laboratory. (b) 3D plots; the top row is the DSD; the data in the other rows are discussed in the text.

Note that the DSD moment differs from the moment of the probability density by a factor of N_t because the integral of a DSD is N_t, rather than 1, which is the integral of the probability density function $p(D) = \dfrac{N(D)}{N_t}$.

The characteristic size of a rain DSD is normally represented by the median volume diameter (D_0), expressed as

$$\int_0^{D_0} D^3 N(D)\,dD = \int_{D_0}^{D_{max}} D^3 N(D)\,dD, \tag{2.6}$$

which means that the total volume for drops of diameter less than D_0 is equal to the total volume for drops of diameter equal to or larger than D_0.

The mean diameter $<D>$ and mass/volume-weighted diameter D_m are also commonly used and can be defined using DSD moments as

$$<D> = M_1/M_0, \qquad D_m = M_4/M_3. \tag{2.7}$$

For remote sensing applications, the effective diameter (EFD or D_{eff}) and radar estimated size (RES) are introduced. In optical remote sensing, the measurements are related to the geometric cross section, which is proportional to the second moment. Hence, EFD is defined as the ratio of the third and the second moments, written as

$$D_{eff} = M_3/M_2. \tag{2.8}$$

A dual-frequency radar measures reflectivity ($\sim M_6$) and attenuation ($\sim M_3$); a dual-polarization radar measures reflectivities close to M_6 for horizontal polarization and $M_{5.4}$ for vertical polarization (Zhang et al. 2001). The RES defined for dual-frequency (D_{df}) and that for dual-polarization (D_{dp}) radar are

$$D_{df} = (M_6/M_3)^{1/3} \quad \text{and} \quad D_{dp} = (M_6/M_{5.4})^{1/0.6}, \tag{2.9}$$

respectively. A general form for a typical size is

$$D_{mn} = (M_m/M_n)^{1/(m-n)} \tag{2.10}$$

with a different choice of m and n for different applications or data sets.

For convenient application, rain DSDs are usually represented by distribution models, such as the exponential distribution, Gamma distribution, and lognormal distribution models. A DSD model usually contains a few free parameters that are easily determined or calculated and are capable of capturing the main physical processes and properties. The exponential distribution with two free parameters is the most commonly used DSD model, and it is given by

$$N(D) = N_0 \exp(-\Lambda D), \tag{2.11}$$

where N_0 (m^{-3} mm^{-1}) is the intercept parameter and Λ (mm^{-1}) is a slope parameter. The two parameters can be determined from two DSD moments or two remote measurements. The fitting procedure is provided in Appendix 2B.

In some cases, only a single moment of the DSD is available from either model prediction (e.g., the bulk microphysics model predicts one of the moments) or single radar measurement. Either the intercept parameter N_0 or the slope Λ is fixed so that the other parameter Λ (or N_0) is uniquely related to the water content, W (g m^{-3}), which in turn is linearly related to the third moment of the DSD. The Marshall–Palmer (M–P; Marshall and Palmer 1948) exponential DSD model with the N_0 value fixed at 8000 m^{-3} mm^{-1} = 8×10^6 m^{-4} is widely used for representing rain (Kessler 1969) and ice (e.g., Lin et al. 1983) microphysics.

In recent years, the three-parameter gamma distribution has been widely used to characterize rain DSDs (Ulbrich 1983):

$$N(D) = N_0 D^\mu \exp(-\Lambda D), \tag{2.12}$$

where N_0 ($\text{mm}^{-\mu-1}\ \text{m}^{-3}$) is a number concentration parameter, μ is a distribution shape parameter, and Λ (mm^{-1}) is a slope term. The three governing parameters of the distribution are determined using the second, fourth, and sixth moments of observed DSDs. There are other moment combinations and methods used to fit the gamma DSD model (Cao and Zhang 2009) (see Appendix 2C). Although the moment method may not be optimal for all applications, it is widely used in radar meteorology because radar measurements are like DSD moments: weighted integrals of DSDs.

Figure 2.5 shows an example of an observed DSD and its fittings to the single-parameter M–P, two-parameter exponential, and three-parameter gamma distribution models. The asterisk (*) represents 2DVD measurements, the solid line represents results from the M–P model, the dashed line represents the exponential

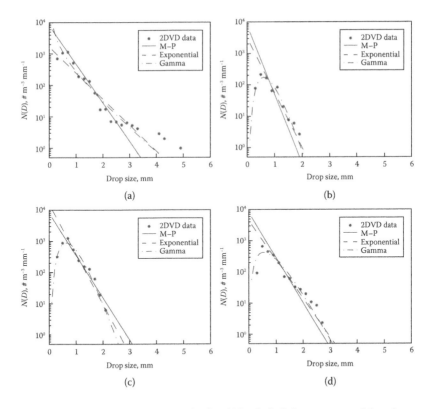

FIGURE 2.5 Example of rain DSD and its fitted Marshall–Palmer, exponential, and gamma distributions. (a) Strong convection at 0755 UTC, (b) weak convection at 0830 UTC, (c) stratiform rain at 0930 UTC, and (d) stratiform rain at 1030 UTC. Asterisks: disdrometer measurements. Solid line: fitted Marshall–Palmer; dashed line: fitted exponential distribution; dash-dotted line: gamma distribution using the second, fourth, and sixth moments.

model, and the dash-dotted line represents the gamma distribution model. The gamma model has the best fit with the data points because it has the largest number of free parameters.

2.1.1.2   Drop Shape

Thus far, we have used an equivolume diameter D_{eq} for raindrop size; in other words, we have used the diameter of a sphere that has the same volume as the raindrop. In fact, the shape of falling raindrops is not spherical; rather, it becomes more oblate as the size increases. It is this nonspherical nature of raindrops that motivated the development of weather radar polarimetry to improve radar rain estimation (Seliga and Bringi 1976). It is still important to correctly model drop shape to simulate polarimetric radar signature and to assist PRD analysis.

The shape of raindrops has been measured through photography (Jones 1959). For example, Pruppacher and Beard (1970) made careful measurements with water drops suspended in the air stream of a vertical wind tunnel, and their results are shown in Figure 2.6. This figure shows that drops are distorted with a flattened base, especially for large drops. In addition to Pruppacher and Beard's measurements, pictures of drops were taken after they had fallen a sufficient distance (>10 m) in stagnant air and had reached their terminal velocity. Thurai and Bringi (2005) made measurements by placing a 2DVD under a bridge. All of these measurements indicate that raindrops larger than 1 mm are nonspherical and can be approximated as oblate spheroids.

Assume an oblate spheroid has a semimajor axis of a and semiminor axis of b; it has a volume of V as follows:

$$V = \frac{4\pi}{3} a^2 b \equiv \frac{\pi}{6} D^3, \tag{2.13}$$

where D is the equivolume diameter, also noted as D_{eq} (the subscript "eq" is omitted in this book for simplicity), the diameter of a sphere that has an equivalent volume to the oblate spheroid. The semimajor axis of drops imaged by the 2DVD (Figure 2.3) is estimated from the geometric mean of two semimajor axes of the two images of a raindrop shown in Figure 2.3 as $a = \sqrt{a_1 a_2}$.

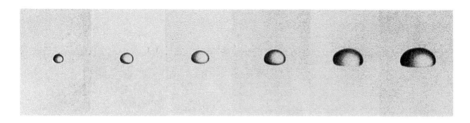

FIGURE 2.6   Photographs of water drops falling at terminal velocities in air. Their equivolume diameters are 2.7, 3.45, 5.3, 5.8, 7.75, and 8.0 mm, respectively. (With kind permission from Springer Science+Business Media: *Microphysics of clouds and precipitation*, vol. 18, 1996, Pruppacher, H., and J. Klett.)

The ratio of the semiminor and semimajor axes: $\gamma = b/a$, called *axis ratio* (or *aspect ratio*), represents the oblateness, that is, the shape of a raindrop. Pruppacher and Beard (1970) introduced an empirical relation in a linear form:

$$\gamma = \begin{cases} 1.0 & (D < 0.48 \text{ mm}) \\ 1.03 - 0.062D & (D > 0.48 \text{ mm}) \end{cases} \tag{2.14}$$

to quantitatively describe the raindrop shape as a function of its equivolume diameter D.

The nonspherical shape of raindrops is the result of the balance of all the forces acting on the water drop when it falls in the air. These forces include the (i) surface tension that tends to keep the drop spherical; (ii) aerodynamic force caused by the drop falling through the air, which yields the flattened base; (iii) hydrostatic pressure gradient; and (iv) internal circulation. Pruppacher and Pitter (1971) formulated the problem for the static equilibrium in which net energy change should be zero for any small displacement of the surface area; they were able to solve the equation to obtain the shape of a raindrop, shown in Figure 2.7 as the dashed line.

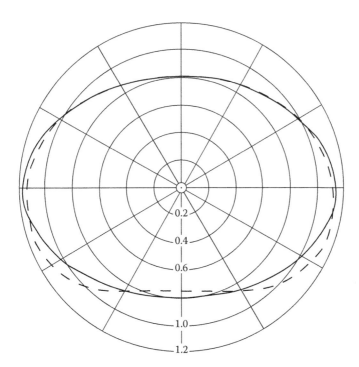

FIGURE 2.7   Shape of a raindrop with an equivolume diameter of 5 mm calculated from the equilibrium models: dashed line for more accurate calculation by Pruppacher and Pitter (1971) and the solid line from Green (1975). (From Green, A. W., 1975. *Journal of Applied Meteorology*, 14, 1578–1583.)

Green (1975) applied the equilibrium equation at the equator of the spheroid, where aerodynamic and hydrostatic pressures vanish, and obtained the simplified equation

$$D^2 = 4[\sigma/(g\rho_w)](\gamma^{-2} - 2\gamma^{-1/3} + 1)\gamma^{-1/3}, \tag{2.15}$$

where the surface tension is $\sigma = 72.75$ (g s^{-2}), $g = 9800$ (mm s^{-2}) is the gravity, and $\rho_w = 1.0 \times 10^{-3}$ (g mm^{-3}). Although the empirical relation in Equation 2.14 provides a reasonable characterization of raindrop shape, there are concerns regarding the real shape of raindrops in nature, where there is vortex shedding, collisions, turbulence, and/or wind shear, which might have an impact on the drop shape and in turn affect the calculation of polarimetric parameters. Beard and Chuang (1987) examined "equilibrium" shapes, that is, the mean shape of drops falling under the influence of gravity and subject to a balance of forces acting at the water–air interface. Other studies revealed the importance of natural environmental effects (Beard and Jameson 1983; Beard and Kubesh 1991; Beard et al. 1983; Pruppacher and Pitter 1971), which cause drops in the free atmosphere to oscillate with axisymmetric and transverse modes and to have mean shapes that differ from equilibrium. The argument for more spherical mean shapes is also supported by drop observations made by aircraft (Bringi et al. 1998; Chandrasekar et al. 1988) and by laboratory experiments (Andsager et al. 1999).

Hence, the observations of Pruppacher and Pitter (1971), Chandrasekar et al. (1988), Beard and Kubesh (1991), and Andsager et al. (1999) are combined to derive (Brandes et al. 2002, 2005)

$$\gamma = 0.9951 + 0.0251D - 0.03644D^2 + 0.005303D^3 - 0.0002492D^4. \tag{2.16}$$

The observations and experimental results are shown in Figure 2.8, along with the empirical linear relation (Equation 2.14), and Green's relation (Equation 2.15). It is clear that Equation 2.16 yields axis ratios that are significantly more spherical than were found by Pruppacher and Beard (1970) and that given by the theory of Green (1975), particularly for drops with $1 < D < 4$ mm. It agrees quite well with the relationship of Andsager et al. (1999, Equation 1), except for large drops ($D > 5$ mm), where that study differs from the observations of Pruppacher and Pitter (1971). Nevertheless, Equation 2.16 is now widely used in the radar meteorology community.

2.1.1.3 Terminal Velocity

As shown in Equation 2.4, rain rate is proportional to the drop's terminal velocity. Near the ground where vertical wind speed is typically negligible, mass flux is proportional to the terminal velocity, which is always positive although it is downward directed. Terminal velocity is the velocity at which the forces acting on the drop are in balance. Ignoring the air-floating force (Pruppacher and Klett 1996), buoyancy, pressure gradient, and inertial forces, the terminal velocity (v) is a result of the counterbalancing forces of drag and gravity (Rogers and Yau 1989), specifically:

$$\frac{\pi}{4}D^2 \frac{1}{2}\rho v^2 C_D = \frac{\pi}{6}D^3\rho_w g, \tag{2.17}$$

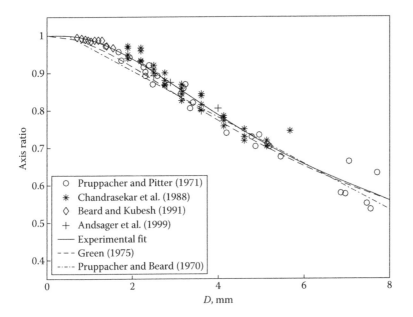

FIGURE 2.8 Comparison of axis ratio as a function of the equivalent volume diameter from theory, observations, and experiments. The dash-dotted line is the empirical relation in Equation 2.14. The dashed line is the result of Equation 2.15. The solid line is the experimental fit (Equation 2.16).

where ρ is the air density, ρ_w is the water density, g is the gravity coefficient, and C_D is the drag coefficient. Using $C_D = 0.45$ and the air density ρ_0 for standard atmosphere at sea level, we obtain the terminal velocity for drops of diameter $D > 1$ mm as

$$v = 4.92 \left(\frac{\rho_0}{\rho} \right)^{1/2} \sqrt{D}, \qquad (2.18)$$

where D is the drop diameter in mm and v is the terminal velocity in m/s.

However, for droplets of diameter <60 μm such that pressure gradients and inertial forces are negligible compared to viscous forces, the theory predicts an exponent of 2, and for drops of 60 μm $< D < 1$ mm, the velocity is modeled as linearly related to the diameter (Rogers and Yau 1989).

Note that Equation 2.18 is derived based on the assumption of a spherical drop with a constant drag coefficient. Due to its simple form, it has been used in model parameterization (Kessler 1969; Lin et al. 1983). In practice, however, large drops become oblate and the drag coefficient increases, yielding a smaller terminal velocity than that by Equation 2.18. Hence, other forms for the terminal velocity at sea level have been proposed and used in the radar meteorology community. They are as follows:

1. The empirical power-law relation proposed by Atlas and Ulbrich (1977)

$$v = 3.78 D^{0.67}. \qquad (2.19)$$

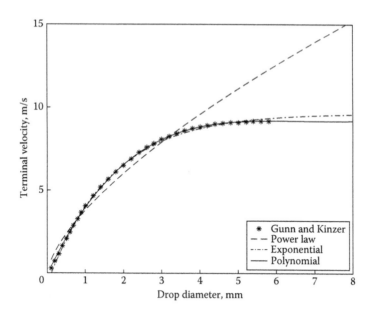

FIGURE 2.9 Comparison of empirical relations for terminal fall velocity of raindrops at sea level.

2. The exponential form used by Atlas et al. (1973)

$$v = 9.65 - 10.3 \exp(-0.6_D).$$ (2.20)

3. The polynomial form derived by Brandes et al. (2002)

$$v = \begin{cases} -0.1021 + 4.932D - 0.9551D^2 + 0.07934D^3 - 0.0023626D^4 & (D < 6\,\text{mm}) \\ 9.17 & (D \geq 6\,\text{mm}) \end{cases}.$$ (2.21)

The above three empirical power-law relations for terminal velocity are plotted together along with the earlier measurements made by Gunn and Kinzer (1949), as shown in Figure 2.9.

Both the exponential and polynomial relations fit with the measurements very well, whereas the power-law relation starts to differ substantially when $D > 4.0$ mm.

2.1.2 SNOW

Snow is common in winter precipitation and can cause serious damage and injury when it is heavy (Brandes et al. 2007; Martner et al. 1992; Zhang et al. 2011b). The common description for snowfall is the snow water equivalent, defined as

$$S = 6\pi \times 10^{-4} \int_0^{D_{max}} \left(\frac{\rho_s}{\rho_w} \right) D^3 v(D) N(D) dD \quad \left[\text{mm hr}^{-1} \right].$$ (2.22)

where ρ_s is the snow density. Although the snow water equivalent S gives the precipitation amount of snowfall, it is important to characterize the microphysics of snowstorms so that snow state parameters can be linked to polarimetric radar measurements. Like rain DSD, snow particle size distribution (PSD) is the fundamental property of microphysics, along with shape and terminal fall velocity. In addition, unlike the constant density of raindrops, the density of snow varies significantly depending on the snow type and climatological location. These microphysical properties are discussed in the following subsections.

2.1.2.1 Snow Particle Size Distribution

Figure 2.10 shows examples of 2DVD measurements of four snowflakes. They vary greatly in terms of size (from ~2.5 mm in Figure 2.10d to ~1.5 cm in Figure 2.10b), shape, and density. Their equivolume diameters are estimated from the maximum widths ($2a_1$ and $2a_2$) and heights ($2b$) of the 2D images measured by the 2DVD: $D = 2(a_1 a_2 b)^{1/3}$. In addition, the oblateness and terminal velocity of snowflakes are also estimated from the 2DVD images. Although the shapes of snowflakes are irregular, they can also be modeled as spheroids, typically with a constant axis ratio of $\gamma = 0.75$. Because snowflakes fall with their major axis aligned mainly in the horizontal direction, the mean canting angle of snow aggregates is normally assumed to be $0°$ and the standard deviation of the canting angle can be assumed to be $20°$.

Snow PSDs are obtained by sorting the 2DVD-measured particles into size bins. Because snowflakes tend to have a slower falling velocity than do raindrops, 5-min segments of the data are used to form a PSD. Figure 2.11 shows an example of snow PSDs collected in Oklahoma by the OU 2DVD on November 30, 2006. Compared with the rain DSDs shown in Figure 2.3, the snow PSDs contain hydrometeors that are larger in size, up to 1.5 cm, and have a lower number concentration, with a typical value of a few hundred snowflakes per meter cubed. This is on approximately the same order as that of stratiform rain DSDs.

2.1.2.2 Snow Bulk Density

As mentioned earlier, there is a large range in snow density, although most numerical weather prediction parameterization schemes use a constant density of 0.1 g cm^{-3} (Lin et al. 1983). Snow density, however, is critical in estimating liquid equivalents and for linking microphysics to polarimetric radar measurements. To determine density, the total volume can be found by summing the volumes of all the snowflakes collected by 2DVD during a 5-min interval, and the precipitation mass can be measured by a snow gauge. Using data collected in Colorado (Brandes et al. 2007), the snow density was determined in this manner to be

$$\rho_s = 0.178 D^{-0.922}, \tag{2.23}$$

where D is in mm and ρ_s is in g cm^{-3}. This expression is very similar to the inverse linear relation $\rho_s = 0.17 D^{-1}$ found by Holroyd (1972) for Great Lakes snow storms and to an intermediate relation $\rho_s = 0.15 D^{-1}$ determined by Fabry and Szyrmer (1999) from a literature survey. It is noted that these are all mean relations, and snow density can vary from event to event, depending on the type of snowstorm.

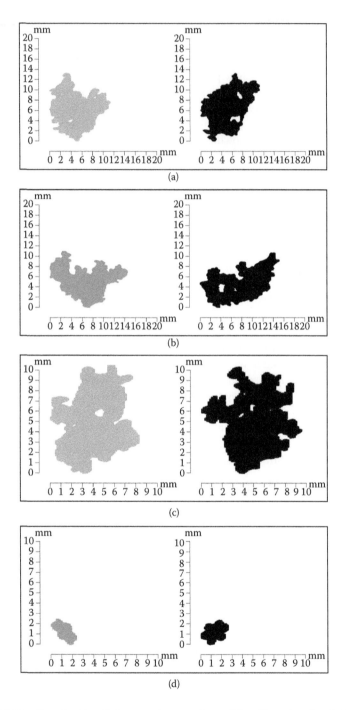

FIGURE 2.10 Sample video disdrometer images of snowflakes. Front (gray) and side (black) profiles are shown. Size increments are (a, b) 2 mm and (c, d) 1 mm. (From Brandes, E. A., et al., 2007. *Journal of Applied Meteorology and Climatology*, 46, 634–650.)

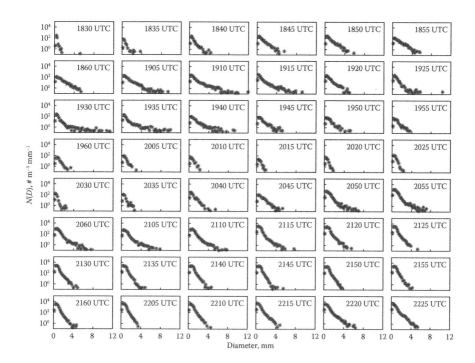

FIGURE 2.11 Snow PSDs measured by OU 2DVD in Oklahoma on November 30, 2006.

Note also that Equation 2.23 is a mean relation derived from disdrometer and gauge measurements of Colorado winter storms. While looking at the disdrometer data collected in other locations such as in Oklahoma, it became apparent that the density relation in Equation 2.23 may not best describe the density of the snowfall during these events. Several factors could affect the snow density that cannot be described by one simple size–density relation: there are both in-cloud processes that affect the formation and growth of snowflakes and sub-cloud processes that affect the flake during its descent to the ground (Roebber et al. 2003).

Figure 2.12 shows a plot of measured fall velocities of winter precipitating particles measured on November 30, 2006. Also plotted is the empirically derived fall speed of raindrops (Equation 2.21), and the lower curve is the empirically derived terminal velocity for snow particles (Brandes et al. 2007):

$$v = 0.768D^{0.142}. \tag{2.24}$$

The curve plotted in the middle is used to separate the liquid and ice phases, which is the geometric mean of Equations 2.21 and 2.24 (Zhang et al. 2011b). Of primary interest are the plotted asterisks that signify particles in the snow portion of the event; most of the snow velocities measured in Oklahoma are larger than the predicted fall velocity using the Brandes density relation determined from the Colorado data. Therefore, the density of these particles should be greater than that predicted by the fixed relation in Equation 2.23.

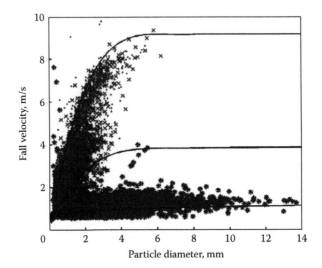

FIGURE 2.12 Plots of snow fall velocity (m/s) vs. diameter (mm) on November 30, 2006, at the Kessler Farm Field Laboratory. Data from the OU 2DVD during the freezing rain period (0000–0800 UTC) are denoted by dots, data from the mixed-phase period (0800–1600 UTC) are denoted by multiplication signs, and data from the frozen precipitation (snow) period (1600–0000 UTC) are denoted with asterisks. Also plotted are Equation 2.21 for raindrop terminal fall speed, Equation 2.24 of a power-law relation for the terminal fall speed of snow, and the velocity function used to separate the rain and snow PSDs. (From Zhang, G., et al., 2011b. *Journal of Applied Meteorology and Climatology,* 50, 1558–1570.)

To have a more realistic density value, a terminal velocity–based modification to the density value was derived from the equation for terminal velocity using Equation 2.17. The value from Equation 2.23 is recast as a baseline density ρ_{sb}, the measured velocity is represented by v_m, and we use the baseline velocity v_{sb} (Equation 2.24) to create an estimate of ρ_s to replace the snow density:

$$\rho_s = \alpha \rho_{sb}, \tag{2.25}$$

where

$$\alpha = \left(\frac{v_m}{v_{sb}}\right)^2 \frac{\rho_{aO}}{\rho_{aC}}. \tag{2.26}$$

Here, α is the adjustment. This adjustment is being used to estimate density variability for all factors rather than for one alone. In the example in Figure 2.13, air densities (ρ_{aC} and ρ_{aO}) are estimated from the pressures at 1742 m MSL for Marshall Station, Colorado, and 344 m MSL for the Washington Mesonet site at the Kessler Farm Field Laboratory in Oklahoma. Figure 2.13 shows the adjusted densities as a function of particle size for the November 30 event. The results for these events were recalculated using Equations 2.24 through 2.26 to determine how the calculated

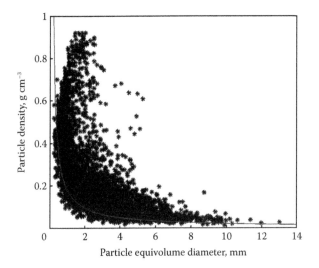

FIGURE 2.13 Velocity-adjusted density vs. diameter for the snow data collected in Oklahoma on November 30, 2006. Also plotted is the baseline density (Equation 2.23). (From Zhang, G., et al., 2011b. *Journal of Applied Meteorology and Climatology*, 50, 1558–1570.)

polarimetric radar variables were affected by this density adjustment. The same approach can be used for the density adjustment for snow data collected in other locations.

2.1.2.3 Melting Models

To derive rain DSDs from snow PSDs for comparison with observed rain DSDs, simple melting models were developed and applied to the PSD data of frozen precipitation measured by the 2DVD disdrometer (Zhang et al. 2011a). Power-law relationships are used for the density and velocity of frozen particles, as well as the velocity of raindrops. Both models assume that the mass of a single particle will be conserved. One model assumes that the total number of drops will be conserved and, thus, the total liquid water content will be conserved. As a result, this model will be referred to as the *mass conservation* (MC) *model.* The other assumes that the number flux (# m^{-2} s^{-1}) of the distribution will be conserved and is thus known as the *flux conservation* (FC) *model.* They are formulated as follows.

Because the mass of a particle will be conserved, we have

$$\rho_s D_s^3 = \rho_r D_r^3, \tag{2.27}$$

assuming that $\rho_s = aD^{-b}$ and $\rho_r = 1$, rearranging gives

$$D_s = a^{-\frac{1}{3-b}} D_r^{\frac{3}{3-b}}, \tag{2.28}$$

$$D_r = a^{\frac{1}{3}} D_s^{\frac{3-b}{3}}. \tag{2.29}$$

Differentiating Equation 2.29 with respect to D_r, we arrive at the following:

$$\frac{dD_s}{dD_r} = \frac{3}{3-b} a^{\frac{-1}{3-b}} D_r^{\frac{b}{3-b}}. \tag{2.30}$$

1. MC model

In the MC model, the mass of each particle and the total number of particles will be conserved, which means that the total liquid water content of a distribution is conserved:

$$N_r(D_r)dD_r = N_s(D_s)dD_s. \tag{2.31}$$

Rearranging leads to

$$N_r(D_r) = N_s(D_s)\frac{dD_s}{dD_r}. \tag{2.32}$$

Substituting Equations 2.28 and 2.30 into Equation 2.32, we arrive at

$$N_r(D_r) = N_s(D_s)\frac{3}{3-b} a^{\frac{-1}{3-b}} D_r^{\frac{b}{3-b}}. \tag{2.33}$$

After melting, the uniform bin size set by the disdrometer no longer applies, and a new bin size must be calculated. Rearranging Equation 2.30 and using $\Delta D_s = 0.2$ mm,

$$\Delta D_r = \frac{3-b}{15} a^{\frac{1}{3-b}} D_r^{\frac{-b}{3-b}}, \tag{2.34}$$

where ΔD_s and D_r are in mm.

2. FC model

The other model, the FC model, also assumes that the mass of a particle is conserved. Thus, this model assumes that the number flux instead of the total number of particles is conserved. Unlike in the MC model, the water content will not be conserved, but the snow water equivalent and rain rates should be equal. The rain DSD from the original frozen PSD can be found in a similar manner to the MC model, but since

$$N_r(D_r)v_r(D_r)dD_r = N_s(D_s)v_s(D_s)dD_s, \tag{2.35}$$

rearrangement gives

$$N_r(D_r) = N_s(D_s)\frac{v_s(D_s)}{v_r(D_r)}\frac{dD_s}{dD_r}. \tag{2.36}$$

We assume a power-law relationship, $v_s = \alpha e D_s^g$, for snow velocity where α is the velocity adjustment factor to density and also assume a power-law relation

(Equation 2.19) for rain velocity, $v_r = cD_r^d$. Inserting the power-law relationships and again substituting Equations 2.29 and 2.30 into Equation 2.36 gives

$$N_r(D_r) = N_s(D_s)\frac{3e\alpha}{c(3-b)}a^{-\frac{g+1}{3-b}}D_r^{\frac{3g+b}{3-b}-d}.$$ (2.37)

The new bin sizes are found as in Equation 2.34.

Figure 2.14 shows an example of four snow PSDs and their melted DSDs using both the MC and FC models. They were selected from the periods of heaviest snow from the snow events on November 30, 2006, and January 27, 2007, in central Oklahoma to ensure that the measured distributions had sufficient numbers of particles. The snow PSDs and rain DSDs are shown as curves and fitted to a gamma distribution (Equation 2.12) with the moment estimator using the second, fourth, and sixth moments as shown in Appendix 2C. It can be observed that they fit the gamma distribution well.

As expected, the measured snow PSDs have long tails containing a few large flakes, and applying the melting models shortens the tails. The shortening of

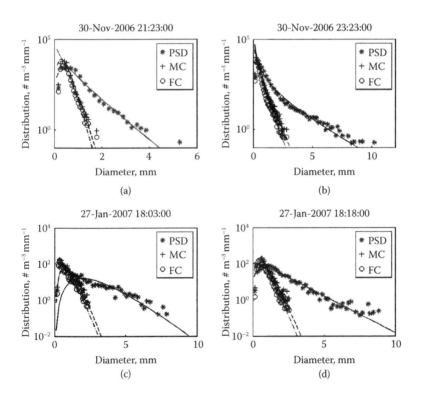

FIGURE 2.14 Examples of melting model applications to a measured snow PSD from (a, b) November 30, 2006, and (c, d) January 27, 2007. Solid lines and asterisks denote the measured snow PSD; dashed lines and plus signs, the mass conservation model; dash-dotted lines and circles, the flux conservation model.

the distributions' tails is apparent in all four panels and is consistent with previous observations (Stewart et al. 1984). This is because the large snowflakes have lower density and hence shrink more in size when they melt. The melting effect on the distribution of very small drops is also apparent in that it increases the number concentration of small drops. The MC model yields a larger increase in the number concentration of small drops than the FC model, in which the increase in number concentration is offset by a reduction caused by the faster fall velocities of raindrops (as compared to snowflakes). Whereas measured snow PSDs can be better described by an exponential distribution or a concave gamma distribution, the melted rain DSDs have more of a convex shape—similar to the observed rain DSDs for stratiform warm precipitation, as shown in Figure 2.5c and d in Subsection 2.1.1.1 and reported in the existing literature (Cao and Zhang 2009; Zhang et al. 2008).

To quantitatively compare the two melting models, microphysical parameters including the mean values for number concentration (N_t), water content (W), median volume diameter (D_0), reflectivity factor (Z), and differential reflectivity (Z_{DR}) were calculated for melting-model-derived DSDs from snow and for observed rain DSDs (Table 2.2). Although there are some differences, the results are close to each other and indicate the similarity of winter precipitation microphysics between rain and snow periods, especially for the median volume diameter and radar reflectivity.

Two melting models were developed and applied to periods of the frozen precipitation shown in Figure 2.14, and the results were compared to periods of rain during an episode of mixed precipitation types. Applying the melting models to measured snow PSDs yields rain DSDs that are similar to those recorded during the rain periods for the same precipitation event. The distributions' tails are shortened, and the number of small drops increases because the large snowflakes become small drops. Compared with the rain period DSDs, melted snow DSDs consistently have more small drops and occasionally have more large drops. The median volume diameter, reflectivity, and differential reflectivity in melted snow DSDs are smaller than those in rain period DSDs. Both melting models yield comparable results, but the FC model yields a slightly smaller number concentration than does the MC model.

TABLE 2.2

Comparison of Microphysical Parameters between Rain and Snow

Variables	Rain	Snow: MC	Snow: FC
$\langle N_t \rangle$, # m^{-3}	90.2	410.1	284.2
$\langle W \rangle$, g m^{-3}	0.050	0.061	0.040
$\langle D_0 \rangle$, mm	1.06	0.79	0.75
$\langle Z \rangle$, dBZ	20.6	17.8	17.3
$\langle Z_{DR} \rangle$, dB	0.467	0.315	0.342

Note: FC, flux conservation; MC, mass conservation.

2.1.3 HAIL AND GRAUPEL

Hail and graupel are another type of solid precipitation that is normally associated with strong supercells. Hailstones, forming mainly from accretion of ice particles or raindrops and having a high density from 0.5 to 0.9 g cm^{-3}, can grow to a very large size. The largest recorded hailstone was found in Nebraska during a thunderstorm on June 22, 2003, having a diameter of 17.8 cm. Figure 2.15 shows a picture of hailstones on the ground after a tornadic supercell passed across northern Oklahoma on May 29, 2004. The big hailstone is about 10 cm in diameter. Although big hailstones are rare, common hailstone sizes range from 5 to 50 mm. The literature does not have many measurements of hail PSDs. Smaller hailstones have been represented by an exponential (Cheng and English 1983) or gamma distribution, but larger hailstones often seem to have narrow distributions symmetrical about a certain size (Ziegler et al. 1983).

Hailstones consist of ice balls and irregular lumps and hence have irregular shapes. For modeling convenience, however, hailstones are modeled as spheres or oblate spheroids with protuberances. Ground observations suggest that the majority of hailstones have axis ratios of 0.8, with spongy hail having a lower axis ratio of 0.6–0.8 (Knight 1986; Matson and Huggins 1980). Dry hailstones tend to tumble while they fall and are therefore considered to have random orientations. By contrast, wet hailstones fall with their major axes aligned horizontally with a mean canting angle of 0°. This is because, as a wet hailstone melts while falling, the outer water layer tends to stabilize tumbling motions. It is commonly accepted that the axis ratio of both dry and wet hailstones is $\gamma = 0.75$. Therefore, the standard

FIGURE 2.15   A picture of hailstones taken on May 29, 2004, after a tornadic supercell passed across northern Oklahoma. (Courtesy of Kumjian and Schenkman.)

deviation of the canting angle is parameterized as a function of the fractional water content as

$$\sigma = 60°(1 - cf_w), \qquad (2.38)$$

where f_w is the water fraction within melting hail (water–ice mixtures) and c is a coefficient of about 0.8 (Jung et al. 2008a).

Graupel, also called *soft hail*, is formed when supercooled water droplets collect and coalesce on a falling snowflake, in which riming is a main process. Similar to hailstones, graupels have irregular shapes, but their density is lower than that of hailstones and higher than that of snowflakes.

2.1.4 CLOUD WATER AND ICE

A cloud is a collection of suspended water droplets or ice crystals, or both. Cloud water droplets form from the condensation of super-saturated water vapor onto cloud condensation nuclei. They are spherical due to surface tension, have a typical size of ~10 μm, and have a typical number concentration of ~10^8 #/m^3 (Pruppacher and Klett 1996). Cloud water content is on the order of 1 g/m^3 or less, and reflectivity factors are usually less than 10 dBZ.

Ice clouds consist of ice crystals that are formed from the deposition of water vapor on ice nuclei. Cloud ice crystals have a typical size (measured along their longest dimension) of ~100 μm, a number concentration of ~10^5 #/m^3, a water content of ~1 g/m^3, and a reflectivity factor of about 10–20 dBZ. Ice crystals have a variety of shapes, which depend on the ambient environment. Figure 2.16 shows the growth habit of ice crystals from laboratory experiments and observations of natural ice crystals. The type of ice crystal depends on the ice super-saturation: low super-saturation tends to yield slow growth and solid plates/columns with high density, whereas high super-saturation allows for fast growth for low-density dendrites. The growth habit also depends on temperature: from columns/needles ($T > -10°C$) to mostly dendrites ($-40°C < T < -10°C$) to columns ($T < -40°C$). Since polarimetric radar signatures are highly dependent on hydrometer shape and density, it is important to know these growth habits in interpreting PRD. For example, the high density dry ice crystals can have very strong polarimetric signatures that can be even stronger than those of melting snow, but low density dry snowflakes have very weak polarimetric signatures even if they are large. This is especially true at the S-band frequency.

2.2 ELECTROMAGNETIC PROPERTIES

The EM properties of a medium or material are its ability to respond to EM fields. These include conductivity, resistivity, insolation, attenuation, phase delay, reflection, refraction, and so forth. These EM properties all affect how radar transmitted/received signals propagate and are scattered in the atmosphere. The EM properties of hydrometeors depend on their phase, composition, and density. Raindrops are composed of water; hailstones are in the form of ice; snowflakes

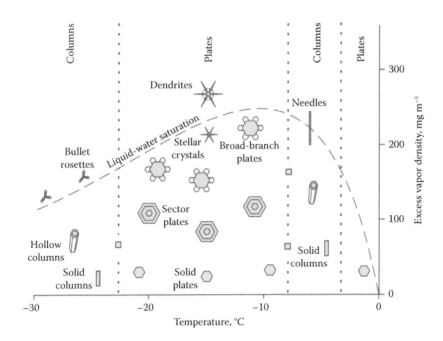

FIGURE 2.16   Ice crystals formed from various growth habits as deduced from laboratory experiments and natural observations. (From Lamb, D., and J. Verlinde, 2011. *Physics and chemistry of clouds.* Cambridge University Press.)

are an air–ice mixture; and melting snowflakes or graupels are mixtures of air, ice, and water. Therefore, it is important to know the EM properties of water, ice, air, and their mixtures.

For nonmagnetic media such as hydrometeors, the EM properties are described by a dielectric constant and/or index of refraction. Whereas a refraction index is easily understood as the ratio of an EM wave's speed in free space to its speed in the given medium, the dielectric constant is the representation of the properties of a dielectric material or medium (another term for electric insulator). The concept of the dielectric constant has been extended to partially conducting media and requires further explanation of EM theory.

2.2.1 Dielectric Constant

The fundamental EM theory is a set of Maxwell equations governing the relationships among the four EM fields: electric field $\vec{E}$ in volts per meter, electric flux density $\vec{D}$ in coulombs per square meter, magnetic field $\vec{H}$ (amperes per meter: A/m), and magnetic flux density $\vec{B}$ (webers per square meter: W/m²) (Maxwell 1873). They are

$$\nabla \times \vec{E} = -\frac{\partial \vec{B}}{\partial t}.$$

(2.39)

$$\nabla \times \vec{H} = \frac{\partial \vec{D}}{\partial t} + \vec{J} \tag{2.40}$$

$$\nabla \cdot \vec{B} = 0 \tag{2.41}$$

$$\nabla \cdot \vec{D} = \rho, \tag{2.42}$$

where $\vec{J}$ is the electric current density (amperes per square meter: A/m²), and ρ is the electric charge density. The first equation (Equation 2.39) is Faraday's law of induction. The second equation is called *Ampere's law* (Equation 2.40). The third (Equation 2.41) and fourth (Equation 2.42) equations are Gauss' laws for magnetic and electric fluxes, respectively.

Assuming that the charge densities are known, this leaves eight equations (as shown in Equations 2.39 through 2.42, with Equations 2.39 and 2.40 each representing three equations), of which only six are independent. Each of the five field vectors $\left(\vec{E}, \vec{D}, \vec{H}, \vec{B}, \vec{J}\right)$ consists of three variables, resulting in a total of 15 unknown variables. This means that other relations or constraints among the fields are needed to resolve the problem. These relations are called *constitute relations*, and they depend on the properties of the medium. For a linear, passive, and isotropic medium, the relationships are

$$\vec{B} = \mu \vec{H} \tag{2.43}$$

$$\vec{D} = \varepsilon \vec{E}, \tag{2.44}$$

where μ is the permeability (henries/meter: H/m), and ε is the permittivity or dielectric constant (farads/meter: F/m). In free space, we have $\mu_0 = 4\pi \times 10^{-7}$ (H/m) and $\varepsilon_0 = 8.854 \times 10^{-12}$ (F/m). Equations 2.43, 2.44, and 2.48 constitute another nine equations to close the formulation, we can solve the fifteen variables because we now have fifteen equations.

For a general nonmagnetic medium ($\mu = \mu_0$) such as water, the dielectric constant is the main concern. Besides the dielectric constant (also called *permittivity*), an alternative description of the dielectric property is to use the electric susceptibility χ_e and the electric polarization $\vec{P}$. The electric flux density is expressed as a sum of the free space part $\varepsilon_0 \vec{E}$ and the electric (or dielectric) polarization $\vec{P}$, written as

$$\vec{D} = \varepsilon_0 \vec{E} + \vec{P}. \tag{2.45}$$

Combining Equations 2.44 and 2.45 and solving for P, we obtain the electric polarization

$$\vec{P} = (\varepsilon - \varepsilon_0)\vec{E} = \chi_e \varepsilon_0 \vec{E}, \tag{2.46}$$

where the electric susceptibility χ_e is a dimensionless constant, indicating the degree of polarization of the dielectric material in response to an applied electric field. The polarization can be produced by one of the following four mechanisms: (i) electric polarization, caused by the displacement of the mass center of electrons from the atomic nuclei, in response to the applied electric field; (ii) atomic polarization, generated by the displacement of differently charged atoms; (iii) molecular (orientation) polarization, produced by existing molecular dipoles aligning to the direction of the applied field; and (iv) interfacial polarization, generated by a spatial separation of charged objects.

Among the four polarization mechanisms, the molecular polarization is especially important in radar meteorology because it is the case for water from which hydrometeors are formed. As shown in Figure 2.17a, a water molecule (H_2O) has an existing dipole formed from the anisotropically arranged positively charged hydrogen and negatively charged oxygen atoms. Figure 2.17b shows the molecular states of water before and after the electric field is applied, in response to the applied field.

To quantify the polarization, the polarization vector is defined as the dipole moments per unit volume of the medium, expressed by

$$\vec{P} = N\vec{p} = Nq\vec{l} = N\alpha\vec{E}',$$ (2.47)

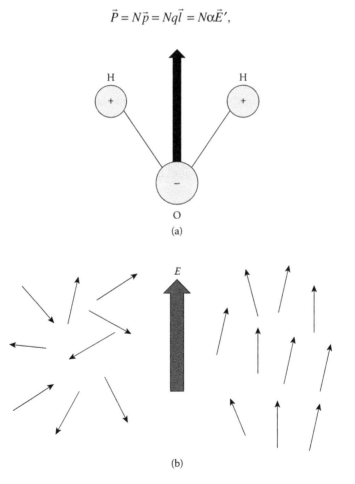

FIGURE 2.17 Molecular polarization mechanism: (a) a water molecule with existing dipole and (b) molecular states before and after an electric field is applied.

where N is the number of dipoles per unit volume, $\vec{P}$ is the dipole moment of each elementary dipole, q is the charge, l is the distance between the pair of the charges, α is the polarizability, and $\vec{E'}$ is the local field, not the applied field. Whereas Equation 2.46 is the macro-description of the electric polarization, Equation 2.47 is the micro-description.

2.2.2 Dielectric Constant for a Lossy Medium

In the last subsection, we focused our discussion on insulate media and ignored losses caused by currents induced in the particles in an otherwise nearly lossless media. The definition of *dielectric constant* discussed above can be extended to that for lossy media. Lossy media (or conductors) are characterized by conductivity σ, in siemens/meter or the reciprocal ohm/meter, through Ohm's law, which is written as

$$\vec{J} = \sigma\vec{E}. \tag{2.48}$$

To differentiate it from source current $\vec{J}_s$, the $\vec{J}$ in Equation 2.48 is the conduction/induced current.

Rewriting Equation 2.40 of Ampere's law for a time-harmonic wave and dropping e^{jwt} yields

$$\nabla \times \vec{H} = j\omega\varepsilon\vec{E} + \sigma\vec{E} + \vec{J}_s$$

$$= j\omega\varepsilon_c\vec{E} + \vec{J}_s, \tag{2.49}$$

where the complex dielectric constant is

$$\varepsilon_c = \varepsilon - \frac{j\sigma}{\omega} \tag{2.50}$$

with the second term, $\dfrac{j\sigma}{\omega}$, representing conductivity contribution. The relationship between the real and imaginary parts can be used to determine if the material is an insulator $\left(\varepsilon > \dfrac{\sigma}{\omega}\right)$ or a conductor $\left(\varepsilon < \dfrac{\sigma}{\omega}\right)$. The relative complex dielectric constant is then

$$\varepsilon_r = \frac{\varepsilon_c}{\varepsilon_0} = \frac{\varepsilon}{\varepsilon_0} - j\frac{\sigma}{\omega\varepsilon_0} = 1 + \chi_e - j\frac{\sigma}{\omega\varepsilon_0} = \varepsilon' - j\varepsilon''. \tag{2.51}$$

In the low frequency limit ($f \to 0$ or $\lambda \to \infty$), the relative dielectric constants and conductivities for typical materials are listed in Table 2.3.

In general, however, the relative dielectric constants (ε', ε'') of water and ice are frequency-dependent. Such dependence is established through models for the field polarization mechanisms discussed in Section 2.2.1. The electric and atomic polarizations are modeled as the harmonic damping oscillation of electrons bound to a heavy nuclei (positive charges). This is also called the *Lorentz model*

TABLE 2.3

Relative Dielectric Constants and Conductivities for Typical Materials

Material	Relative Dielectric Constant (ε')	Conductivity (σ) (S/m)
Air	1.0006	10^{-12}
Freshwater	81	10^{-3}
Seawater	81	4
Ice	3	10^{-7}
Grass	2–10	10^{-3}–10^{0}
Soil	5–10	10^{-3}–10^{0}
Copper	1	5.8×10^{7}
Aluminum	1	3.5×10^{7}

because it was first developed by H.A. Lorentz at the beginning of the last century. Molecular and interfacial polarizations, however, are modeled differently. A model for water molecular (H_2O) polarization is described next.

2.2.3 DEBYE FORMULA

In the case of water or ice, molecules have permanent dipole moments (shown in Figure 2.17a). Instead of electrons oscillating, water molecules/dipoles rotate in response to the applied wave field. There is no overshooting or oscillation when the randomly oriented dipoles align toward the direction of the applied field. This process is referred to as the *Debye relaxation model* (Debye 1929).

When a constant field E_0 is suddenly applied at time t_0, the polarization can be illustrated as a function of time (Figure 2.18).

Assuming an exponential form to describe its approach to an equilibrium state, the polarization can be written as follows:

$$P(t) = P_h + (P_s - P_h)\left\{1 - e^{-(t-t_0)/\tau}\right\}, \tag{2.52}$$

where $P_h = \varepsilon_0 \chi_\infty E$ is the polarization at the high frequency limit, which is a state describing the dipoles after they are instantaneously induced by the applied field; $P_s = \varepsilon_0 \chi_0 E$ is the polarization at the low frequency limit (the equilibrium state); and τ is the scale (or relaxation) time. Replacing the polarizations at the high and low frequency limits with the abovementioned relations, Equation 2.52 becomes

$$P(t) = \varepsilon_0 \chi_\infty E + \varepsilon_0 \left(\chi_0 - \chi_\infty\right)\left\{1 - e^{-(t-t_0)/\tau}\right\}E$$

$$= \varepsilon_0 \chi_\infty E + \varepsilon_0 \left(\chi_0 - \chi_\infty\right)\int_{t_0}^{t} E e^{-(t-t')/\tau}\, dt'/\tau. \tag{2.53}$$

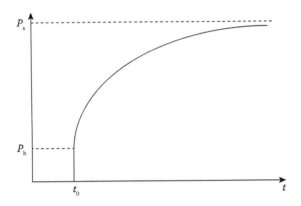

FIGURE 2.18 Sketch of the polarization as a function of time in Debye relaxation model.

To obtain the frequency dependence of the polarization, we change the constant field to the time harmonic field $E \rightarrow E_0 e^{jwt}$ and have

$$P(t) = \varepsilon_0 \chi_\infty E_0 e^{j\omega t} + \varepsilon_0 \left(\chi_0 - \chi_\infty \right) \int_{t0}^{t} E_0 e^{j\omega t' - (t-t')/\tau} \, dt'/\tau$$

$$= \varepsilon_0 \chi_\infty E_0 e^{j\omega t} + \varepsilon_0 \left(\chi_0 - \chi_\infty \right) E_0 e^{j\omega t} \frac{1}{1 + j\omega\tau}. \tag{2.54}$$

From Equation 2.46 and the right-hand side of Equation 2.54, we obtain the polarizability

$$\chi_e = \chi_\infty + \frac{\left(\chi_0 - \chi_\infty \right)}{1 + j\omega\tau} \tag{2.55}$$

and the relative dielectric constant with the conduction term is

$$\varepsilon_r = \varepsilon_\infty + \frac{\left(\varepsilon_s - \varepsilon_\infty \right)}{1 + j\omega\tau} - j\frac{\sigma}{\omega\varepsilon_0}. \tag{2.56}$$

Replacing the angular frequency with wavelength $\omega = \dfrac{2\pi c}{\lambda}$ and relaxation time with relaxation wavelength $\tau = \dfrac{\lambda_s}{2\pi c}$, Equation 2.56 becomes

$$\varepsilon_r = \varepsilon_\infty + \frac{\left(\varepsilon_s - \varepsilon_\infty \right)}{1 + j\lambda_s / \lambda} - j\frac{\sigma\lambda}{2\pi c\varepsilon_0} \tag{2.57}$$

with real and imaginary parts

$$\varepsilon' = \varepsilon_\infty + \frac{\left(\varepsilon_s - \varepsilon_\infty \right)}{1 + (\lambda_s / \lambda)^2},$$

$$\varepsilon'' = \frac{(\varepsilon_s - \varepsilon_\infty)(\lambda_s/\lambda)}{1 + (\lambda_s/\lambda)^2} + \frac{\sigma\lambda}{2\pi c \varepsilon_0}. \tag{2.58}$$

Note that the real and imaginary parts are not independent of each other, but are related through the Kramers–Kronig relations (Toll 1956). The Debye formulas (Equation 2.58) were modified by Cole and Cole (1941) to include spread effects to better match with experimental results as follows:

$$\varepsilon' = \varepsilon_\infty + \frac{(\varepsilon_s - \varepsilon_\infty)\left[1 + (\lambda_s/\lambda)^{1-\alpha}\sin(\alpha\pi/2)\right]}{1 + 2(\lambda_s/\lambda)^{1-\alpha}\sin(\alpha\pi/2) + (\lambda_s/\lambda)^{2(1-\alpha)}}$$

$$\varepsilon' = \frac{(\varepsilon_s - \varepsilon_\infty)\left[(\lambda_s/\lambda)^{1-\alpha}\cos(\alpha\pi/2)\right]}{1 + 2(\lambda_s/\lambda)^{1-\alpha}\sin(\alpha\pi/2) + (\lambda_s/\lambda)^{2(1-\alpha)}} + \frac{\sigma\lambda}{2\pi c \varepsilon_0}. \tag{2.59}$$

The parameters ε_s, ε_∞, α, λ_s, σ were found through a least-squares fitting of experiment data as a function of temperature (Ray 1972) and are converted to SI units as follows:

$$\varepsilon_s = 78.54[1.0 - 4.579 \times 10^{-3}(T - 25)$$
$$+ 1.19 \times 10^{-5}(T - 25)^2 - 2.8 \times 10^{-8}(T - 25)^3]$$
$$\varepsilon_\infty = 5.27137 + 2.16474 \times 10^{-2}\,T - 1.31198 \times 10^{-3}T^2$$
$$\alpha = -16.8129/(T + 273) + 6.09265 \times 10^{-2} \tag{2.60}$$
$$\lambda_s = 3.3836 \times 10^{-6}\exp[2513.98/(T + 273)] \quad [\text{m}]$$
$$\sigma = 1.1117 \times 10^{-4} \quad [\text{S/m}],$$

where temperature, T, is in Celsius. The model result is plotted as a function of wavelength in Figure 2.19.

In the case of ice, the parameters in the dielectric constant expression (Equation 2.59) are

$$\varepsilon_s = 203.168 + 2.5T + 0.15T^2$$
$$\varepsilon_\infty = 3.168$$
$$\alpha = 0.288 + 5.2 \times 10^{-3}T + 2.3 \times 10^{-4}T^2 \tag{2.61}$$
$$\lambda_s = 9.990288 \times 10^{-6}\exp[6643.5/(T + 273)]$$
$$\sigma = 1.1156 \times 10^{-13}\exp[-6291.2/(T + 273)].$$

Sample calculations of the dielectric constant and the refraction index of ice are shown in Figure 2.20. It can be observed that the real part of the dielectric constant is a constant for the wavelength range at a value of 3.17, and the imaginary part is very small and less than 0.01 for microwave frequencies. Both the real and imaginary parts of the dielectric constant of ice are much smaller than those of the dielectric constant of water, which means that ice particles result in less scattering for the radar to detect than do same-size water drops and that ice particles cause less attenuation for wave propagation.

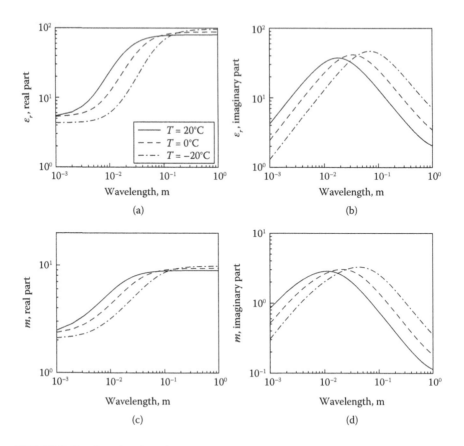

FIGURE 2.19 Sample calculations of the dielectric constant and refraction index for water as a function of wavelength.

2.2.4 DIELECTRIC CONSTANT FOR A MIXTURE OF TWO AND THREE MATERIALS

Besides rain, most hydrometeors are in the form of a mixture of two or three media. For examples, dry snowflakes are a mixture of air and ice; melting hailstones are a mixture of ice and water; and melting snowflakes/graupels are a mixture of air, ice, and water. Vegetation and biological objects are a mixture of water and other organic materials, in which the water content plays a very important role in determining dielectric constants.

Let us start with a mixture of two materials, shown in Figure 2.21. There are N spheres with a relative dielectric constant of ε_2 (or ε_1) embedded in the background medium with a relative dielectric constant of ε_1 (or ε_2).

As described earlier, a dielectric constant is the measure of polarizability of a given material in response to an electric field. A dielectric constant is related to polarizability through the Clausius–Mossotti equation (Ishimaru 1991):

$$\varepsilon_r = \frac{1 + 2N\alpha / 3\varepsilon_0}{1 - N\alpha / 3\varepsilon_0}. \tag{2.62}$$

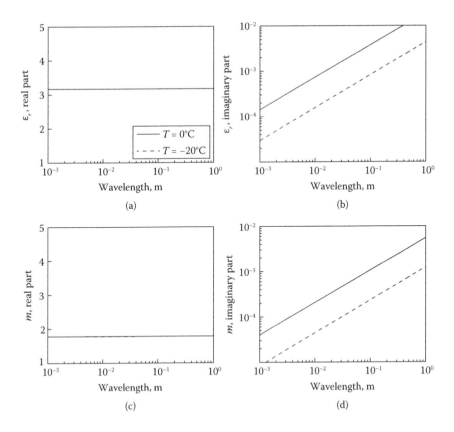

FIGURE 2.20   Dielectric constant and refraction index for ice as a function of wavelength.

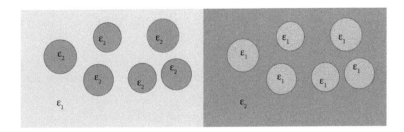

FIGURE 2.21   Dielectric constant of a mixture of two media.

In the case of Figure 2.21, we have the background medium with a dielectric constant of $\varepsilon_1\varepsilon_0$ instead of ε_0 in Equation 2.62. Hence, the effective dielectric constant of ε_e for the mixture can be obtained by replacing ε_0 with $\varepsilon_1\varepsilon_0$:

$$\frac{\varepsilon_e}{\varepsilon_1} = \frac{1 + 2N\alpha/3\varepsilon_1\varepsilon_0}{1 - N\alpha/3\varepsilon_1\varepsilon_0} \qquad (2.63)$$

and the polarizability of the sphere can also be found by extending that in free space as

$$\alpha = \frac{\varepsilon_2 - \varepsilon_1}{\varepsilon_2 + 2\varepsilon_1} 3\varepsilon_1 \varepsilon_0 V,\tag{2.64}$$

where V is the volume of the sphere.

Substituting Equation 2.64 into Equation 2.63 and noting the fractional volume $f = NV$ that all spheres occupy, we have

$$\frac{\varepsilon_e}{\varepsilon_1} = \frac{1 + 2fy}{1 - fy}\tag{2.65}$$

with

$$y = \frac{\varepsilon_2 - \varepsilon_1}{\varepsilon_2 + 2\varepsilon_1}.\tag{2.66}$$

This is the Maxwell-Garnett (M-G) mixing formula from which the effective dielectric constant can be easily calculated. However, it is valid only in the case of $fy \ll 1$ because it was obtained under the assumption that included spheres contribute less to the effective dielectric constant than does the background. Therefore, when using the M-G mixing formula it is important to make the most appropriate choice of background and inclusions, which cannot simply depend on the fractional volume, but must also take into consideration the dielectric constants.

In practice, it is not necessarily true that one material acts as background (and contributes relatively more to the dielectric constant) while another material acts as an inclusion (and contributes relatively less to the dielectric constant). In such cases, it is reasonable to imagine that both materials are inclusions and embedded in an effective medium. The polarizabilities in both materials are represented in terms of the effective medium

$$\alpha_1 = \frac{\varepsilon_1 - \varepsilon_e}{\varepsilon_1 + 2\varepsilon_e} 3\varepsilon_e \varepsilon_0 V_1\tag{2.67}$$

$$\alpha_2 = \frac{\varepsilon_2 - \varepsilon_e}{\varepsilon_2 + 2\varepsilon_e} 3\varepsilon_e \varepsilon_0 V_2.$$

The polarization vector is the sum of both contributions and should be zero with reference to the effective medium:

$$\vec{P} = (N_1 \alpha_1 + N_2 \alpha_2)\vec{E}_e = 0.\tag{2.68}$$

Substituting Equation 2.67 into Equation 2.68 and noting $f_1 = N_1 V_1$, $f_2 = N_2 V_2$ ($f_1 + f_2 = 1$), we obtain the equation for the effective dielectric constant:

$$f_1 \frac{\varepsilon_1 - \varepsilon_e}{\varepsilon_1 + 2\varepsilon_e} + f_2 \frac{\varepsilon_2 - \varepsilon_e}{\varepsilon_2 + 2\varepsilon_e} = 0.\tag{2.69}$$

This is the Polder–van Santern (P–S) mixing formula; ε_e is not in apparent form as in Equation 2.65 but can be solved through a quadratic form.

Figure 2.22 shows sample calculations of the dielectric constant for dry snow (air–ice mixture) and melting hail (ice–water mixture), which can also be modeled

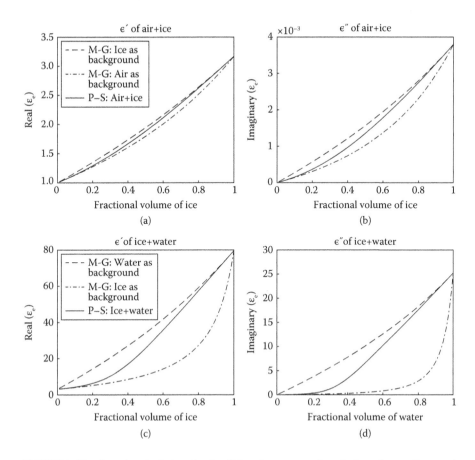

FIGURE 2.22 Sample calculations for the dielectric constant of an air–ice mixture (dry snow) and melting hail using the M-G formula with air as the background, the M-G formula with ice as the background, and the P–S mixing formula: (a, b) show real and imaginary parts of the dielectric constant of air–ice mixture, respectively and (c, d) show these of ice–water mixture.

as two-layer particles (not discussed), using the M-G and P–S mixing formulas. In modeling with the M-G mixing formula, we consider both the case with air as the background and that with ice as the background. It is evident that the dielectric constant calculated with ice as the background is larger than that with air as the background, whereas the results of the P–S mixing formula fall between the M-G results. This is because the calculated effective dielectric constants are affected more by the background than by inclusion in the case of the M-G mixing formula.

A mixture of three materials is also important in polarimetric weather radar applications. For example, melting snowflakes and graupels are composed of air, ice, and water. During the melting process, the relative increase in water content causes an increase in the dielectric constant. This results in enhanced radar reflectivity, namely bright band and other polarimetric radar signatures (an increase in differential reflectivity and reduction in correlation coefficient). The calculation of the effective dielectric constant for the three-material mixture can be extended from the M-G and P–S two-material mixture formulas discussed above.

Using the snowflakes or graupels as an example, the three components of air, ice, and water have relative dielectric constants of ε_a, ε_i, and ε_w, respectively. Their fractional volumes are f_a, f_i, and f_w and $f_a + f_i + f_w = 1$. A two-step procedure is used to calculate the effective dielectric constant for the mixture as follows:

Step 1a: Let $\varepsilon_1 = \varepsilon_a$ and $\varepsilon_2 = \varepsilon_i$; calculate their fractional volumes in the two-medium mixture with $f_1 = f_a / (f_a + f_i)$ and $f_2 = 1 - f_1$.

Step 1b: Calculate the effective dielectric constant $\varepsilon_{ai} = \varepsilon_e$ for the air–ice mixture (dry snow or graupel) using the M-G formula (Equations 2.65 and 2.66) with $f = f_2$ or the P–S formula (Equation 2.69).

Step 2a: Let $\varepsilon_1 = \varepsilon_w$ and $\varepsilon_2 = \varepsilon_{ai}$; determine their fractional volumes of water and dry snow or graupel with $f_1 = f_w$ and $f_2 = 1 - f_w$.

Step 2b: Repeat Step 1b to obtain the effective dielectric constant for the three-medium mixture.

Most of the time, however, the fractional volumes for each component of melting snow or graupel are not known, but can be calculated using the dry snow density ρ_{ds} and the percentage of melting γ_w. In this case, the wet-snow density ρ_{ws} needs to be modeled, and fractional volumes can then be calculated. At the very early stage of melting, the size of the snow aggregate does not change much with increasing γ_w so that the density increases slowly. As melting progresses, γ_w further increases, the snow particle collapses, inducing the shrinkage of the particle, and the density increases more rapidly. To model this melting process as the snow aggregate particles descend, the density of the melting snow aggregate is represented by a quadratic function of γ_w as

$$\rho_{ws} = \rho_{ds}\left(1 - \gamma_w^2\right) + \rho_w \gamma_w^2. \tag{2.70}$$

Once the density of wet snow is known, the fractional volumes for each component are calculated as

$$f_w = \gamma_w \, \rho_{ws} / \rho_w$$
$$f_i = (1 - \gamma_w) \, \rho_{ws} / \rho_i \tag{2.71}$$
$$f_a = 1 - f_i - f_w.$$

The results of the dielectric constant for wet snow are shown as functions of the percentage of melting in Figure 2.23.

This continuously melting model is needed to accurately interpret bright-band radar signatures (Brandes and Ikeda 2004) and to accurately simulate polarimetric radar signatures (Jung et al. 2008a). Figure 2.24 shows vertical profiles of reflectivity and differential reflectivity measured the NCAR S-pol radar for a stratiform rain event on September 17, 1998. There is 7–10 dB reflectivity enhancement at the bright band caused by snow melting, and the peak location of the differential reflectivity is a little lower than that of reflectivity. (This result is explained in Chapter 3.)

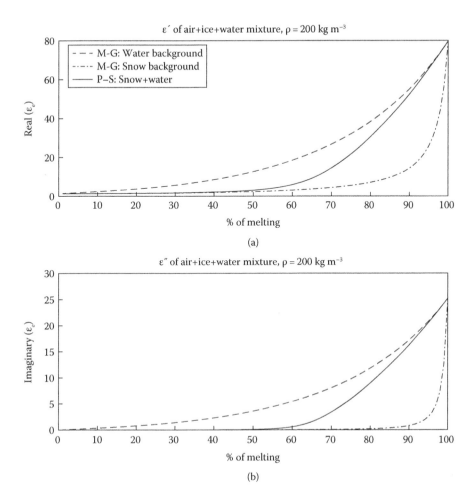

FIGURE 2.23 Effective dielectric constant for an air–ice–water mixture (melting snow) as a function of the percentage of melting (a) real part and (b) imaginary part.

The vertical changes of the reflectivity factor during melting are attributed to (i) an increase in the dielectric constant, (ii) a shrinking of the particle size due to snowflake collapse, and (iii) a reduction in the particle number concentration due to faster falling. This is simply expressed by the reflectivity

$$\eta = \frac{\pi^5}{\lambda^4}\left|\frac{\varepsilon_e - 1}{\varepsilon_e + 2}\right|^2 \int D^6 N(D)dD \equiv \frac{\pi^5}{\lambda^4}|K|^2 Z \tag{2.72}$$

with two terms: (i) a dielectric factor of

$$K = \frac{\varepsilon_e - 1}{\varepsilon_e + 2} \tag{2.73}$$

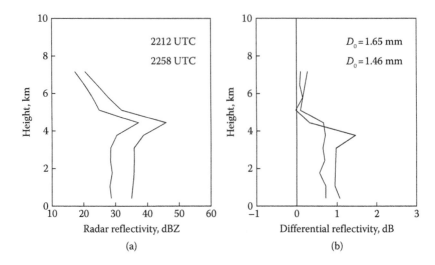

FIGURE 2.24 Profiles of reflectivity and differential reflectivity for a stratiform rain event observed by the S-pol radar at 2212 and 2258 UTC on September 17, 1998. Median volume drop diameters (D_0) from 2DVD measures are also shown. (From Brandes et al., 2004b. *Journal of Applied Meteorology*, 43, 461–475.)

and (ii) a reflectivity factor of

$$Z = \int D^6 N(D) dD,$$ (2.74)

contributing to the measured radar reflectivity.

Although the dielectric factor K for wet snow can be calculated from the effective dielectric constant for the melting snow that was described earlier, the reflectivity factor Z also varies due to the change in particle size and the PSD. The effective dielectric constant is calculated in the two-step procedure: (i) the dielectric constant of dry snow is calculated using the P–S mixing formula first, and then (ii) that of wet snow is calculated using the M-G formula with water as the background.

From Equation 2.18, we have a falling velocity proportional to the square-root of the particle density. Using mass conservation similar to Equation 2.27 and number flux conservation as in Equation 2.35, we have the reflectivity factor for wet snow as follows:

$$Z_{ws} = \int D_{ws}^6 N(D_{ws}) dD_{ws} = \int \left(\frac{\rho_{ds}}{\rho_{ws}}\right)^2 D_{ds}^6 N(D_{ds}) \left(\frac{v_{ds}}{v_{ws}}\right) dD_{ds} = \left(\frac{\rho_{ds}}{\rho_{ws}}\right)^{2.5} Z_{ds}.$$

Hence, we obtain the reflectivity ratio between wet and dry snow as follows:

$$\frac{Z_{ws}}{Z_{ds}} = \left(\frac{\rho_{ds}}{\rho_{ws}}\right)^{2.5}.$$ (2.75)

The results of the dielectric and reflectivity factors for melting snow are shown in Figure 2.25, with the top plot in linear units and the bottom plot in decibel units.

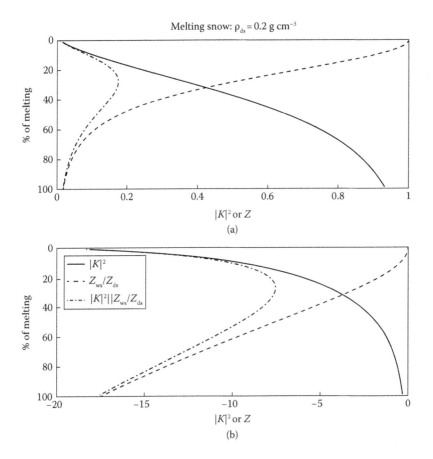

FIGURE 2.25 Dependence of the radar reflectivity of melting snow on the percentage of melting: (a) in linear units and (b) in decibel units.

As expected, the dielectric factor $|K|^2$, represented by the solid line, increases from 0.016 (−18 dB) for dry snow to 0.92 (−0.36 dB) for rain during the whole melting procedure. Although there is a large dynamic change (over 17 dB) in the dielectric factor $|K|^2$, $|K|^2$ has been assumed to be proportional to snow density for dry snow and a constant of water for wet snow in traditional cloud analysis (Pan et al. 2016; Smith et al. 1975). By taking into account the dynamic change in $|K|^2$, the continuous melting model described here is an improvement upon traditional cloud analysis.

For modelers' convenience in using the continuous melting results, the dielectric factors $|K|^2$ are parameterized as a function of snow density, represented by Equation 2.70 for wet snow, and as a function of melting percentage for wet hail as follows.

$$|K_{ws}|^2 = \left(-0.5158 + 6.1449\rho_{ws} - 10.533\rho_{ws}^2 + 8.6207\rho_{ws}^3 - 2.7187\rho_{ws}^4\right)|K_w|^2 \qquad (2.76)$$

$$|K_{wh}|^2 = \left(0.2069 - 0.0378\gamma_w + 3.977\gamma_w^2 - 4.939\gamma_w^3 + 1.791\gamma_w^4\right)|K_w|^2 \qquad (2.77)$$

where the wet snow density is represented by (2.70) and the dry snow density is the 100 kg /m³ that was used in Lin et al. (1983).

The reflectivity factor, shown by the dashed line, for wet snow decreases to 0.018 times (1.8% or –17.5 dB) that for dry snow due to the decreasing snow particle size and number concentration. The radar reflectivity, plotted by the dash-dotted line, would be the result of the dielectric (EM), particle size (physics), and DSD changes. The reflectivity sharply increases (resulting from the sharp increase in the dielectric constant) as snow starts to melt, reaches a maximum at ~30% melting; it then decreases as snow particles start to collapse to become raindrops.

A recent study (Andrić et al. 2013) revealed more interesting polarimetric features and understanding of a winter storm in central Oklahoma, as shown in Figure 2.26. Range-height indicator (RHI) images of reflectivity (Z_H), differential reflectivity (Z_{DR}), co-polar correlation coefficient (ρ_{hv}), and specific differential phase (K_{DP}) are shown

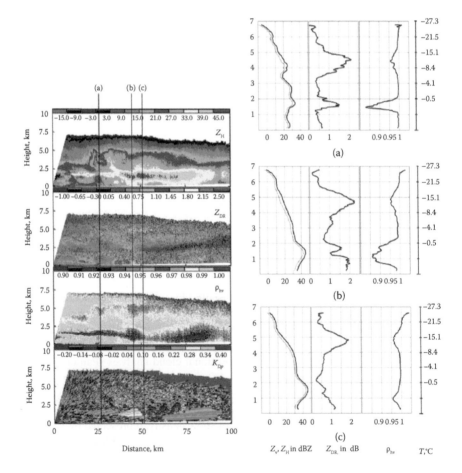

FIGURE 2.26 Range-height indicator display of polarimetric parameters for a winter storm observed by KOUN radar on January 19, 2009, at 2317 UTC at an azimuth of 181°. (From Andrić, J., et al., 2013. *Journal of Applied Meteorology and Climatology*, 52, 682–700.)

on the left, and three profiles of polarimetric measurements, except for specific differential phase, are plotted on the right. As expected, radar reflectivity increases from cloud top to melting layer as ice crystals increase in size through deposition growth, as further growth occurs through aggregation, and as the dielectric constant increases through melting. Differential reflectivity and co-polar correlation coefficients have different features, indicating different information content. In addition to the melting layer, there is a strong enhancement of Z_{DR} at the layer where the temperature is about $-10°C$, where deposition growth is dominant and can produce high density plates or columns, as shown in Figure 2.16. Corresponding to the Z_{DR} enhancement, there is a reduction in ρ_{hv}, which could be due to the random orientation of ice crystals. Further understanding of these features requires knowledge regarding wave scattering, which is discussed in later chapters.

APPENDIX 2A: FITTING PROCEDURES FOR DSD MODELS: MARSHALL–PALMER DISTRIBUTION

The nth moment of the exponential DSD (Equation 2.11) is

$$M_n = \int D^n N(D)\,dD = N_0 \Lambda^{-(n+1)} \Gamma(n+1). \tag{2A.1}$$

Because N_0 is fixed as a known parameter, the only unknown parameter Λ can be determined from any moment estimated from DSD data:

$$\Lambda = \left(\frac{N_0 \Gamma(n+1)}{M_n} \right)^{\frac{1}{n+1}}. \tag{2A.2}$$

APPENDIX 2B: FITTING PROCEDURES FOR DSD MODELS: EXPONENTIAL DISTRIBUTION

In the case of exponential distribution, the DSD parameters, N_0 and Λ, can be determined from any two moments (M_n, M_m) as

$$\Lambda = \left(\frac{M_n \Gamma(m+1)}{M_m \Gamma(n+1)} \right)^{\frac{1}{m-n}}, \tag{2B.1}$$

$$N_0 = \frac{M_n \Lambda^{n+1}}{\Gamma(n+1)}. \tag{2B.2}$$

APPENDIX 2C: FITTING PROCEDURES FOR DSD MODELS: GAMMA DISTRIBUTION

The nth moment of the gamma DSD (Equation 2.12) is

$$M_n = \int D^n N(D)\,dD = N_0 \Lambda^{-(\mu+n+1)} \Gamma(\mu+n+1). \tag{2C.1}$$

The gamma distribution parameters using five different moment estimators are described as follows.

1. M012:

$$\eta = \frac{M_1^2}{M_0 M_2}, \quad \mu = \frac{1}{(1-\eta)} - 2,$$

$$\Lambda = \frac{M_0}{M_1}(\mu + 1), \quad N_0 = \frac{M_0 \Lambda^{\mu+1}}{\Gamma(\mu+1)}. \tag{2C.2}$$

2. M234:

$$\eta = \frac{M_3^2}{M_2 M_4}, \quad \mu = \frac{1}{(1-\eta)} - 4,$$

$$\Lambda = \frac{M_2}{M_3}(\mu + 3), \quad N_0 = \frac{M_2 \Lambda^{\mu+3}}{\Gamma(\mu+3)}. \tag{2C.3}$$

3. M246:

$$\eta = \frac{M_4^2}{M_2 M_6}, \quad \mu = \frac{(7-11\eta) - \left(\eta^2 + 14\eta + 1\right)^{0.5}}{2(\eta - 1)},$$

$$\Lambda = \left[\frac{M_2}{M_4}(\mu + 3)(\mu + 4)\right]^{0.5}, \quad N_0 = \frac{M_2 \Lambda^{(\mu+3)}}{\Gamma(\mu+3)}. \tag{2C.4}$$

4. M346:

$$\eta = \frac{M_4^3}{M_3^2 M_6}, \quad \mu = \frac{(8-11\eta) - \left(\eta^2 + 8\eta\right)^{0.5}}{2(\eta - 1)}$$

$$\Lambda = \frac{M_3}{M_4}(\mu + 4), \quad N_0 = \frac{M_3 \Lambda^{(\mu+4)}}{\Gamma(\mu+4)}. \tag{2C.5}$$

5. M456:

$$\eta = \frac{M_5^2}{M_4 M_6}, \quad \mu = \frac{1}{(1-\eta)} - 6,$$

$$\Lambda = \frac{M_4}{M_5}(\mu + 5), \quad N_0 = \frac{M_4 \Lambda^{(\mu+5)}}{\Gamma(\mu+5)}. \tag{2C.6}$$

In addition to the moment estimators, the L-moment method (LM: Kliche et al. 2008) and maximum-likelihood (ML) approach are also used. Suppose the number concentration of each size category D_i ($i = 1 \ldots 41$)

has an integer number N_i, and the summation of N_i is N. The LM and ML estimators are described as follows.

6. LM:

$$b_0 = \frac{1}{N}\sum_{k=1}^{N} D_k, \quad b_1 = \frac{1}{N(N-1)}\sum_{k=1}^{N-1} kD_{(k+1)},$$

$$l_1 = b_0, \quad l_2 = 2b_1 - b_0,$$

$$\frac{l_2}{l_1} = \frac{\Gamma(\mu+1.5)}{\sqrt{\pi}\Gamma(\mu+2)}, \tag{2C.7}$$

where l_1 and l_2 are the first two L-moments; $D_{(k)}$ is the kth size category, with N_i from small to large in sequence. The estimate of μ is calculated by nonlinear iteration. After obtaining the estimate of μ, Λ can be calculated by

$$\Lambda = \frac{\mu+1}{l_1}. \tag{2C.8}$$

7. ML:

The estimate of μ is calculated by iteration from the following formula:

$$\ln(\mu+1) - \Psi(\mu+1) = \ln\left[\frac{\frac{1}{N}\sum_{k=1}^{N} D_k}{\left(\prod_{i=1}^{N} D_i\right)^{1/N}}\right], \tag{2C.9}$$

where Ψ is the psi function defined by $\Psi(x) = \dfrac{\Gamma'(x)}{\Gamma(x)}$. The estimate of Λ has the same form as Equation 2C.8:

$$\Lambda = \frac{\mu+1}{\frac{1}{N}\sum_{k=1}^{N} D_k}. \tag{2C.10}$$

As described above, ML and LM estimators give estimates of μ and Λ. The third DSD parameter, N_0, is estimated from the 0th moment here. The estimate of N_0 is given by

$$N_0 = \frac{M_0\Lambda^{(\mu+1)}}{\Gamma(\mu+1)}. \tag{2C.11}$$

Problems

2.1 Assume that a rain DSD follows the M–P model and has a water content of 1 g m^{-3}. Calculate the slope parameter, total number concentration, mean diameter, and mass-weighted mean diameter.

2.2 Given two estimated moments of M_m and M_n, show the exponential DSD parameters that can be calculated using Equations 2B.1 and 2B.2. Assume that M_2 and M_4 are given; find the expressions of N_0 and Λ, in terms of the moments (M_2 and M_4).

2.3 Given three estimated moments of M_2, M_3, and M_4 of a rain DSD, show the gamma DSD parameters that can be calculated using Equation 2C.3.

2.4 Using the provided rain DSD data, do the following:

   Plot the DSDs, total surface distributions $a(D) = \pi D^2 N(D)$, mass distributions, and reflectivity distributions $z(D) = D^6 N(D)$; comment on their differences.

   Calculate and plot the total number concentrations, rain water contents, rain rates, and radar reflectivity factors.

   Fit the measured DSDs to the M–P, exponential, and gamma DSD models using M_3, (M_2, M_4) and (M_2, M_4, M_6), respectively. Plot the DSD parameters as a function of the DSD index. In addition, plot the 35th DSD and the 115th DSD separately, along with their model fittings. Discuss the goodness of fit for each.

   Calculate the number concentrations, water contents, rain rates, and reflectivity factors from your DSD parameters and compare with those from DSD data.

2.5 Using the Debye theory (Ray 1972), calculate the relative dielectric constants at S-band (2.8 GHz), C-band (5.6 GHz), X-band (10 GHz), and Ka-band (35 GHz) for the following media/cases:

   a. Water at 0°, 10°, and 20°C
   b. Ice at –20°, –10°, and 0°C
   c. Snow at 0°C using the M-G mixing formula at S-band. Plot the dielectric constant of snow at S-band as a function of density changing from 100 kg m⁻³ to 913 kg m⁻³. Perform the calculation using two approaches: (i) mixing air and ice with air as the background and (ii) mixing air and ice with ice as the background. Discuss the difference in your results between the two approaches and under what conditions each approach is more appropriate.

   Repeat Problem 2.5c, again calculating the dielectric constant of snow, but this time using the P–S formula. Compare the results with those from Problem 2.5c.

2.6 Calculate and plot the dielectric constant of wet snow as a function of percentage of melting during the melting process from dry snowflakes to raindrops at S-band (2.8 GHz). Assume the of dry snow is 100 kg m⁻³, use the formula $\rho_{ws} = \rho_{ds}(1 - \gamma_w^2) + \rho_w \gamma_w^2$ (γ_w, percentage of melting) for the density of wet snow and use the M-G formula with water as the background. Use your results to explain bright band.

3 Wave Scattering by a Single Particle

Weather radars measure electromagnetic (EM) wave scattering by the hydrometeors that are in clouds and precipitation. To understand polarimetric radar signatures as they relate to the shape, orientation, and composition of these hydrometeors for specific weather conditions, it is important to understand EM wave scattering by various types of hydrometeors. This chapter deals with wave scattering by a single particle, focusing on the sphere and the spheroid shapes on which many hydrometeors can be modeled. This chapter describes the basic physics behind and mathematical representation of EM waves and wave scattering. The approaches for scattering calculations, including the Rayleigh approximation, Mie theory, and T-matrix method, are provided. The basic scattering characteristics of hydrometeors are also described.

3.1 WAVE AND ELECTROMAGNETIC WAVE

A wave is the propagation of a vibration or oscillation and is an efficient way of transporting energy and information. Waves are omnipresent in our daily lives. We hear each other through sound, which are acoustic waves, and see our environment by light, which are EM waves at extremely high frequencies. An acoustic wave is a longitudinal wave, which vibrates air molecules along the direction of the wave propagation. As such, only a one-dimensional variable of the displacement from its reference is needed to describe wave characteristics. A longitudinal wave is also called a *scalar wave*. An EM wave, however, is a transverse wave, which vibrates electrons (if present) in a direction perpendicular to the direction of propagation. Since a two-dimensional vector is needed to define the vibration direction, a transverse wave is also called a *vector wave*.

3.1.1 VIBRATION AND WAVE

To mathematically describe a wave, we must start with its vibration. A vibration is a back-and-forth movement that is caused by a force, $f = -\kappa z$, in the opposite direction of the displacement z (κ is the restoring coefficient). Examples of vibration include a spring and a pendulum, as shown in Figure 3.1.

Based on Newton's second law, the equation for the displacement is written as

$$f = m\frac{d^2 z}{dt^2} = -\kappa z, \qquad (3.1)$$

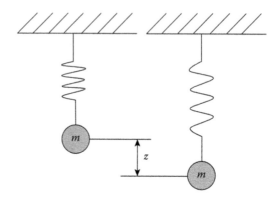

FIGURE 3.1 Conceptual illustration of a spring vibration. The restoring force is opposite to the displacement.

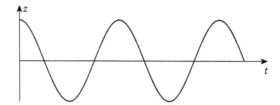

FIGURE 3.2 Sketch of a vibration with instantaneous displacement z as a function of time.

where m is the mass. We can rewrite Equation 3.1 to yield

$$\frac{d^2z}{dt^2} + \omega^2 z = 0,\tag{3.2}$$

where $\omega^2 = \kappa/m$.

Solving Equation 3.2 yields

$$z(t) = A\cos(\omega t + \phi_0),\tag{3.3}$$

where A is the amplitude, which is defined as the maximal displacement of vibration from its equilibrium position, $\omega = 2\pi f$ is the angular frequency ($f = 1/T$ is the frequency, with T as the period), and ϕ_0 in Equation 3.3 is the initial phase at $t = 0$. The phase is the argument of the cosine function, which is zero in Figure 3.2.

As a wave is the propagation of a vibration, its mathematical expression can be obtained by changing the time t to $t - x/v$ in Equation 3.3, yielding

$$z(x,t) = A\cos[\omega(t - x/v) + \phi_0],\tag{3.4}$$

where v is the phase velocity of the wave, and the wavelength is $\lambda = vT$ (as shown from peak to peak in Figure 3.3).

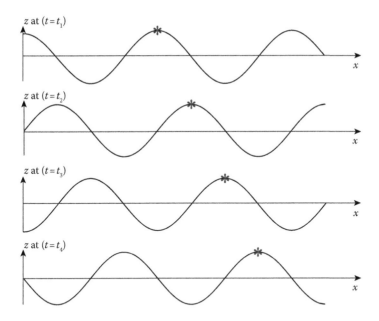

FIGURE 3.3 Sketch of wave propagation: asterisk (*) propagates in the x-direction.

3.1.2 PHASOR REPRESENTATION OF TIME-HARMONIC WAVES

Although a time-harmonic wave can be simply represented by a sinusoidal function like Equation 3.4, it can be more conveniently represented by a complex quantity. For example, Equation 3.4 can be rewritten as

$$z(x,t) = A\cos\left[\omega t + \left(\phi_0 - \omega x/v\right)\right] = \mathrm{Re}\left[Ae^{j(\omega t + \phi_0 - \omega x/v)}\right] = \mathrm{Re}\left[z(x)e^{j\omega t}\right], \quad (3.5)$$

where $z(x) = Ae^{j(\phi_0 - \omega x/v)}$ is a complex quantity, which is equivalent to the real quantity $z(x,t)$ in representing the wave, except for the omission of $[e^{j\omega t}]$ in writing. The complex quantity $z(x)$ is called a *phasor* because it is a complex quantity with phase information, but omits the time dependence term. The advantage of introducing the phasor in the complex representation is the convenience it allows in mathematical manipulation. For example, a derivative with respect to time t is simply a multiplication of $j\omega$: $\dfrac{\partial z(x,t)}{\partial t} \leftrightarrow j\omega z(x,t)$.

3.1.3 EM WAVE

As mentioned earlier, an EM wave is a transverse wave that needs a vector in the form of an electric field $\vec{E}(\vec{r},t)$ to represent it. An EM wave propagates in space or a medium such that a changing electric field causes a magnetic field. This magnetic field is also changing, which in turn yields a varying electric field, and so on. Using the analogy of the phasor for a scalar wave, the phasor representation of the electric field is $\vec{E}(\vec{r},t) = \text{Re}\left[\vec{E}(\vec{r})e^{j\omega t}\right]$. In the case of the time-harmonic wave, the Maxwell equations (Equations 2.39–2.42) for the phasors of the electric and magnetic fields become

$$\nabla \times \vec{E} = -j\omega\mu\vec{H} \tag{3.6}$$

$$\nabla \times \vec{H} = j\omega\varepsilon\vec{E} \tag{3.7}$$

$$\nabla \cdot \vec{E} = 0 \tag{3.8}$$

$$\nabla \cdot \vec{H} = 0. \tag{3.9}$$

Taking the curl of Equation 3.6 and using the vector identity of $\nabla \times \nabla \times \vec{E} = \nabla\left(\nabla \cdot \vec{E}\right) - \nabla^2\vec{E}$, and using Equation 3.7, we have

$$\nabla^2\vec{E} + k^2\vec{E} = 0 \tag{3.10}$$

with the wave number

$$k = \omega\sqrt{\mu\varepsilon} = \omega\sqrt{\mu_0\varepsilon_0}\sqrt{\varepsilon_r} = k_0 m = \frac{2\pi}{\lambda_0}m. \tag{3.11}$$

In the uniform medium where the permittivity and permeability are constant, a plane wave is the solution of Equation 3.10, specifically:

$$\vec{E} = \hat{e}E_0 e^{-j\vec{k}\cdot\vec{r}} \tag{3.12}$$

and

$$\vec{H} = \hat{k} \times \hat{e}\frac{E_0}{\eta}e^{-j\vec{k}\cdot\vec{r}}, \tag{3.13}$$

where $\eta = \sqrt{\dfrac{\mu}{\varepsilon}}$ is the intrinsic impedance of the medium, $\hat{e}$ is the unit vector for polarization, $\vec{k} = k\hat{k}$ is the wave vector, and $\hat{k}$ is the unit vector of wave propagation (Figure 3.4).

In the case of a spherical wave, the electric field can be expressed by

$$\vec{E} = \hat{e}E_0 \frac{e^{-jkr}}{r} \tag{3.14}$$

and it is illustrated as in Figure 3.5, where the equal phase constitutes spherical surfaces.

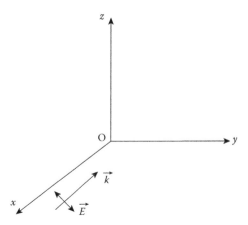

FIGURE 3.4 A sketch of a uniform plane wave propagation, where phases of the wave are equal on a plane that is perpendicular to the wave vector.

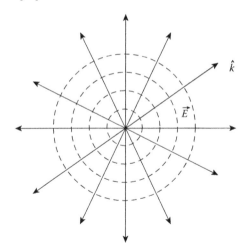

FIGURE 3.5 A sketch of a spherical wave propagation, where equal phase surfaces are perpendicular to the wave vector.

3.1.4 Wave Polarization and Representation

An EM wave is a transverse wave, and the electric field $\vec{E}$ varies sinusoidally with time (vibration) at a given location. The wave polarization is a description of the vibration direction, which is the locus of the tip of the $\vec{E}$ vector as time progresses.

Figure 3.6 shows examples of the polarization of an electric wave field from a coherent source or sources. If the locus of the $\vec{E}$ vector is a straight line, as shown in Figure 3.6a, the wave is said to be linearly polarized (or in linear polarization); if the loci form a circle as in Figure 3.6b, it is circularly polarized (circular polarization); and if the locus is an ellipse as in Figure 3.6c, it is elliptically polarized (elliptical polarization). If the locus is random, however, the wave is unpolarized. Sunlight or

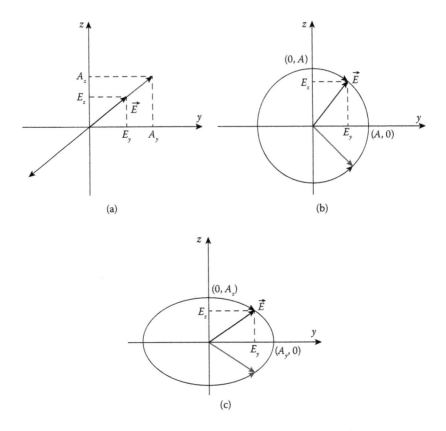

FIGURE 3.6 Polarization of an electric wave field. (a) Linear polarization, (b) circular polarization, and (c) elliptical polarization.

light reflected from a Lambertian surface (Born and Wolf 1999) is an example of unpolarized waves. A wave can also be partially polarized if it is from both coherent and incoherent sources.

A mathematical representation of a polarized wave can be made by describing the wave field $\vec{E}$ vector in the polarization plane. Assuming that a plane wave propagates along the x-axis, the $\vec{E}$ vector will be in the y–z plane:

$$\vec{E}(x,t) = E_y(x,t)\hat{y} + E_z(x,t)\hat{z} = \mathrm{Re}(\vec{E}e^{j\omega t}) \qquad (3.15)$$

and each component follows the time-harmonic solution (Equation 3.5), which is rewritten as follows:

$$E_y(x,t) = A_y\cos(\omega t - kx + \phi_{01}) \qquad (3.16)$$

$$E_z(x,t) = A_z\cos(\omega t - kx + \phi_{02}), \qquad (3.17)$$

where $(\omega t - kx)$ is the time–space variable phase term; A_y and A_z are the amplitudes for the y and z components, respectively and ϕ_{01} and ϕ_{02} are their corresponding initial phases.

Because the variable phase $(\omega t - kx)$ in Equations 3.16 and 3.17 is a common term, the different types of polarization, such as those illustrated in Figure 3.6, are then represented by Equations 3.15 through 3.17 with different amplitude ratios and phase differences as follows:

3.1.4.1 Linear Polarization

Let the phase difference be

$$\delta = \phi_{02} - \phi_{01} = 0 \text{ or } \pi. \tag{3.18}$$

Substituting Equations 3.16 and 3.17 into Equation 3.15 and using Equation 3.18, we obtain

$$\vec{E}(x,t) = \left(A_y \hat{y} \pm A_z \hat{z}\right) \cos\left(\omega t - kx + \phi_{01}\right), \tag{3.19}$$

which shows that the two components have the same phase term, and the locus is a straight line with a slope of A_z/A_y, as shown in Figure 3.6a. Hence, the wave represented by Equation 3.19 is described as being *linearly polarized*.

3.1.4.2 Circular Polarization

Let the phase difference be

$$\delta = \phi_{02} - \phi_{01} = \pm\frac{\pi}{2} \tag{3.20}$$

and the amplitude ratio is unity, specifically, $A_z/A_y = 1$. Substituting Equations 3.16 and 3.17 into Equation 3.15 and using these conditions, we have

$$\vec{E}(x,t) = A\cos\left(\omega t - kx + \phi_{01}\right)\hat{y} \pm A\sin\left(\omega t - kx + \phi_{01}\right)\hat{z}. \tag{3.21}$$

Equation 3.21 is clearly an equation for a circle in a 2D plane, hence representing the circular polarization illustrated in Figure 3.6b. Because the wave is coming out of the page (the x direction), $\delta = \pi/2$ represents left-hand circular polarization and $\delta = -\pi/2$ represents right-hand circular polarization.

3.1.4.3 Elliptical Polarization

The wave represented by the general equations of (3.11) and (3.12) without further conditions is elliptically polarized. Combining Equations 3.15 through 3.17 yields

$$\vec{E}(x,t) = A_y \cos(\omega t - kz + \phi_{01})\hat{y} + A_z \cos(\omega t - kz + \phi_{02})\hat{z}. \tag{3.22}$$

The tip of the vector forms an ellipse, which can also be represented by cancelling the varying phase term $(\omega t - kx)$ as

$$\frac{E_y^2}{A_y^2} + \frac{E_z^2}{A_z^2} - 2\frac{E_y E_z}{A_y A_z}\cos\delta = \sin^2\delta, \tag{3.23}$$

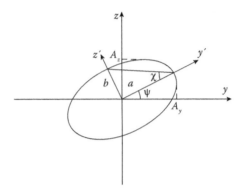

FIGURE 3.7 Representation of an elliptically polarized wave by the semimajor (a) and semiminor (b) axes of the ellipse and the orientation of ψ.

where $\delta = \phi_{02} - \phi_{01}$ is the phase difference. Three parameters (A_y, A_z, δ) describe an elliptically polarized wave field. The ellipse represented by Equation 3.23 is in general a tilted ellipse, which can also be described by the semimajor (a) and semiminor (b) axes of the ellipse and the orientation of ψ, as shown in Figure 3.7, where the ellipticity angle is defined as $\tan\chi = \pm\dfrac{b}{a}$ with the positive sign for left-hand polarization and the negative sign for right-hand polarization (Shen and Kong 1983).

Another way to represent an elliptically polarized wave is to use four parameters with the same dimension, which has been widely used in the remote sensing community but has not received much attention in the radar meteorology community. This parameterization was introduced by G. G. Stokes (1852a, 1852b) and is called the *Stokes parameters*:

$$I = |E_y|^2 + |E_z|^2 = A_y^2 + A_z^2 \tag{3.24}$$

$$Q = |E_y|^2 - |E_z|^2 = A_y^2 - A_z^2 = I \cos 2\chi \cos 2\psi \tag{3.25}$$

$$U = 2\,\mathrm{Re}(E_y^* E_z) = 2A_y A_z \cos\delta = I \cos 2\chi \sin 2\psi \tag{3.26}$$

$$V = 2\,\mathrm{Im}(E_y^* E_z) = 2A_y A_z \sin\delta = I \sin 2\chi. \tag{3.27}$$

Note that there is a relationship among the four parameters, which is

$$I^2 = Q^2 + U^2 + V^2 \tag{3.28}$$

for a fully polarized wave. Hence, the four-parameter Stokes notation of $[I,Q,U,V]$ is equivalent to the representation by the three parameters (A_y, A_z, δ) in this case.

In the case of a partially polarized or unpolarized wave, the amplitudes and the phase difference are random variables. Then, the definitions of the Stokes parameters

(Equations 3.24 through 3.27) are replaced by their respective averages. In this case, it can be shown that Equation 3.28 is replaced by

$$\langle I^2 \rangle > \langle Q^2 \rangle + \langle U^2 \rangle + \langle V^2 \rangle, \tag{3.29}$$

The ratio between the right-hand and left-hand sides, $p = (\langle Q^2 \rangle + \langle U^2 \rangle + \langle V^2 \rangle)/\langle I^2 \rangle$, is the degree of polarization. For elliptically polarized waves, $p = 1$; $0 < p < 1$ for partially polarized waves, and $p = 0$ for unpolarized waves. In weather radar polarimetry, we usually deal with a high degree of polarization (except for those from clutter or biological objects)—as shown in Chapter 4, the degree of polarization is closely related to the co-polar cross correlation coefficient (ρ_{hv}).

3.2 SCATTERING FUNDAMENTALS

Scattering is a physical process in which an object or objects, called *scatterers* (e.g., hydrometeors), redirect the wave incidence on them into all directions. Figure 3.8 shows a sketch of wave scattering by a single particle. The question is how to conceptually understand and quantitatively represent the scattering process by a single particle, which is discussed in this section.

3.2.1 SCATTERING AMPLITUDE, SCATTERING MATRIX, AND SCATTERING CROSS SECTIONS

When the wave is incident on the particle, a part of the wave power is absorbed by the particle, and the other part is scattered out to all directions. It is expected that the scattering and absorption properties depend on the wave properties and the physical and EM characteristics of the particle, as described in Chapter 2.

Let the particle have an arbitrary shape that will be specified later. Its EM properties are represented by permittivity ε and permeability μ as

$$\varepsilon(\vec{r}) = \varepsilon_r(\vec{r})\varepsilon_0 = (\varepsilon'(\vec{r}) - j\varepsilon''(\vec{r}))\varepsilon_0 \tag{3.30}$$

$$\mu(\vec{r}) = \mu_0, \tag{3.31}$$

where ε_0 and μ_0 are free space permittivity and permeability, respectively; ε_r is the relative dielectric constant with ε' as its real part and ε'' as its imaginary part.

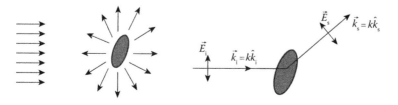

FIGURE 3.8 Conceptual sketch of wave scattering by a hydrometeor particle.

Assume that the incident wave is a linearly polarized plane wave propagating in free space with a permittivity and permeability of (ε_0, μ_0). Hence, the phasor representation of the plane wave is Equation 3.12, which is

$$\vec{E}_i = \hat{e}_i E_0 e^{-j\vec{k}_i \bullet \vec{r}}, \qquad (3.32)$$

where $\hat{e}_i$ is the unit vector for the incident wave polarization; E_0 is the amplitude; and $\vec{k}_i = k\hat{k}_i$ is the incident wave vector, with k as the wave number for the background medium and the unit vector $\hat{k}_i$ for the propagation direction.

With the incident wave impinging on the particle, charges inside the particle are excited and vibrate, yielding a reradiated wave or waves in all directions. If the observed point is far away from the particle, it is called the *far field* if it meets the condition: $r > 2D^2/\lambda$. In this case, the scattered field behaves as a spherical wave and can be represented by an equation similar to Equation 3.14:

$$\vec{E}_s = \vec{s}\left(\hat{k}_s, \hat{k}_i\right) E_0 \frac{e^{-jkr}}{r}, \qquad (3.33)$$

where $\vec{s}\left(\hat{k}_s, \hat{k}_i\right) = s\left(\hat{k}_s, \hat{k}_i\right)\hat{e}_s$ is the scattering amplitude, representing the amplitude, phase, and polarization of the scattered wave field for a unit plane wave incidence on the particle. The scattering amplitude is, in general, complex and contains both magnitude and phase information.

As discussed in Section 3.1.4, wave polarization is fully described by two orthogonal components. Assuming that $(\hat{e}_{i1}, \hat{e}_{i2})$ are the reference unit vectors for the incident wave field and $(\hat{e}_{s1}, \hat{e}_{s2})$ are the reference unit vectors for the scattered wave field, as shown in Figure 3.9, we have

$$\vec{E}_i = \hat{e}_i E_0 e^{-j\vec{k}_i \bullet \vec{r}} = \left(E_{i1}\hat{e}_{i1} + E_{i2}\hat{e}_{i2}\right)e^{-j\vec{k}_i \bullet \vec{r}} \qquad (3.34)$$

$$\vec{E}_s = \hat{e}_s E_{0s} \frac{e^{-jkr}}{r} = \left(E_{s1}\hat{e}_{s1} + E_{s2}\hat{e}_{s2}\right). \qquad (3.35)$$

The scattering equation (Equation 3.33) can then be written in matrix form as

$$\begin{bmatrix} E_{s1} \\ E_{s2} \end{bmatrix} = \frac{e^{-jkr}}{r} \begin{bmatrix} s_{11}\left(\hat{k}_s, \hat{k}_i\right) & s_{12}\left(\hat{k}_s, \hat{k}_i\right) \\ s_{21}\left(\hat{k}_s, \hat{k}_i\right) & s_{22}\left(\hat{k}_s, \hat{k}_i\right) \end{bmatrix} \begin{bmatrix} E_{i1} \\ E_{i2} \end{bmatrix}. \qquad (3.36)$$

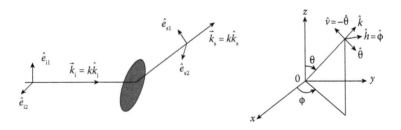

FIGURE 3.9 Coordinate systems for the scattering matrix of a hydrometeor particle.

The matrix $[S] = \begin{bmatrix} s_{11} & s_{12} \\ s_{21} & s_{22} \end{bmatrix}$ is called the *scattering matrix*, with its diagonal terms s_{11} and s_{22} representing the wave scattering for co-polar components and the off-diagonal terms s_{12} and s_{21} representing cross-polar scattering. There are alternate ways to choose the orthogonal unit vectors $(\hat{e}_1, \hat{e}_2)$ to define the scattering matrix—two commonly used methods are based on (i) the scattering plane and (ii) the horizontal and vertical directions.

The scattering plane is the plane that contains the incident and scattered wave vectors. The vector perpendicular to both the wave vector and the scattering plane is called the *perpendicular vector*, defined as $\hat{e}_i = \hat{e}_\perp = \dfrac{\hat{k}_s \times \hat{k}_i}{\left| \hat{k}_s \times \hat{k}_i \right|}$. The vectors on the scattering plane are defined as $\hat{e}_{i\parallel} = \hat{k}_i \times \hat{e}_{i\perp}$ and $\hat{e}_{s\parallel} = \hat{k}_s \times \hat{e}_{s\perp}$ for the incident and scattered waves, respectively. In this case, Equation 3.36 becomes

$$\begin{bmatrix} E_{s\perp} \\ E_{s\parallel} \end{bmatrix} = \frac{e^{-jkr}}{r} \begin{bmatrix} s_{\perp\perp}\left(\hat{k}_s, \hat{k}_i\right) & s_{\perp\parallel}\left(\hat{k}_s, \hat{k}_i\right) \\ s_{\parallel\perp}\left(\hat{k}_s, \hat{k}_i\right) & s_{\parallel\parallel}\left(\hat{k}_s, \hat{k}_i\right) \end{bmatrix} \begin{bmatrix} E_{i\perp} \\ E_{i\parallel} \end{bmatrix}. \tag{3.37}$$

The other coordinate system to define the scattering matrix is based on the horizontal and *vertical* (more precisely described as longitudinal) directions with

$$\hat{h} = \hat{\phi} = \frac{\hat{z} \times \hat{k}}{\left| \hat{z} \times \hat{k} \right|} \quad \text{and} \quad \hat{v} = -\hat{\theta} = \hat{k} \times \hat{h}, \tag{3.38}$$

where

$$\hat{k} = \sin\theta\cos\phi\hat{x} + \sin\theta\sin\phi\hat{y} + \cos\theta\hat{z} \tag{3.39}$$

$$\hat{h} = -\sin\phi\hat{x} + \cos\phi\hat{y} \tag{3.40}$$

$$\hat{v} = -\cos\theta\cos\phi\hat{x} - \cos\theta\sin\phi\hat{y} + \sin\theta\hat{z}. \tag{3.41}$$

Hence, the scattered wave field is related to the incident wave field by

$$\begin{bmatrix} E_{sh} \\ E_{sv} \end{bmatrix} = \frac{e^{-jkr}}{r} \begin{bmatrix} s_{hh}\left(\hat{k}_s, \hat{k}_i\right) & s_{hv}\left(\hat{k}_s, \hat{k}_i\right) \\ s_{vh}\left(\hat{k}_s, \hat{k}_i\right) & s_{vv}\left(\hat{k}_s, \hat{k}_i\right) \end{bmatrix} \begin{bmatrix} E_{ih} \\ E_{iv} \end{bmatrix} \tag{3.42}$$

and is generally elliptically polarized. The scattering matrix elements are discussed further in the following sections.

Whereas a wave field is represented by its complex scattering amplitude and scattering matrix, the wave power is characterized by the power flux density—the power passing through a unit area, which is the magnitude of the power density flux vector.

The power flux density is also called the *Poynting vector*. For a time-harmonic wave, we have the mean Poynting vectors

$$\vec{S}_i = \frac{1}{2}\text{Re}\left(\vec{E}_i \times \vec{H}_i^*\right) = \frac{|E_0|^2}{2\eta_0}\hat{k}_i \qquad (3.43)$$

for the incident wave and

$$\vec{S}_s = \frac{1}{2}\text{Re}\left(\vec{E}_s \times \vec{H}_s^*\right) = \frac{|E_s|^2}{2\eta_0}\hat{k}_s = \frac{\left|s(\hat{k}_s,\hat{k}_i)\right|^2}{r^2}\frac{|E_0|^2}{2\eta_0}\hat{k}_s \qquad (3.44)$$

for the scattered wave. The factor of 1/2 in Equations 3.43 and 3.44 accounts for the difference between the averaged power and the peak power for time-harmonic waves (Shen and Kong 1983).

Hence, the differential power dP_s for the scattered wave through a differential area $da = r^2 d\Omega$ is

$$dP_s = S_s da = \frac{\left|s\left(\hat{k}_s,\hat{k}_i\right)\right|^2}{r^2}\frac{|E_0|^2}{2\eta_0}r^2 d\Omega, \qquad (3.45)$$

where $d\Omega = \sin\theta d\theta d\phi$ is the differential solid angle. Using $S_i = \dfrac{|E_0|^2}{2\eta_0}$ in Equation 3.45, we have the scattered differential power normalized by the incident power flux density

$$\frac{dP_s}{S_i} = \left|s\left(\hat{k}_s,\hat{k}_i\right)\right|^2 d\Omega = \sigma_d\left(\hat{k}_s,\hat{k}_i\right)d\Omega, \qquad (3.46)$$

where $\sigma_d\left(\hat{k}_s,\hat{k}_i\right) = \left|s\left(\hat{k}_s,\hat{k}_i\right)\right|^2$ is called the *differential cross section* because it represents scattered wave power in a unit solid angle for an incident wave having a unit of power density and has a unit of area (section). In radar applications, a factor of 4π is multiplied to obtain a bistatic and backscattering radar cross section

$$\sigma_{bi}\left(\hat{k}_s,\hat{k}_i\right) = 4\pi\sigma_d\left(\hat{k}_s,\hat{k}_i\right), \quad \sigma_b = 4\pi\sigma_d\left(-\hat{k}_i,\hat{k}_i\right), \qquad (3.47)$$

so that it can be compared with the total scattered power in the 4π steradians of a solid angle.

The scattering cross section that represents power loss for the scattered power going to all the angles is the integral of Equation 3.46, specifically

$$\sigma_s = \int_{4\pi}\sigma_d\left(\hat{k}_s,\hat{k}_i\right)d\Omega. \qquad (3.48)$$

Besides the scattered power loss, another part of the wave power is absorbed by the particle and dissipated. This part can be calculated by the volume integral of the power loss inside the particle. Consider that the differential power is a product of different current and voltage:

$$dP = dI \times dV = (Jda) \times (Ed\ell) = \sigma E^2 dad\ell,$$ (3.49)

where the conductivity is $\sigma = \omega \varepsilon'' \varepsilon_0$. Taking the factor of $\dfrac{1}{2}$ into account for the difference between the mean and peak powers, the absorption power is normalized by the incident power flux density to obtain the absorption cross section as

$$\sigma_a = P_a/S_i = \int \frac{1}{2} \omega \varepsilon'' \varepsilon_0 \left| E_{int}(\vec{r}') \right|^2 d\vec{r}'/S_i.$$ (3.50)

The sum of the scattering and absorption represents the total power loss due to the scattering process; then the total cross section is the sum of the scattering cross section and absorption cross section:

$$\sigma_t = \sigma_s + \sigma_a = \frac{4\pi}{k} \text{Im}\left[s\left(\hat{k}_i, \hat{k}_i \right) \right],$$ (3.51)

where $s\left(\hat{k}_i, \hat{k}_i \right)$ is the forward scattering amplitude; $\sigma_t \equiv \sigma_e$ is also called the *extinction cross section*. The second equity of Equation 3.51 shows that the extinction cross section is proportional to the imaginary part of the forward scattering amplitude; it is known as the *optical theorem* (see Appendix 3A). The ratio between the scattering cross section (Equation 3.48) and the extinction cross section, $w = \sigma_s/\sigma_t$, is called *albedo*.

For convenience of comparison, the efficiency factors for extinction, scattering, and absorption are defined (Bohren and Huffman 1983) as the cross sections normalized as follows:

$$Q_t = \sigma_t/\sigma_g, \ Q_s = \sigma_s/\sigma_g, \ Q_a = \sigma_a/\sigma_g,$$ (3.52)

where σ_g is the geometric cross section, and $\sigma_g = \pi a^2$ for a sphere of radius a.

Figure 3.10 shows a plot of the efficiency factors for water spheres, specifically, the normalized cross sections as functions of the electric size (ka). When a particle is very small ($ka \ll 1$), the efficiency factors monotonically increase as the size increases. This is called the *Rayleigh scattering regime*. When a particle is about the size of the wavelength ($ka \sim 1$), the normalized cross sections oscillate, which is called *resonance* or the *Mie scattering regime*. When the particle size is very large compared with the wavelength ($ka \gg 1$), each of the efficiency factors approaches a constant, where the theory of geometric optics applies. It is interesting to note that, in the geometric optical regime, the total (or extinction) cross section is twice that of the geometric cross section for very large particles. This is contrary to our intuition in which an object blocks and makes extinct an area that exactly corresponds to its geometric cross section. This difference between what our intuition tells us and the wave theory is called the *extinction paradox* (Van De Hulst 1957). Next, we discuss the scattering calculations in the aforementioned regimes.

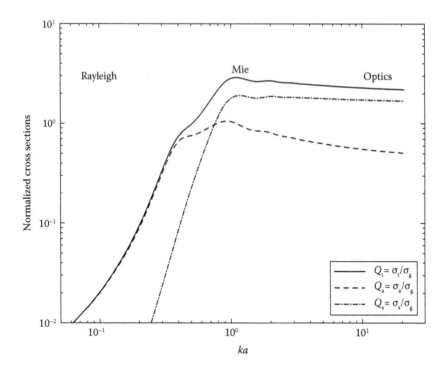

FIGURE 3.10 Normalized extinction cross section Q_t, absorption cross section Q_a, and total scattering cross section Q_s of water spheres.

3.3   RAYLEIGH SCATTERING

As mentioned earlier, wave scattering by a small particle whose size is much smaller than a wavelength is known as *Rayleigh scattering*. Rayleigh scattering is discussed based on the conceptual statement and mathematical representation that follow.

3.3.1   ORIGINAL STATEMENT

The law of Rayleigh scattering was initially developed by Rayleigh (1871) through dimension matching, which was cited by Bohren and Huffman (1983) in the following statement: "When light is scattered by particles which are very small compared with any of the wavelengths, the ratio of the amplitudes of the vibrations of the scattered and incident light varies inversely as the square of the wavelength and the intensity of the lights themselves as the inverse fourth power."

The law of Rayleigh scattering was derived by dimension matching for the amplitude ratio because it was suspected to be related to the physical parameters as follows:

$$\frac{A_s}{A_i} : V, r, \lambda, c, \rho_e, \tag{3.53}$$

where V is the volume of the particle, r is the range from the particle, λ is the wavelength, c is the velocity of the EM wave, and ρ_e is the density of the ether medium.[*] In Equation 3.53, the velocity c has a dimension of [L T^{-1}], but no other term contains the dimension of time [T]. Hence, because the amplitude ratio is dimensionless and there is no other term to cancel out the time dimension in c, c is not a component of $\frac{A_s}{A_i}$. Similarly, the ether density ρ_e contains the dimension of mass [M]. However, there is no other term that contains the dimension of mass; therefore, the dependence on ρ_e should be removed. This leaves V, r, and λ on the right-hand side of Equation 3.53. If the ratio between the scattered wave amplitude and the incident wave amplitude is proportional to the volume of particle V and is inversely proportional to the range r, the wavelength squared is needed in the denominator, yielding

$$\frac{A_s}{A_i} \propto \frac{V}{\lambda^2 r}. \tag{3.54}$$

Taking the square of both sides of Equation 3.54 gives an intensity ratio of

$$\frac{I_s}{I_i} \propto \frac{V^2}{\lambda^4 r^2}. \tag{3.55}$$

Hence, Equations 3.54 and 3.55 constitute the mathematical expression for the law of Rayleigh scattering: the intensity ratio is inversely proportional to the fourth power of the wavelength. This law provides, among other things, the explanation for why the sky is blue and the sun is orange: When looking to the sky, because the wave scattering is stronger for short wavelength (blue) light, the sky is perceived as blue; when looking directly at the sun, because the short wavelength light is removed from the direct line of sight, what remains the long wavelength light and the sun therefore appears orange to our eyes.

Although the law of Rayleigh scattering derived from dimension-matching was successful in explaining natural phenomena, there were unresolved issues such as what the angular and polarization dependences were, and what the absorption was. These issues were addressed through a more rigorous formulation of wave scattering based on the Maxwell equations.

3.3.2 SCATTERING AS DIPOLE RADIATION

When the particle is small compared with the wavelength, the Rayleigh scattering can be understood as a dipole radiation. As illustrated in Figure 3.11, when a wave is incident on a small spherical particle (represented by the dashed circle), an electric field is applied to the particle and causes positive charges to move to one end and negative charges to move to the other end. The wave then changes phases (i.e., the direction of the electric field changes), and the charges move in the opposite directions. This process continues, causing the particle to act as a dipole antenna and radiate an EM wave after it is excited by the incident wave (Ishimaru 1991).

---

[*] At that time, it was not yet understood that the EM wave can propagate in free space. It was rather thought to propagate within a medium called "ether," like a sound wave propagates in air.

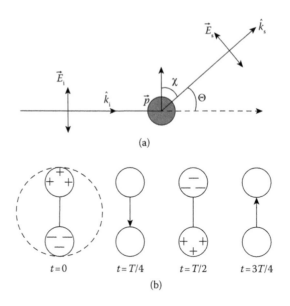

(a)

(b)

FIGURE 3.11 Illustration of wave scattering from a small sphere as a dipole radiation; (a) scattering configuration and (b) sketched process for dipole vibration and radiation.

Therefore, the scattered wave field can be represented by the vector potential $\vec{A}$ (Ishimaru 1978, 1997) as

$$\vec{E}_s\left(\vec{r}\right) = \frac{1}{j\omega\mu_0\varepsilon_0}\nabla \times \nabla \times \vec{A}\left(\vec{r}\right) \tag{3.56}$$

and

$$\vec{A} = \mu_0\int_V G\left(\vec{r},\vec{r}'\right)\vec{J}_{eq}\left(\vec{r}'\right)d\vec{r}', \tag{3.57}$$

where r is the observation point and r' is the source location inside the particle. The free space Green's function is

$$G\left(r,r'\right) = \exp\left(-jk\left|\vec{r} - \vec{r}'\right|\right)/\left(4\pi\left|\vec{r} - \vec{r}'\right|\right), \tag{3.58}$$

and the equivalent current source is

$$J_{eq}\left(\vec{r}'\right) = -j\omega\varepsilon_0\left[\varepsilon_r\left(\vec{r}'\right) - 1\right]\vec{E}_{int}, \tag{3.59}$$

which exists inside the particle only due to the excitation of the charges inside the particle.

The internal field can be represented by an incident wave field as follows (Stratton 1941, 205):

$$\vec{E}_{int} = \frac{3}{\varepsilon_r + 2}\vec{E}_i. \tag{3.60}$$

Substituting Equations 3.58 and 3.59 into Equations 3.56 and 3.57, assuming a unit wave of incidence, and using far-field approximation ($r \gg r'$), we have

$$\vec{E}_s(\vec{r}) = \frac{e^{-jkr}}{r} E_0 \vec{s}\left(\hat{k}_s, \hat{k}_i\right) \tag{3.61}$$

with the scattering amplitude as

$$\vec{s}\left(\hat{k}_s, \hat{k}_i\right) = \frac{k^2}{4\pi} \frac{3(\varepsilon_r - 1)}{\varepsilon_r + 2} V\left[-\hat{k}_s \times \left(\hat{k}_s \times \hat{e}_i\right)\right]$$

$$= k^2 a^3 \frac{\varepsilon_r - 1}{\varepsilon_r + 2} \sin \chi \hat{e}_s, \tag{3.62}$$

which represents the amplitude of the scattered wave field with the unit wave of incidence on a small sphere. As expected, the scattering amplitude depends on the particle's physical property, volume V, electric property ε_r, both the scattering direction and incident direction $\left(\hat{k}_s, \hat{k}_i\right)$, and the polarization direction of the incident wave field $\hat{e}_i$. It is noted that χ is the angle between the scattering direction $\hat{k}_s$ and the incident wave polarization $\hat{e}_i$, as shown in Figure 3.11. The $\sin \chi$ term can be understood as the unit vector of incident wave polarization $\hat{e}_i$ projected onto the scattered wave polarization of $\hat{e}_s$, which is perpendicular to the scattered wave propagation direction of $\hat{k}_s$, demonstrating the transverse nature of EM waves.

Whereas the scattering amplitude represents the property of the scattered wave field, the scattered wave power distribution is represented by the differential scattering cross section as

$$\sigma_d\left(\hat{k}_s, \hat{k}_i\right) = \left(k^2 a^3\right)^2 \left|\frac{\varepsilon_r - 1}{\varepsilon_r + 2}\right|^2 \sin^2 \chi. \tag{3.63}$$

The scattering field patterns and scattering power patterns are plotted in Figure 3.12. Whereas the scattering pattern appears in the $\sin \chi$ distribution in the electric field plane (E-plane), the scattering is isotropic in the magnetic field plane (H-plane). This is why a single polarization monostatic radar tends to use horizontal polarization, but a bistatic radar like BINET (bistatic receiver network) uses vertical polarization (Wurman et al. 1993).

Once the scattering amplitude (Equation 3.62) and the differential scattering cross section (Equation 3.63) are known, the total scattering cross section is the integral over the 4π steradians of a solid angle, yielding

$$\sigma_s = \int_{4\pi} \sigma_d(\hat{k}_s, \hat{k}_i) d\Omega$$

$$= k^4 a^6 \left|\frac{\varepsilon_r - 1}{\varepsilon_r + 2}\right|^2 \int_0^\pi \int_0^{2\pi} \sin^2 \chi \sin \chi \, d\chi \, d\phi = \frac{8\pi k^4 a^6}{3} \left|\frac{\varepsilon_r - 1}{\varepsilon_r + 2}\right|^2. \tag{3.64}$$

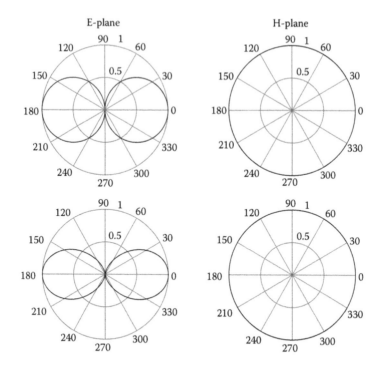

FIGURE 3.12 Normalized scattering field patterns (top row) and power pattern (bottom row). The left column is in the plane containing the incident wave polarization (E-plane), and the right column is the plane perpendicular to the incident wave polarization (H-plane).

The backscattering radar cross section is then

$$\sigma_b = 4\pi\sigma_d\left(-\hat{k}_i,\hat{k}_i\right) = 4\pi k^4 a^6 \left|\frac{\varepsilon_r - 1}{\varepsilon_r + 2}\right|^2. \tag{3.65}$$

Note that the backscattering radar cross section (Equation 3.65) is larger than the total scattering cross section (Equation 3.64). This is because the radar cross section is defined as 4π times the differential scattering cross section, for which both the maximum and the backscattering direction occur at the equator.

Following Equations 3.49 and 3.50, the absorption cross section is

$$\sigma_a = \frac{\int \frac{1}{2}\omega\varepsilon_0\varepsilon''\left|\vec{E}(\vec{r}')\right|^2 d\vec{r}'}{S_i} = \frac{4}{3}\pi a^3 k\varepsilon''\left|\frac{3}{\varepsilon_r + 2}\right|^2. \tag{3.66}$$

Whereas the scattering cross sections (Equations 3.64 and 3.65) are proportional to the volume squared ($\sim a^6 = (D/2)^6$), the absorption cross section (Equation 3.66) is proportional linearly to the volume ($a^3 = (D/2)^3$) of the particle, in the case of Rayleigh scattering. Because the internal wave field has been assumed to be constant in the derivations of the scattering amplitude (Equation 3.62) and cross sections (Equations 3.63 through 3.66), the derived formulas apply only to the electrically small scatterers/particles ($ka \ll 1$, i.e., $a \ll \lambda$). The valid regime for Rayleigh scattering can be seen more clearly in comparison with the Mie theory discussed in Section 3.4.

3.4 MIE SCATTERING THEORY

3.4.1 CONCEPTUAL DESCRIPTION

The exact solution for wave scattering by a sphere was developed in 1908 by Gustav Mie and is called the *Mie theory*. The Mie theory allows for the accurate calculation of wave scattering by a uniform sphere of any size and dielectric constant; these calculations have been well documented in multiple textbooks (Bohren and Huffman 1983; Kerker 1969). The results of the Mie scattering calculation also allow for a determination of the valid regime of the Rayleigh scattering approximation.

In the Rayleigh scattering approximation, the internal wave field is assumed to be constant and the wave scattering is treated as if it were radiation from a dipole antenna. By contrast, the Mie theory allows for the wave field to vary inside the sphere and for not only the dipole mode, but the quadrupole, hexapole, and higher-order pole modes as well (a dipole, a quadrupole, and a hexapole are illustrated in Figure 3.13).

In general, the wave scattering by a sphere is the superposition of all order-pole modes that are resonating in the scatterer, and hence its radiation field has a more complex pattern than that of Rayleigh scattering.

3.4.2 MATHEMATICAL EXPRESSION AND SAMPLE RESULTS

The purpose of the mathematical representation for Mie scattering is to solve the EM boundary problem. That is, under the wave incidence of $\vec{E}_i$, the internal wave field inside the sphere is $\vec{E}_{int}$, and the wave field outside is the summation of the incident wave field $\vec{E}_i$ and the scattered wave field $\vec{E}_s$, specifically $\vec{E}_i + \vec{E}_s$. The first and second Maxwell equations require that the tangential components of the electric and magnetic fields must be continuous. Let us use a spherical coordinate system (r, θ, ϕ) with the sphere center as the origin. We have

$$\hat{r} \times \vec{E}_{int}\Big|_{r=a} = \hat{r} \times \left[\vec{E}_i + \vec{E}_s \right]_{r=a} \tag{3.67}$$

$$\hat{r} \times \vec{H}_{int}\Big|_{r=a} = \hat{r} \times \left[\vec{H}_i + \vec{H}_s \right]_{r=a}. \tag{3.68}$$

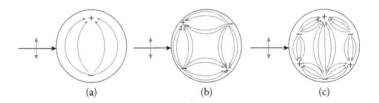

(a)   (b)   (c)

FIGURE 3.13 Conceptual sketch of the electric field inside a scatterer excited by the incident wave field as dipole (a), quadrupole (b), and hexapole (c) radiation.

Each part of Equations 3.67 and 3.68 represents two equations if it is expressed in component form in the θ and ϕ directions, totaling a set of four equations.

Because the wave fields can be expanded in vector spherical harmonics, the incident, scattered, and internal wave fields are expressed as

$$\vec{E}_i(\vec{r}) = \sum_{n=1}^{\infty} C_n \left[\vec{M}_{o1n}^{(1)}(kr,\theta,\phi) + j\vec{N}_{e1n}^{(1)}(kr,\theta,\phi) \right] \tag{3.69}$$

$$\vec{E}_s(\vec{r}) = \sum_{n=1}^{\infty} C_n \left[-b_n \vec{M}_{o1n}^{(4)}(kr,\theta,\phi) - ja_n \vec{N}_{e1n}^{(4)}(kr,\theta,\phi) \right] \tag{3.70}$$

$$\vec{E}_{int}(\vec{r}) = \sum_{n=1}^{\infty} C_n \left[c_n \vec{M}_{o1n}^{(1)}(k'r,\theta,\phi) + d_n \vec{N}_{e1n}^{(1)}(k'r,\theta,\phi) \right], \tag{3.71}$$

where $C_n = (-j)^n E_0 (2n+1)/[n(n+1)]$ are the expansion coefficients for the incident field, and (a_n, b_n) and (c_n, d_n) are those for the scattered and internal wave fields, respectively. $\vec{M}_{mn}$ and $\vec{N}_{mn}$ are the vector spherical harmonics (see their expressions in Appendix 3B), and the subscripts e and o indicate the even and odd modes, respectively.

Using Equations 3.69 through 3.71 and their derived expressions for magnetic fields from Equations 3.67 and 3.68 and solving the equations yields the scattering coefficients:

$$a_n = \frac{\psi_n(ka)\psi_n'(kma) - m\psi_n(kma)\psi_n'(ka)}{\zeta_n(ka)\psi_n'(kma) - m\psi_n(kma)\zeta_n'(ka)} \tag{3.72}$$

$$b_n = \frac{m\psi_n(ka)\psi_n'(kma) - \psi_n(kma)\psi_n'(ka)}{m\zeta_n(ka)\psi_n'(kma) - \psi_n(kma)\zeta_n'(ka)}, \tag{3.73}$$

where $\psi_n(x) = xj_n(x) = \sqrt{\pi x / 2} J_{n+1/2}(x)$ is the Riccati–Bessel function, and $j_n(x)$ is the spherical Bessel function. Then, the scattered wave fields are

$$\begin{bmatrix} E_{s\perp} \\ E_{s\parallel} \end{bmatrix} = \frac{e^{-jkr}}{r} \begin{bmatrix} s_{\perp\perp}(\theta) & 0 \\ 0 & s_{\parallel\parallel}(\theta) \end{bmatrix} \begin{bmatrix} E_{i\perp} \\ E_{i\parallel} \end{bmatrix}, \tag{3.74}$$

where

$$s_{\perp\perp}(\theta) = \frac{1}{jk} \sum_{n=1}^{\infty} \frac{2n+1}{n(n+1)} \left[a_n \pi_n(\cos\theta) + b_n \tau_n(\cos\theta) \right], \tag{3.75}$$

$$s_{\parallel\parallel}(\theta) = \frac{1}{-jk} \sum_{n=1}^{\infty} \frac{2n+1}{n(n+1)} \left[a_n \tau_n(\cos\theta) + b_n \pi_n(\cos\theta) \right], \tag{3.76}$$

and

$$\pi_n(\cos\theta) = \frac{P_n^1(\cos\theta)}{\sin\theta} \text{ and } \tau_n(\cos\theta) = \frac{d}{d\theta} P_n^1(\cos\theta) \tag{3.77}$$

with $P_n^1(\cos\theta) = \dfrac{d}{d\theta} P_n(\cos\theta)$ and $P_n(\cos\theta)$ as the nth order Legendre polynomial.

The sample calculation results for the amplitude patterns at the X-band of two water spheres with diameters of 2 mm (top row) and 2 cm (bottom row) are shown in Figure 3.14. It is apparent that the pattern for the sphere with a diameter of 2 mm is stronger in the forward direction than the pattern in the backward direction (the pattern magnitude is ~10% different between forward and backward scattering), but is somewhat similar to those patterns of Rayleigh scattering shown in Figure 3.12. (For hydrometeors of diameters larger than 2 mm, the Rayleigh scattering approximation starts to become invalid.) As is evident, for the sphere with a diameter of 2 cm, the forward scattering pattern is dominant and is of an order larger than that of backscattering. Note also that the backscattering by the 2-cm sphere is still two orders larger than the backscattering by the 2-mm sphere, although the main scattered energy is in the forward direction. This is because the total scattering cross section increases by many orders when the sphere increases in size from 2 mm to 2 cm.

Using the scattering amplitude (Equations 3.75 and 3.76) in the expressions in Equations 3.47, 3.48, and 3.51, we have the cross sections for Mie scattering as follows:

The radar backscattering cross section is

$$\sigma_b = \frac{\pi}{k^2}\left|\sum_{n=1}^{\infty}(2n+1)(-1)^n\left(a_n - b_n\right)\right|^2, \tag{3.78}$$

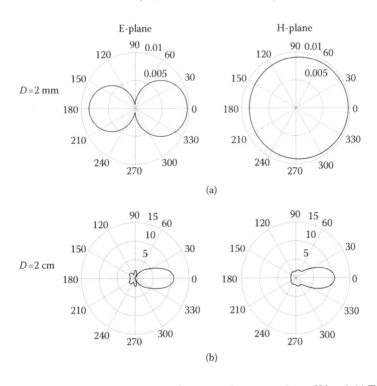

(a)

(b)

FIGURE 3.14 Sample amplitude patterns for water sphere scattering at X-band: (a) Top row for a sphere with a 2-mm diameter and (b) bottom row for a sphere with a 2-cm diameter.

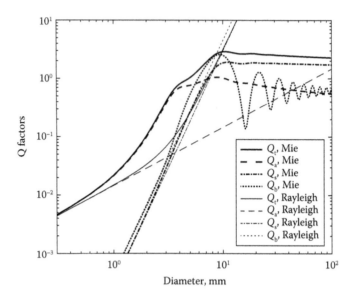

FIGURE 3.15 Dependence of efficiency factors on water sphere diameter at X-band.

The total scattering cross section is

$$\sigma_s = \frac{2\pi}{k^2} \sum_{n=1}^{\infty} (2n+1)\left(|a_n|^2 + |b_n|^2\right) \tag{3.79}$$

and the extinction cross section is

$$\sigma_t = \frac{4\pi}{k} \text{Im}\left(\frac{s(0)}{jk}\right) = \frac{2\pi}{k^2} \sum_{n=1}^{\infty} (2n+1)\text{Re}[a_n + b_n] \tag{3.80}$$

Now, we have cross sections from both the Rayleigh scattering approximation given in Equations 3.64 through 3.66 and the rigorous calculation given by the Mie theory (Equations 3.78 through 3.80). The efficiency factors, cross sections normalized by the geometric cross section, are calculated using the two aforementioned approaches and compared in Figure 3.15.

The calculations are performed for water spheres at X-band (wavelength: $\lambda = 3$ cm) with a dielectric constant of $(44 - j43)$. It is evident that the scattering efficiency factors of Rayleigh scattering and those of Mie theory agree well up to a diameter of 2 mm. That means Rayleigh scattering approximation is valid for a sphere whose diameter $D < \lambda/16$, which is equivalent to $2kD < \pi/4$, meaning that the wave field inside the particle can be treated as a constant. However, the results of absorption and extinction start to differ at smaller sizes, $D < 1$ mm. Typically, there is a more stringent requirement for Rayleigh scattering to be valid, specifically $D < \lambda/50$. This value is close to $k(m' - 1)D < \pi/4$ (with $m' = 7$), which requires that the phase difference between a wave passing through the particle and a wave in free space is small.

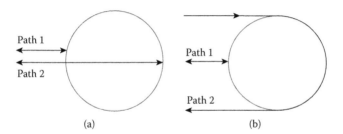

FIGURE 3.16 Conceptual sketches for resonance effect: (a) reflection model and (b) reflection and creeping wave model.

At the other end of the extreme, when a sphere is much larger than the wavelength ($D > 100\lambda$), geometric optics is a good approximation. However, the extinction cross section is twice as large as the geometric cross section, which is counterintuitive. This unexpected result is called the *extinction paradox*, as noted in Section 3.2. The extinction paradox can be resolved if we note that (i) anything removed from the forward direction is considered part of scattering and (ii) the extinction cross section is defined in the far field, and any scatterer, no matter how big it is, has an edge that can alter wave propagation and can cause diffraction that geometric optics does not take into account. Hence, there is a difference between our intuition and the full wave theory.

As for the resonance effect of the up and down changes in the normalized backscattering cross section, the intuitive explanation is constructive and destructive summation of scattered waves going through different paths. Figure 3.16 shows two conceptual models: (a) the reflection model and (b) the reflection and creeping wave model. The reflection model is shown in Figure 3.16a: one wave is reflected back from the front edge, the other penetrates into the sphere and is then reflected from the back edge. In Figure 3.16b, Path 1 is the same as that in Figure 3.16a, but the second wave propagates around the half-sphere and then comes out in the opposite direction.

In either case, we can expect a maximum if the two waves are in phase and constructively add, meaning the path difference is an integer of the wavelength, but a minimum if the difference is an odd number of a half of a wavelength. Using the reflection and creeping wave model in Figure 3.16b, the first maximum that satisfies the condition of $k(\ell_2 - \ell_1) = k(D + \pi D/2) = 2\pi$, results in $D = 0.39\lambda$. In the case of the result shown in Figure 3.15 for an X-band with $\lambda = 3$ cm, we have the maximum at $D = 1.17$ cm, which is close to that of ~1.0 cm, shown in the figure.

3.5 SCATTERING CALCULATIONS FOR A NONSPHERICAL PARTICLE

As discussed earlier, backscattering by a homogenous sphere exhibits no difference in polarization, meaning that the backscattering amplitudes and the radar cross sections are the same for horizontally and vertically polarized waves. However, most hydrometeors are not spherical, meaning that they give different radar returns and allow for additional information from polarimetric radar measurements. Wave scattering by a single nonspherical particle is fundamental to understanding and interpreting the polarimetric measurements. Shape and orientation are the two key factors that determine wave-scattering characteristics. Once the shape and orientation of a scatterer are determined, the scattering amplitudes can be calculated using the Rayleigh scattering approximation, T-matrix method, or other numerical approach, depending on its electric size and the complexity of its shape.

3.5.1 Basic Nonspherical Shape: Spheroid

As described in Chapter 2, there are a variety of shapes for hydrometeors: from a spherical shape for cloud droplets and small rain drops, to an oblate spheroidal shape with a flattened base for large raindrops, to an irregular shape for snowflakes and hailstones. Furthermore, an ice crystal in a snowflake can be in the shape of a needle, disk, plate, cylinder, dendrite, or a combination of these shapes. It is difficult to rigorously model the irregular shape of snowflakes and hailstones with a few parameters. Such rigorous modeling is unnecessary in radar polarimetry because there are, after all, only a few types of radar measurements. These measurements can capture the main statistical properties of the hydrometeors without requiring detailed structural information about the hydrometeors. Hence, a simple spheroidal shape model, as long as it is not based on a perfect sphere, is usually sufficient for interpreting polarimetric radar data. This is one reason the spheroid model is often used. Another reason that the spheroid is used to model hydrometeors is because it is one of the basic shapes for which the EM scattering problem can be analytically solved for small scatterers and numerically solved (with relatively low computation costs) for large ($D \sim \lambda$) scatterers.

A spheroid is a special case of an ellipsoid; its two semidiameters are equal, which can be obtained by rotating an ellipse around its principal z-axis. When the ellipse rotates around its minor axis, we obtain an oblate spheroid ($a_1 = a_2 > b$); when the ellipse rotates around its major axis, we have a prolate spheroid ($a_1 = a_2 < b$). Both are illustrated in Figure 3.17.

Oblate spheroids are used to model raindrops, snowflakes, hailstones, and ice crystal plates and dendrites, whereas prolate spheroids can be used to represent needles and columns of ice crystals.

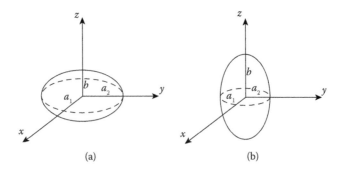

(a)   (b)

FIGURE 3.17 Basic shape models for hydrometeor scattering: (a) oblate spheroid; (b) prolate spheroid.

3.5.2 RAYLEIGH SCATTERING APPROXIMATION FOR SPHEROIDS

The mathematical representation of Rayleigh scattering by a sphere was provided in Section 3.3.2. The expression of scattering amplitude provided there (Equation 3.62) can be rewritten as

$$
\vec{s}\left(\hat{k}_s,\hat{k}_i\right) = \frac{k^2}{4\pi\varepsilon_0}\int_V\left\{-\hat{k}_s\times\left[\hat{k}_s\times\vec{P}(\vec{r}')\right]\right\}e^{j\hat{k}_s\cdot\vec{r}'}d\vec{r}'
$$

$$
\approx\frac{k^2V(\varepsilon_r-1)}{4\pi}\left[-\hat{k}_s\times\left(\hat{k}_s\times\vec{E}_{int}\right)\right],
$$

(3.81)

where the internal field is

$$
\vec{E}_{int} = \vec{E}_i + \vec{E}_p.
$$

(3.82)

Although the internal field for a sphere can be easily represented by the incident wave field (Equation 3.60), the internal field for an ellipsoid (spheroid) is more difficult to find because it depends on the incident wave polarization and the particle's orientation. However, the polarized wave field for each component can be expressed by the following (Stratton 1941):

$$
E_p = \int_s\frac{dq}{4\pi\varepsilon_0 r^2}(-\hat{r}\cdot\hat{p}) = -\frac{L}{\varepsilon_0}P = -L(\varepsilon_r-1)E_{int}.
$$

(3.83)

Combining Equations 3.82 and 3.83 and solving for the internal field for each polarization component on its symmetry axis, we obtain the following (Van De Hulst 1957):

$$
E_{int\,x} = \frac{E_{ix}}{1+L_x(\varepsilon_r-1)}; E_{int\,y} = \frac{E_{iy}}{1+L_y(\varepsilon_r-1)}; \text{ and } E_{int\,z} = \frac{E_{iz}}{1+L_z(\varepsilon_r-1)},
$$

(3.84)

where

$$
L_x = \int_0^\infty\frac{a_1 a_2 b}{2(s+a_1^2)\left[(s+a_1^2)(s+a_2^2)(s+b^2)\right]^{1/2}}ds,
$$

(3.85)

$$
L_y = \int_0^\infty\frac{a_1 a_2 b}{2(s+a_2^2)\left[(s+a_1^2)(s+a_2^2)(s+b^2)\right]^{1/2}}ds,
$$

(3.86)

$$
L_z = \int_0^\infty\frac{a_1 a_2 b}{2(s+b^2)\left[(s+a_1^2)(s+a_2^2)(s+b^2)\right]^{1/2}}ds
$$

(3.87)

are factors that depend on the shape of the scatterer. It can be shown (Stratton 1941) that the shape factors follow an equality of

$$
L_x + L_y + L_z = 1.
$$

(3.88)

In general, the shape factors are inversely proportional to the corresponding dimension. If the scatterer is sufficiently spheroidal, that is, the axis ratio does not differ from the unity too much ($0.5 < b/a < 2$), an approximate relation exists

$$L_x : L_y : L_z \approx 1/a_1 : 1/a_2 : 1/b. \tag{3.89}$$

In the case of spheroids, the shape factors can be calculated by using Equation 3.88 and performing the integral of Equation 3.87 with ($a_1 = a_2 = a$), giving

$$L_x = L_y = \frac{1}{2}(1 - L_z) \tag{3.90}$$

and $$L_z = \frac{1+g^2}{g^2}\left(1 - \frac{1}{g}\arctan g\right) \text{ and } g^2 = \left(\frac{a}{b}\right)^2 - 1 = \frac{1}{\gamma^2} - 1 \tag{3.91}$$

for an oblate spheroid ($a > b$) and

$$L_z = \frac{1-e^2}{e^2}\left(-1 + \frac{1}{2e}\ln\frac{1+e}{1-e}\right) \text{ and } e^2 = 1 - \left(\frac{a}{b}\right)^2 \tag{3.92}$$

for a prolate spheroid. Once the shape factors (L) and the dielectric constant are known, the scattering amplitude can be calculated. Consider a wave incident on a spheroid in the x-direction: the scattering amplitudes in the scattering plane for wave polarization aligned at the major and minor axes are obtained by substituting Equation 3.84 into Equations 3.81 through 3.83 and letting $L_y = L_a$ and $L_z = L_b$ as follows:

$$\vec{s}_{a,b}\left(\hat{k}_s, \hat{k}_i\right) = k^2 a^2 b \frac{\varepsilon_r - 1}{3[1 + L_{a,b}(\varepsilon_r - 1)]}\sin\chi\hat{e}_s. \tag{3.93}$$

It is evident that the scattering amplitude of Equation 3.93 for a spheroid is similar to the scattering amplitude of Equation 3.62 for a sphere. The normalized amplitude patterns are the same with the term $\sin\chi$ as shown in Figure 3.11. The magnitude is different with $\dfrac{\varepsilon_r - 1}{3[1 + L_{a,b}(\varepsilon_r - 1)]}$, instead of $\dfrac{\varepsilon_r - 1}{\varepsilon_r + 2}$, which depends on the polarization. The backward and forward scattering amplitudes are equal.

Using the axis ratio of Equation 2.16 for raindrops (Equations 3.90 and 3.91), the forward $\left(\hat{k}_s = \hat{k}_i; \Theta = 0\right)$ or backward $\left(\hat{k}_s = -\hat{k}_i; \Theta = \pi\right)$ scattering amplitudes for polarizations on the major and minor axes are plotted in Figure 3.18. The effective shapes of a few raindrops are also shown in the figure. As expected, the scattering amplitude with polarization on the major axis (s_a: solid line) is larger than for polarization on the minor axis (s_b: dashed line); the larger dimension of the scatterer causes stronger scattering because a larger dipole moment is formed. The difference in the scattering amplitudes between the two polarizations increases as raindrops become more oblate with size increases. The Rayleigh scattering approximation for spheroids is valid for clouds, rain, and dry snow at the S-band, but is not applicable for hail and melting snow, which require more accurate calculations for scattering amplitudes. At C-band and X-band or higher frequencies, wave scattering from all precipitation particles require more accurate calculation than is provided by Rayleigh scattering approximations.

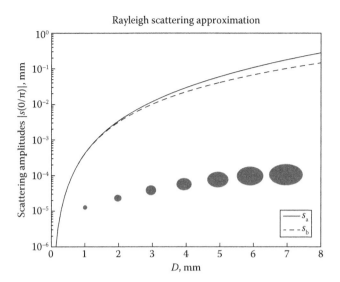

FIGURE 3.18 Scattering amplitudes of raindrops as a function of equivolume diameter for polarization at major (s_a) and minor (s_b) axes, respectively.

3.5.3 T-Matrix Method

When a hydrometeor's size is comparable to its wavelength, the Rayleigh scattering approximation introduced above becomes invalid, and an analytical solution does not exist. There is thus a need for a rigorous method to numerically calculate wave scattering. The T-matrix method is a numerical method that has been successfully developed and widely used in the radar meteorology community (Barber and Yeh 1975; Seliga and Bringi 1976; Vivekanandan et al. 1991; Waterman 1965). The idea of the T-matrix method is (i) to expand the incident, scattered, and internal wave fields in terms of their vector spherical harmonics and (ii) to use extended boundary conditions to determine the expansion coefficients through a transition matrix, which is briefly described in the following.

Wave scattering by an irregularly shaped particle is illustrated in Figure 3.19. Under the wave incidence $\vec{E}_i(\vec{r})$, there is the scattered wave field $\vec{E}_s(\vec{r})$ outside the particle and the internal wave field $\vec{E}_{int}(\vec{r})$ inside the particle. These wave fields are expanded in vector spherical harmonics

$$\vec{E}_s(\vec{r}) = \sum_{m,n}\left[a_{mn}\vec{M}_{mn}^{(4)}(kr,\theta,\phi) + b_{mn}\vec{N}_{mn}^{(4)}(kr,\theta,\phi)\right] \qquad (3.94)$$

$$\vec{E}_{int}(\vec{r}) = \sum_{m,n}\left[c_{mn}\vec{M}_{mn}^{(1)}(kr,\theta,\phi) + d_{mn}\vec{N}_{mn}^{(1)}(kr,\theta,\phi)\right] \qquad (3.95)$$

$$\vec{E}_i(\vec{r}) = \sum_{m,n}\left[e_{mn}\vec{M}_{mn}^{(1)}(k'r,\theta,\phi) + f_{mn}\vec{N}_{mn}^{(1)}(k'r,\theta,\phi)\right]. \qquad (3.96)$$

Because the incident wave is known, the expansion coefficients (e_{mn}, f_{mn}) are also known. That leaves four sets of expansion coefficients, including the scattering

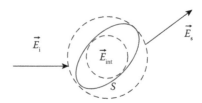

FIGURE 3.19 Illustration of wave scattering by a nonspherical scatterer and the concept of extended boundary conditions.

coefficients (a_{mn}, b_{mn}) and the internal coefficients (c_{mn}, d_{mn}), to be determined using the boundary conditions. This problem is similar to the problem set up by the Mie theory, but is more difficult to solve because we cannot apply the continuity conditions at $r = $ constant as we could in the case of a sphere, which complicates the angular dependence of the wave fields.

To address the difficulty in solving the boundary problem, the concept of extended boundary conditions was introduced by Waterman (1965, 1969), and the extinction theorem and Huygens principle are used. As illustrated in Figure 3.19, two spheres are drawn in dashed lines: the inner sphere and the outer sphere circumscribing the scatterer. Inside the inner sphere, the scattered wave field cancels out the incident wave field, which is written as

$$\vec{E}_i(\vec{r}) + \int_S d\bar{a}\left[i\omega\mu\hat{n} \times \vec{H}_{int}(\vec{r}') \cdot \bar{\bar{G}}(\vec{r},\vec{r}') + \hat{n} \times \vec{E}_{int}(\vec{r}') \cdot \nabla \times \bar{\bar{G}}(\vec{r},\vec{r}')\right] = 0, \quad (3.97)$$

where $\bar{\bar{G}}(\vec{r},\vec{r}')$ is the dyadic Green's function that can be represented by the vector spherical harmonics. Substitution of Equations 3.95 and 3.96 into Equation 3.97 yields

$$\begin{bmatrix} e_{mn} \\ f_{mn} \end{bmatrix} = [A]\begin{bmatrix} c_{mn} \\ d_{mn} \end{bmatrix}, \quad (3.98)$$

which links the expansion coefficients between the incident wave field and the internal field.

Outside the scatter, by contrast, the scattered wave field is caused by the surface wave field. From the Huygens principle, we have

$$\vec{E}_s(\vec{r}) = \int_S d\bar{a}\left[i\omega\mu\hat{n} \times \vec{H}_{int}(\vec{r}') \cdot \bar{\bar{G}}(\vec{r},\vec{r}') + \hat{n} \times \vec{E}_{int}(\vec{r}') \cdot \nabla \times \bar{\bar{G}}(\vec{r},\vec{r}')\right]. \quad (3.99)$$

Using Equations 3.94 and 3.95 in Equation 3.99, we obtain

$$\begin{bmatrix} a_{mn} \\ b_{mn} \end{bmatrix} = [B]\begin{bmatrix} c_{mn} \\ d_{mn} \end{bmatrix}. \quad (3.100)$$

Solving (c_{mn}, d_{mn}) from Equation 3.98 and substituting them into Equation 3.100, we have

$$\begin{bmatrix} a_{mn} \\ b_{mn} \end{bmatrix} = [B][A]^{-1}\begin{bmatrix} e_{mn} \\ f_{mn} \end{bmatrix} = [T]\begin{bmatrix} e_{mn} \\ f_{mn} \end{bmatrix}, \quad (3.101)$$

where $[T] = [B][A]^{-1}$ is the transition matrix that relates the coefficients of the scattered wave field to the coefficients of the incident wave field. It is called the *T-matrix*, and this method is known as the *T-matrix method*. Once the coefficients (a_{mn}, b_{mn}) are known, the scattered wave field and the scattering amplitude/matrix can be found.

Figure 3.20 shows the T-matrix calculations for backscattering amplitudes in magnitude (left column) and phase (right column) for wave scattering by spheroid

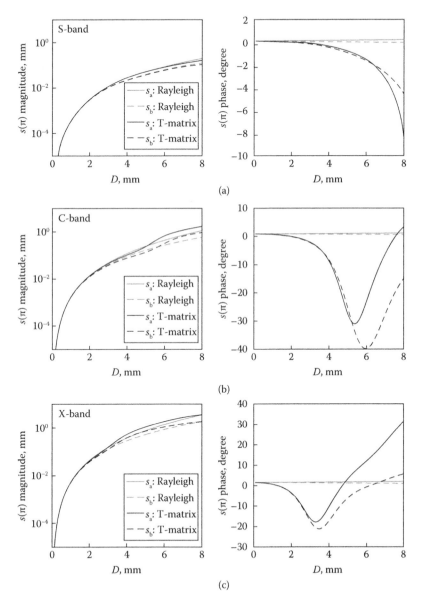

FIGURE 3.20 Magnitudes and phases of scattering amplitudes as a function of raindrop size at S-band (a), C-band (b), and X-band (c) frequencies.

raindrops at different frequencies, as compared with the backscattering amplitudes obtained with the Rayleigh scattering approximation. A temperature of 10°C was used for the calculation. At S-band, the T-matrix-calculated magnitudes agree well with their corresponding Rayleigh approximation results until the drop diameter increases to 6 mm. The phases of the scattering amplitudes are very small. At C-band, however, the results between the T-matrix calculations and the Rayleigh approximation start to differ at diameters of approximately 3 mm, because the electric size (ka) at C-band is double that at S-band. The T-matrix calculations show the resonance effect and substantial phase for the scattering amplitudes. At X-band, as the wavelength becomes even shorter, the resonance effect and scattering phase appear at an even smaller sizes ($D \sim 2$ mm); Rayleigh approximation then becomes invalid even for median-sized raindrops. The forward scattering amplitudes have similar properties; interested readers can download the T-matrix results and compare their own plots.

Figure 3.21 shows a summary of the results given in Figure 3.20 by plotting the normalized backscattering cross section (upper left), backscattering magnitude ratios

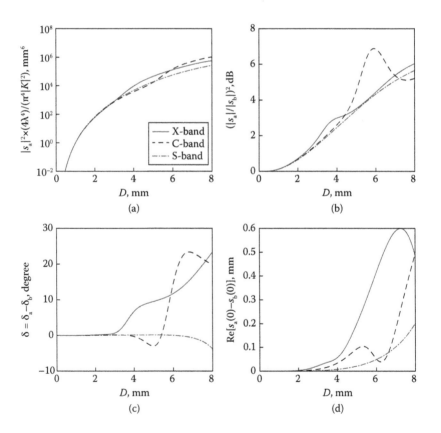

FIGURE 3.21 Normalized backscattering cross section (a), backscattering magnitude ratios (b), backscattering phase differences (c), and the real part of the forward scattering amplitudes (d) as a function of raindrop size at S-band, C-band, and X-band frequencies.

(upper right), backscattering phase differences (bottom left), and the real part of the forward scattering amplitudes (bottom right). Chapter 4 shows that the normalized backscattering cross section and magnitude ratio are the reflectivity factor and differential reflectivity, respectively, for monodispersion drop size distributions. There is a peak in the C-band magnitude ratio, which is caused by the resonance effect. In general, the scattering phase difference increases as drop size and/or frequency increases. The phase difference is a main factor that causes signal decorrelation (ρ_{hv}) between the dual-polarizations. The real part of the forward scattering amplitudes is associated with the specific differential phase (K_{DP}). These are described in Chapter 4.

The scattering amplitudes of snowflakes and hailstones were also calculated using the T-matrix method, and the results are shown in Figure 3.22. Both snowflakes and hailstones are assumed to be oblate spheroids with an axis ratio of $\gamma = 0.75$. The difference between the snow and hail modeling is the density: the density of snowflakes is assumed to be 0.1 g/cm³, whereas hailstones have a density of 0.92 g/cm³. The cases for both dry and wet snow/hail are calculated. For wet snow/hail, 20%

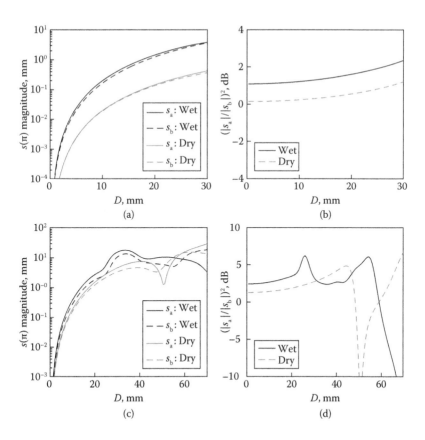

FIGURE 3.22 Magnitudes (left column) and ratios (right column) of scattering amplitudes as a function of particle size at S-band for snow (top row) and hail (bottom row).

melting is used. A melting hailstone can be treated as a water-coated ice (two-layer) particle. At low frequencies, it can be shown that the water-coated ice is equivalent to an effective particle of water–ice mixture from the Maxwell-Garnett formula with water as background (see Problem 3.4). As in Figure 3.20, the left column shows the magnitudes of the backscattering amplitudes with polarizations aligned with the major and minor axes. A substantial difference is evident in the scattering amplitude between the wet and dry cases: there is almost an order of difference in the magnitudes between wet and dry snow. The difference between the polarizations in the major and minor axes is very small for dry snow. To better show the polarization difference, the right column contains the ratio of the scattering amplitude magnitudes squared in decibels. It is clear that the ratio between the two polarizations is much larger for wet snow than that for dry snow.

The scattering characteristics of hail are more complicated: there is a greater difference between the two polarizations, but a smaller difference between the wet and dry cases. The resonance effect is very pronounced for hail with a diameter greater than 2 cm, which could yield backscattering for the polarization on the major axis that is smaller than that on the minor axis. At higher frequencies (C- and X-bands), these differences are even greater.

3.5.4 Other Numerical Methods for Scattering Calculations

Besides T-matrix methods, other numerical methods have also been developed for scattering calculations. These include physical optics (Born and Wolf 1999), method of moment (Harrington 1968), and discrete dipole approximation (Goodman et al. 1991; Purcell and Pennypacker 1973). The numerical methods can be used to calculate wave scattering by irregularly shaped objects such as terrain, vegetation, and biological objects (Zhang 1998; Zhang et al. 1996).

3.6 SCATTERING FOR ARBITRARY ORIENTATIONS

The scattering amplitudes s_a and s_b represent wave scattering for polarizations in the major and minor axes of a spheroid (Figure 3.23). Natural hydrometeors, however, occur in random orientations, and their major and minor axes don't necessarily align with the radar polarization base directions (typically horizontal and vertical). Whereas raindrops fall with their major axis aligned mostly horizontally, hailstones tumble as they fall, yielding random orientations. Hence, the scattering amplitudes for arbitrary orientation are needed. In general, the problem needs to be solved by the EM boundary condition for such scattering configurations. In special cases, wave scattering by a canted particle can be derived with the simple approaches discussed below.

3.6.1 Scattering Formulation through Coordinate Transformation

When a scatterer cants in the polarization plane with an angle of φ, wave scattering can be represented by either local (body) polarization reference $\left(\hat{h}', \hat{v}' \right)$ (symmetry

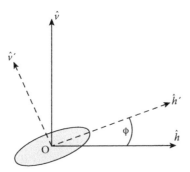

FIGURE 3.23 Coordinate systems for wave scattering by a scatterer canted in polarization plane with an orientation angle φ; $(\hat{h}', \hat{v}')$ is the local polarization reference on its major and minor axis, and $(\hat{h}, \hat{v})$ is the radar polarization reference.

axis of the scatterer) or global (radar) polarization reference $(\hat{h}, \hat{v})$. As shown earlier, with the local polarization reference, the wave scattering can be expressed by

$$\begin{bmatrix} E'_{sh} \\ E'_{sv} \end{bmatrix} = \frac{e^{-jkr}}{r} \begin{bmatrix} s_a & 0 \\ 0 & s_b \end{bmatrix} \begin{bmatrix} E'_{ih} \\ E'_{iv} \end{bmatrix} = \frac{e^{-jkr}}{r} \bar{\bar{S}}^{(b)} \bar{E}'_i. \tag{3.102}$$

There is no cross-polarization in this case because the polarization reference is on the scatterer's symmetry axis.

Because the local reference is a rotation of the global reference system by an angle of φ, the relation of the wave field represented in the two polarization references is

$$\begin{bmatrix} E_h \\ E_v \end{bmatrix} = \begin{bmatrix} \cos\varphi & -\sin\varphi \\ \sin\varphi & \cos\varphi \end{bmatrix} \begin{bmatrix} E'_h \\ E'_v \end{bmatrix} = \bar{\bar{R}}(\varphi)\bar{E}'. \tag{3.103}$$

Using Equations 3.103 and 3.102, we have the scattered wave field in the global reference system as

$$\bar{E}_s = \bar{\bar{R}}(\varphi)\bar{E}'_s = \frac{e^{-jkr}}{r}\bar{\bar{R}}(\varphi)\bar{\bar{S}}^{(b)}\bar{E}'_i = \frac{e^{-jkr}}{r}\bar{\bar{R}}(\varphi)\bar{\bar{S}}^{(b)}\bar{\bar{R}}^{-1}(\varphi)\bar{E}_i \equiv \frac{e^{-jkr}}{r}\bar{\bar{S}}\bar{E}_i, \tag{3.104}$$

where

$$\bar{\bar{S}} = \bar{\bar{R}}^{-1}(\varphi)\bar{\bar{S}}^{(b)}\bar{\bar{R}}(\varphi) = \begin{bmatrix} s_a \cos^2\varphi + s_b \sin^2\varphi & (s_a - s_b)\sin\varphi\cos\varphi \\ (s_a - s_b)\sin\varphi\cos\varphi & s_a \sin^2\varphi + s_b \cos^2\varphi \end{bmatrix} \tag{3.105}$$

is the scattering matrix with the canting angle taken into account.

3.6.2 GENERAL EXPRESSION FOR RAYLEIGH SCATTERING

In the case of the Rayleigh approximation, wave scattering for an arbitrary orientation can be described through the projection of a dipole radiation as follows.

As shown in Figure 3.24, a scatterer has its principal axis in the z_b direction of the scattering body coordinate system (x_b, y_b, z_b), which has an orientation (θ_b, ϕ_b) in the global coordinate system (x, y, z). The two coordinate systems are related by

$$
\begin{bmatrix} \hat{x}_b \\ \hat{y}_b \\ \hat{z}_b \end{bmatrix} = \begin{bmatrix} \cos\theta_b\cos\phi_b & \cos\theta_b\sin\phi_b & -\sin\theta_b \\ -\sin\phi_b & \cos\phi_b & 0 \\ \sin\theta_b\cos\phi_b & \sin\theta_b\sin\phi_b & \cos\theta_b \end{bmatrix} \begin{bmatrix} \hat{x} \\ \hat{y} \\ \hat{z} \end{bmatrix}.
\tag{3.106}
$$

According to the Rayleigh scattering approximation, as expressed in Equations 3.81 through 3.84, the scattering amplitude can be written as

$$
\vec{s}\left(\hat{k}_s, \hat{k}_i\right) = \frac{k^2}{4\pi\varepsilon_0}\left[\vec{p} - \hat{k}_s\left(\hat{k}_s \cdot \vec{p}\right)\right],
\tag{3.107}
$$

and the scatterer polarization $\vec{p}$ can be expressed in the scatterer body coordinate system as

$$
\begin{bmatrix} p_x \\ p_y \\ p_z \end{bmatrix} = \begin{bmatrix} \alpha_x & 0 & 0 \\ 0 & \alpha_y & 0 \\ 0 & 0 & \alpha_z \end{bmatrix} \begin{bmatrix} E_{ix} \\ E_{iy} \\ E_{iz} \end{bmatrix},
\tag{3.108}
$$

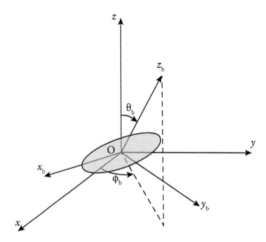

FIGURE 3.24 Coordinate systems for wave scattering by a spheroidal scatterer with orientation angle (θ_b, ϕ_b); (x, y, z) is the global coordinate system, and (x_b, y_b, z_b) is the scatterer body coordinate system.

where

$$\alpha_j = V\varepsilon_0 \frac{\varepsilon_r - 1}{1 + L_j(\varepsilon_r - 1)}. \tag{3.109}$$

Projecting the scattering amplitude to the reference (horizontal and vertical) polarization directions $(\hat{h}, \hat{v})$ defined in Equation 3.38 and then writing them in matrix form, we obtain the scattering matrix

$$\overline{\overline{S}} = \frac{k^2}{4\pi\varepsilon_0} \overline{\overline{P}}_s \overline{\overline{\alpha}}^{(b)} \overline{\overline{P}}_i, \tag{3.110}$$

where the projection matrix

$$\overline{\overline{P}}_s = \begin{bmatrix} \hat{h}_s \cdot \hat{x}_b & \hat{h}_s \cdot \hat{y}_b & \hat{h}_s \cdot \hat{z}_b \\ \hat{v}_s \cdot \hat{x}_b & \hat{v}_s \cdot \hat{y}_b & \hat{v}_s \cdot \hat{z}_b \end{bmatrix} \tag{3.111}$$

for the scattered wave field, and the projection matrix

$$\overline{\overline{P}}_i = \begin{bmatrix} \hat{h}_i \cdot \hat{x}_b & \hat{v}_i \cdot \hat{x}_b \\ \hat{h}_i \cdot \hat{y}_b & \hat{v}_i \cdot \hat{y}_b \\ \hat{h}_i \cdot \hat{z}_b & \hat{v}_i \cdot \hat{z}_b \end{bmatrix} \tag{3.112}$$

for the incident wave field. It can be shown that each element is

$$S_{hh} = \frac{k^2}{4\pi\varepsilon_0}\left(\alpha_x\left(\hat{h}_s \cdot \hat{x}_b\right)\left(\hat{h}_i \cdot \hat{x}_b\right) + \alpha_y\left(\hat{h}_s \cdot \hat{y}_b\right)\left(\hat{h}_i \cdot \hat{y}_b\right) + \alpha_z\left(\hat{h}_s \cdot \hat{z}_b\right)\left(\hat{h}_i \cdot \hat{z}_b\right)\right) \tag{3.113}$$

$$S_{hv} = \frac{k^2}{4\pi\varepsilon_0}\left(\alpha_x\left(\hat{h}_s \cdot \hat{x}_b\right)\left(\hat{v}_i \cdot \hat{x}_b\right) + \alpha_y\left(\hat{h}_s \cdot \hat{y}_b\right)\left(\hat{v}_i \cdot \hat{y}_b\right) + \alpha_z\left(\hat{h}_s \cdot \hat{z}_b\right)\left(\hat{v}_i \cdot \hat{z}_b\right)\right) \tag{3.114}$$

$$S_{vh} = \frac{k^2}{4\pi\varepsilon_0}\left(\alpha_x\left(\hat{v}_s \cdot \hat{x}_b\right)\left(\hat{h}_i \cdot \hat{x}_b\right) + \alpha_y\left(\hat{v}_s \cdot \hat{y}_b\right)\left(\hat{h}_i \cdot \hat{y}_b\right) + \alpha_z\left(\hat{v}_s \cdot \hat{z}_b\right)\left(\hat{h}_i \cdot \hat{z}_b\right)\right) \tag{3.115}$$

$$S_{vv} = \frac{k^2}{4\pi\varepsilon_0}\left(\alpha_x\left(\hat{v}_s \cdot \hat{x}_b\right)\left(\hat{v}_i \cdot \hat{x}_b\right) + \alpha_y\left(\hat{v}_s \cdot \hat{y}_b\right)\left(\hat{v}_i \cdot \hat{y}_b\right) + \alpha_z\left(\hat{v}_s \cdot \hat{z}_b\right)\left(\hat{v}_i \cdot \hat{z}_b\right)\right). \tag{3.116}$$

3.6.3 BACKSCATTERING MATRIX FOR A SPHEROID

In the case of a spheroid, we have $\alpha_x = \alpha_y = \alpha_a$ and $\alpha_z = \alpha_b$. To simplify the expressions in Equations 3.113 through 3.116 for backscattering with horizontal incidence, let $\theta_i = \pi/2$, $\phi_i = \pi$, $\theta_s = \pi/2$, and $\phi_s = 0$; we have

$$\hat{h}_i = -\hat{y}; \hat{v}_i = \hat{z} \tag{3.117}$$

$$\hat{h}_s = \hat{y}; \hat{v}_s = \hat{z}. \tag{3.118}$$

From Equation 3.106, we have the body orientation vectors

$$\hat{x}_b = \cos\theta_b \cos\phi_b \hat{x} + \cos\theta_b \sin\phi_b \hat{y} - \sin\theta_b \hat{z} \qquad (3.119)$$

$$\hat{y}_b = -\sin\phi_b \hat{x} + \cos\phi_b \hat{y} \qquad (3.120)$$

$$\hat{z}_b = \sin\theta_b \cos\phi_b \hat{x} + \sin\theta_b \sin\phi_b \hat{y} + \cos\theta_b \hat{z}. \qquad (3.121)$$

Applying Equations 3.117 through 3.121 to Equations 3.113 through 3.116, we obtain the scattering amplitude elements as follows:

$$s_{hh} = -\left(s_a(\cos^2\theta_b \sin^2\phi_b + \cos^2\phi_b) + s_b \sin^2\theta_b \sin^2\phi_b\right) \qquad (3.122)$$

$$s_{hv} = -s_{vh} = -(s_a - s_b)\sin\theta_b \cos\theta_b \sin\phi_b \qquad (3.123)$$

$$s_{vv} = \left(s_a(\sin^2\theta_b + \cos^2\phi_b) + s_b \cos^2\theta_b\right). \qquad (3.124)$$

3.6.4 FORWARD SCATTERING ALIGNMENT VERSUS BACKSCATTERING ALIGNMENT

Thus far, we have used the scatterer location as the origin of the coordinate system and with reference vectors of $\left(\hat{h}_i, \hat{v}_i, \hat{k}_i\right)$ and $\left(\hat{h}_s, \hat{v}_s, \hat{k}_s\right)$, which is convenient in theoretical scattering calculations. In monostatic radar applications, however, this is not convenient because it requires a change in the reference polarization and wave vector (Bringi and Chandrasekar 2001; Ulaby and Elach 1990), as shown in Figure 3.25. This is called the *forward scattering alignment* (FSA: $\left(\hat{h}_s, \hat{v}_s, \hat{k}_s\right)$) convention.

In the case of FSA, both $\hat{h}_s$ and $\hat{k}_s$ have directions opposite to that of the incident wave in the back direction. To avoid the changing the directions of the reference vectors, a set of radar-based reference vectors $\left(\hat{h}_r, \hat{v}_r, \hat{k}_r\right) = \left(-\hat{h}_s, \hat{v}_s, -\hat{k}_s\right)$ are used, and

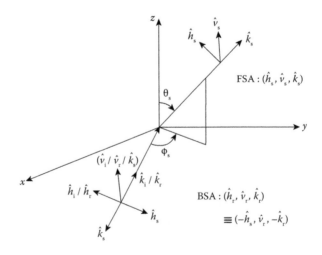

FIGURE 3.25 Diagram of coordinate systems for forward scattering alignment (FSA) and back scattering alignment (BSA) conventions.

it follows $\left(\hat{h}_r, \hat{v}_r, \hat{k}_r\right) = \left(\hat{h}_i, \hat{v}_i, \hat{k}_i\right)$ in backscattering, called the *back scattering alignment* (BSA) convention. Hence, the scattering amplitude/matrix in BSA is related to that in FSA by

$$\begin{bmatrix} s_{hh} & s_{hv} \\ s_{vh} & s_{vv} \end{bmatrix}_B = \begin{bmatrix} -1 & 0 \\ 0 & 1 \end{bmatrix}_B \begin{bmatrix} s_{hh} & s_{hv} \\ s_{vh} & s_{vv} \end{bmatrix}_B = \begin{bmatrix} -s_{hh} & -s_{hv} \\ s_{vh} & s_{vv} \end{bmatrix}. \tag{3.125}$$

Using Equations 3.122 through 3.124 in Equation 3.125, we obtain

$$\overline{\overline{S}}_B$$

$$= \begin{bmatrix} s_a(\cos^2\theta_b \sin^2\phi_b + \cos^2\phi_b) + s_b \sin^2\theta_b \sin^2\phi_b & (s_a - s_b)\sin\theta_b \cos\theta_b \sin\phi_b \\ (s_a - s_b)\sin\theta_b \cos\theta_b \sin\phi_b & s_a(\sin^2\theta_b + \cos^2\phi_b) + s_b \cos^2\theta_b \end{bmatrix}. \tag{3.126}$$

It is worth discussing two extreme cases when the scatterer is oriented in a specific direction. If the scatterer is canted in the polarization $(y\text{–}z)$ plane, meaning $\phi_b = \pi/2$. Let the canting angle in the polarization plane be $\theta_b \equiv \varphi$; Equation 3.126 reduces to

$$\overline{\overline{S}}_B = \begin{bmatrix} s_a\cos^2\varphi + s_b\sin^2\varphi & (s_a - s_b)\sin\varphi\cos\varphi \\ (s_a - s_b)\sin\varphi\cos\varphi & s_a\sin^2\varphi + s_b\cos^2\varphi \end{bmatrix}. \tag{3.127}$$

This shows that both the horizontal and vertical scattering amplitudes change from those amplitudes based on the polarizations on the major and minor axes, and there is cross-polarization due to canting. This is expected, because the projected dimensions in the horizontal and vertical directions change from those of the major and minor axes, and the electric wave field inside the scatterer is no longer along the symmetry axis, hence yielding cross-polarization components. The co-polar ratio and cross-to-co-polar ratio are plotted in Figure 3.26.

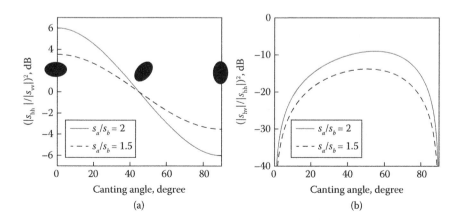

FIGURE 3.26 Dependences of polarization ratio on canting angle: (a) co-polar ratio and (b) cross-to-co-polar ratio.

By contrast, if the scatterer is canted in the scattering (z–x) plane, $\phi_b = 0$. Let the canting in the scattering plane be $\theta_b \equiv \vartheta$; Equation 3.126 becomes

$$\overline{\overline{S}}_B = \begin{bmatrix} s_a & 0 \\ 0 & s_a \sin^2 \vartheta + s_b \cos^2 \vartheta \end{bmatrix}. \qquad (3.128)$$

This makes physical sense, because the scatterer's horizontal dimension does not change when it is canted in the scattering plane; hence, there is no change in scattering for the horizontal polarization, and there is no cross-polarization component.

3.6.5 SCATTERING MATRIX BY A SPHEROID WITH ANY ORIENTATION

In the last subsection, we obtained the scattering matrix for an arbitrary orientation expressed by either Equation 3.126 or by combining Equations 3.127 and 3.128. To combine Equations 3.127 and 3.128, we replace "s_b" in Equation 3.127 with "$s_a \sin^2 \vartheta + s_b \cos^2 \vartheta$" based on Equation 3.128. Because we are using the backscattering alignment in this book, we can omit the subscript B and have

$$
\begin{aligned}
\overline{\overline{S}} &= \begin{bmatrix} s_a \cos^2\varphi + (s_a \sin^2\vartheta + s_b \cos^2\vartheta)\sin^2\varphi & (s_a - s_b)\cos^2\vartheta \sin\varphi\cos\varphi \\ (s_a - s_b)\cos^2\vartheta \sin\varphi\cos\varphi & s_a \sin^2\varphi + (s_a \sin^2\theta + s_b \cos^2\vartheta)\cos^2\varphi \end{bmatrix} \\
&= \begin{bmatrix} s_a(\cos^2\varphi + \sin^2\vartheta \sin^2\varphi) + s_b \cos^2\vartheta \sin^2\varphi & (s_a - s_b)\cos^2\vartheta \sin\varphi\cos\varphi \\ (s_a - s_b)\cos^2\vartheta \sin\varphi\cos\varphi & s_a(\sin^2\varphi + \sin^2\vartheta \cos^2\varphi) + s_b \cos^2\vartheta \cos^2\varphi \end{bmatrix} \\
&= \begin{bmatrix} As_a + Bs_b & (s_a - s_b)\sqrt{BC} \\ (s_a - s_b)\sqrt{BC} & Cs_b + Ds_a \end{bmatrix} \equiv \begin{bmatrix} s_{hh} & s_{hv} \\ s_{vh} & s_{vv} \end{bmatrix},
\end{aligned} \qquad (3.129)
$$

where the canting angle dependent factors are

$$A = \cos^2\varphi + \sin^2\vartheta\sin^2\varphi \qquad (3.130)$$

$$B = \cos^2\vartheta\sin^2\varphi \qquad (3.131)$$

$$C = \cos^2\vartheta\cos^2\varphi \qquad (3.132)$$

$$D = \sin^2\varphi + \sin^2\vartheta\cos^2\varphi. \qquad (3.133)$$

Once the canting angles (ϑ, φ) are known, the scattering matrix can be calculated using Equations 3.129 through 3.133 for a particle with any orientation. In reality, however, hydrometeors are randomly orientated; the canting angles (ϑ, φ) should be treated as random variables, and their statistics should be characterized—these points are addressed in Chapter 4.

APPENDIX 3A: DERIVATION OF OPTICAL THEOREM (FORWARD SCATTERING THEOREM)

Consider a plane wave E_i incident on a particle (Figure 3A.1); the scattered wave field is E_s, and the total wave field is

$$\vec{E} = \vec{E}_i + \vec{E}_s \quad \text{and} \quad \vec{H} = \vec{H}_i + \vec{H}_s. \tag{3A.1}$$

Then, the total absorbed power is

$$P_a = -\oint_A \frac{1}{2} \text{Re}\left(\vec{E} \times \vec{H}^*\right) \cdot d\vec{a}. \tag{3A.2}$$

Substitution of Equation 3A.1 into Equation 3A.2 yields

$$P_a = -\oint_A \frac{1}{2} \text{Re}\left[\left(\vec{E}_i + \vec{E}_s\right) \times \left(\vec{H}_i + \vec{H}_s\right)^*\right] \cdot d\vec{a}$$

$$= -\oint_A \frac{1}{2} \text{Re}\left(\vec{E}_i \times \vec{H}_i^* + \vec{E}_s \times \vec{H}_s^* + \vec{E}_i \times \vec{H}_s^* + \vec{E}_s \times \vec{H}_i^*\right) \cdot d\vec{a}. \tag{3A.3}$$

Noting $\vec{S}_i = \frac{1}{2}\text{Re}\left(\vec{E}_i \times \vec{H}_i^*\right)$ and $\vec{S}_s = \frac{1}{2}\text{Re}\left(\vec{E}_s \times \vec{H}_s^*\right)$, we have

$$P_i = \oint_A \vec{S}_i \cdot d\vec{a} = \frac{1}{2}\oint_A \text{Re}\left(\vec{E}_i \times \vec{H}_i^*\right) \cdot d\vec{a}$$

$$= \frac{1}{2\eta_0}|E_i|^2 \oint_S \hat{k}_i \cdot d\vec{a} \tag{3A.4}$$

$$= \frac{1}{2\eta_0}|E_i|^2 \int_0^\pi \int_0^{2\pi} \cos\theta \cdot r^2 \sin\theta\, d\theta\, d\varphi = 0$$

$$P_s = \oint_A \vec{S}_s \cdot d\vec{a} = \frac{1}{2}\oint_A \text{Re}\left(\vec{E}_s \times \vec{H}_s^*\right) \cdot d\vec{a}. \tag{3A.5}$$

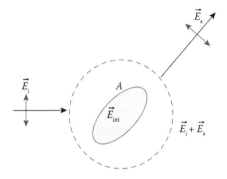

FIGURE 3A.1 A sketch of wave scattering by a nonspherical scatterer.

Using Equations 3A.4 and 3A.5 in Equation 3A.3, we have

$$\begin{aligned}
P_{\mathrm{a}} + P_{\mathrm{s}} &= -\frac{1}{2} \oint_A \mathrm{Re}\left(\vec{E}_{\mathrm{i}} \times \vec{H}_{\mathrm{s}}^* + \vec{E}_{\mathrm{s}} \times \vec{H}_{\mathrm{i}}^* \right) \cdot d\vec{a} \\
&= -\frac{1}{2} \oint_A \mathrm{Re}\left(\vec{E}_{\mathrm{i}} \times \vec{H}^* + \vec{E} \times \vec{H}_{\mathrm{i}}^* \right) \cdot d\vec{a} \\
&= -\frac{1}{2} \oint_A \mathrm{Re}\left(\vec{E}_{\mathrm{i}}^* \times \vec{H} + \vec{E} \times \vec{H}_{\mathrm{i}}^* \right) \cdot d\vec{a} \\
&= -\frac{1}{2} \int_V \mathrm{Re}\, \nabla \cdot \left(\vec{E}_{\mathrm{i}}^* \times \vec{H} + \vec{E} \times \vec{H}_{\mathrm{i}}^* \right) \cdot dv.
\end{aligned} \tag{3A.6}$$

Using the vector formula

$$\nabla \cdot (\vec{A} \times \vec{B}) = \vec{B} \cdot \nabla \times \vec{A} - \vec{A} \cdot \nabla \times \vec{B}, \tag{3A.7}$$

we have

$$\begin{aligned}
\nabla \cdot \left(\vec{E}_{\mathrm{i}}^* \times \vec{H} \right) &= \vec{H} \cdot \nabla \times \vec{E}_{\mathrm{i}}^* - \vec{E}_{\mathrm{i}}^* \cdot \nabla \times \vec{H} \\
&= \vec{H} \cdot \left(j\omega\mu_0 \vec{H}_{\mathrm{i}} \right)^* - \vec{E}_{\mathrm{i}}^* \cdot \left(-j\omega\varepsilon \vec{E} \right) \\
&= -j\omega\mu_0 \vec{H} \cdot \vec{H}_{\mathrm{i}}^* + j\omega\varepsilon_r\varepsilon_0 \vec{E} \cdot \vec{E}_{\mathrm{i}}^*
\end{aligned} \tag{3A.8}$$

$$\nabla \cdot \left(\vec{E} \times \vec{H}_{\mathrm{i}}^* \right) = j\omega\mu_0 \vec{H} \cdot \vec{H}_{\mathrm{i}}^* - j\omega\varepsilon_0 \vec{E} \cdot \vec{E}_{\mathrm{i}}^*. \tag{3A.9}$$

Combining Equations 3A.8 and 3A.9, taking the real part, we have

$$\begin{aligned}
\mathrm{Re}\left[\nabla \cdot \left(\vec{E}_{\mathrm{i}}^* \times \vec{H} + \vec{E} \times \vec{H}_{\mathrm{i}}^* \right) \right] &= \mathrm{Re}\left[0 - j\omega\varepsilon_0 (\varepsilon_r - 1)\vec{E} \cdot \vec{E}_{\mathrm{i}}^* \right] \\
&= -\mathrm{Im}\left[\omega\varepsilon_0 (\varepsilon_r - 1)\vec{E} \cdot \vec{E}_{\mathrm{i}}^* \right].
\end{aligned} \tag{3A.10}$$

Substitution of Equation 3A.10 into Equation 3A.6 yields

$$P_{\mathrm{a}} + P_{\mathrm{s}} = \mathrm{Im} \int_V \frac{1}{2} \omega_0 \varepsilon_0 (\varepsilon_r - 1)\vec{E} \cdot \vec{E}_{\mathrm{i}}^* \cdot dv. \tag{3A.11}$$

From Equation 2.20 (Ishimaru 1997),

$$\hat{e}_{\mathrm{s}} s\left(\hat{k}_{\mathrm{s}}, \hat{k}_{\mathrm{i}} \right) = \frac{k^2}{4\pi E_0} \int_V \left[-\hat{k}_{\mathrm{s}} \times \hat{k}_{\mathrm{s}} \times \vec{E}(\vec{r}) \right] (\varepsilon_r - 1)e^{-j\hat{k}_{\mathrm{s}} \cdot \vec{r}} \cdot dv. \tag{3A.12}$$

In the forward direction, Equation 3A.12 becomes

$$\hat{e}_{\mathrm{i}} s\left(\hat{k}_{\mathrm{i}}, \hat{k}_{\mathrm{i}} \right) = \frac{k^2}{4\pi E_0} \int_V (\varepsilon_r - 1)\vec{E}(\vec{r}) \cdot dv$$

Hence,

$$\int_V (\varepsilon_r - 1)\vec{E} \cdot \vec{E}_i^* \cdot dv = \frac{4\pi |E_0|^2}{2k^2} s\left(\hat{k}_i, \hat{k}_i\right) \hat{e}_i \cdot \hat{e}_i. \tag{3A.13}$$

Substitution of Equation 3A.13 into 3A.11 leads to

$$P_a + P_s = \frac{4\pi \omega \varepsilon_0 |E_0|^2}{2k^2} \text{Im}\left[s\left(\hat{k}_i, \hat{k}_i\right)\right]$$

$$= \frac{4\pi}{k} \frac{|E_0|^2}{2\eta_0} \text{Im}\left[s\left(\hat{k}_i, \hat{k}_i\right)\right] \tag{3A.14}$$

$$P_t = S_i \sigma_t = S_i \frac{4\pi}{k} \text{Im}\left[s\left(\hat{k}_i, \hat{k}_i\right)\right]. \tag{3A.15}$$

Hence,

$$\sigma_t = \frac{4\pi}{k} \text{Im}\left[s\left(\hat{k}_i, \hat{k}_i\right)\right]. \tag{3A.16}$$

APPENDIX 3B: VECTOR SPHERICAL WAVE HARMONICS

Let the eigen solution of a scalar wave equation in spherical coordinates be

$$\psi_{mn} = z_n(kr)P_n^m(\cos\theta)e^{jm\phi}, \tag{3B.1}$$

where $z_n(kr)$ is one of the four spherical Bessel functions: j_n, y_n, $h_n^{(1)}$, and $h_n^{(2)}$, depending on whether the wave is incoming or outgoing. Because we use the phasor of $e^{j\omega t}$, $z_n^{(1)} = j_n$ represents the incident and internal wave, and $z_n^{(4)} = h_n^{(2)}$ represents the scattered wave. The vector spherical wave harmonics are defined based on ψ_{mn} as

$$\vec{M}_{mn} = \nabla \times \left(\psi_{mn}\vec{r}\right) \tag{3B.2}$$

$$\vec{N}_{mn} = \frac{1}{k}\nabla \times \vec{M}_{mn}. \tag{3B.3}$$

Problems

3.1 As discussed in this chapter, a plane wave with any polarization can be represented by

$$\vec{E}(x,t) = A_y \cos(\omega t - kx + \delta_1)\hat{y} + A_z \cos(\omega t - kx + \delta_2)\hat{z}. \tag{3.134}$$

Let $A_y = A_z$, plot the trace of the tip of the electric field in the y–z plane and give the name for each polarization for the following phase differences:

a. $\delta_2 - \delta_1 = 0$ and $\delta_2 - \delta_1 = \pi$

b. $\delta_2 - \delta_1 = \pm\dfrac{\pi}{2}$

c. $\delta_2 - \delta_1 = \pm\dfrac{\pi}{4}$

d. $\delta_2 - \delta_1 = \pm\dfrac{3\pi}{4}$.

What polarization (of the wave described above) is the polarimetric WSR-88D radar likely to transmit?

3.2 Using the given cross sections (hw2p2.dat) calculated from the Mie theory, plot the Q factors as functions of the particle diameter. The four columns are the particle diameter in millimeters, extinction cross section, scattering cross section, and backscattering cross section, respectively. It is known that the relative dielectric constant is (41, 41) and the wavelength is 3 cm. Calculate the Q factors using the Rayleigh scattering approximation and compare them with the results from the Mie theory. In addition, calculate and show the scattering albedo for both the Mie theory and the Rayleigh approximation. Discuss the valid regime for Rayleigh scattering, and explain the extinction paradox.

3.3 Assuming raindrops to be oblate spheroids with the following axis ratio (represented in polynomial form):

$$\gamma = b/a = 0.9951 + 0.0251D - 0.03644D^2 + 0.005303D^3 - 0.0002492D^4.$$

and assuming the relative dielectric constant to be (80, 17), do the following:

a. Using the Rayleigh scattering approximation, calculate the scattering amplitudes of the raindrops at S-band (2.8 GHz) and plot them as a function of equivolume diameter ranging from 0.1 to 8 mm. Using the provided results (hm4.dat, rigorously calculated with the T-matrix method), verify your calculations. Show the scattering magnitudes of both your results and the provided results in terms of real and imaginary parts as well as magnitude and phase and explain the similarities and differences.

b. Write down the formula used in your calculation and discuss how the ratio of the scattering amplitudes $|s_a/s_b|$ changes when the dielectric constant is replaced by that of dry snow with a fractional volume of 10%. Explain why this change occurred.

3.4 Show that the polarizability of a coated sphere (with the expression below) is equal to that of a mixture sphere calculated using the Maxwell-Garnett mixing formula with medium 2 as its background (the Maxwell-Garnett formula is discussed in Chapter 2). The expression of the coated sphere polarizability is

$$\alpha = V\varepsilon_0 \times 3 \frac{(\varepsilon_2 - 1)(\varepsilon_1 + 2\varepsilon_2) + f_v(\varepsilon_1 - \varepsilon_2)(1 + 2\varepsilon_2)}{(\varepsilon_2 + 2)(\varepsilon_1 + 2\varepsilon_2) + f_v(2\varepsilon_2 - 2)(\varepsilon_1 - \varepsilon_2)},$$

where ε_1 and ε_2 are the relative dielectric constant of the inner sphere and outer shell, respectively. The fractional volume is $f_v = a^3/a'^3$, where a and a' are the radius for the inner and outer spheres, respectively. Use the expression $\alpha = V\varepsilon_0 \dfrac{3(\varepsilon_e - 1)}{(\varepsilon_e + 2)}$ for the effective polarizability of the coated sphere, where ε_e is the effective dielectric constant for the mixture sphere.

3.5 Assume that the backscattering amplitudes for polarizations on the major and minor axes are s_a and s_b, respectively. Show that the backscattering matrix in the reference of horizontal and vertical polarizations for a scatterer with a canting angle of φ in the polarization plane follows

$$\begin{bmatrix} s_{hh} & s_{hv} \\ s_{vh} & s_{vv} \end{bmatrix}_B = \begin{bmatrix} s_a \cos^2\varphi + s_b \sin^2\varphi & (s_a - s_b)\sin\varphi\cos\varphi \\ (s_a - s_b)\sin\varphi\cos\varphi & s_a \sin^2\varphi + s_b \cos^2\varphi \end{bmatrix}.$$

4 Scattering and Propagation in Clouds and Precipitation

What a weather radar measures are the collective effects of wave scattering and propagation from clouds and/or precipitation containing numerous hydrometeors. Not only do the hydrometeors scatter waves sampled by the radar to yield Doppler radar measurements such as reflectivity, Doppler velocity, and spectrum width, as well as measurements of polarimetric variables such as differential reflectivity, linear depolarization ratio, co-polar correlation coefficient, and cross-polar correlation coefficients, but hydrometeors along the beam also cause attenuation, phase delay, and depolarization of wave propagation. The wave scattering and propagation properties depend on both hydrometeor characteristics and wave properties such as frequency and transmitted wave polarization. It is important to know the wave statistics and how polarimetric radar variables are related to the microphysical and statistical properties of hydrometeors. This chapter deals with wave scattering and propagation in clouds and precipitation. It describes the concepts of scattering models, coherent and incoherent scattering, coherent wave propagation, wave statistics, and polarimetric radar variables.

4.1 SCATTERING MODELS

Hydrometeors are randomly distributed in clouds and precipitation, and their positions and orientations randomly change in space and time. Hence, the signal received by the radar, which is the total scattered wave resulting from all the hydrometeor particles in the radar resolution volume, is also random, depending on the constructive or destructive contributions from individual particles. The randomness of the medium and the radar signal depends on the nature of hydrometeor distribution and the random relative displacements of particles from sample to sample, as well as radar wavelength. Because radar uses time samples (pulse to pulse) to represent ensembles, the medium can be considered deterministic if the relative random motion is much less than half a wavelength, partially random if the random motion is within half a wavelength, and completely random if the random motion exceeds half a wavelength. To study the wave statistics, scattering models are needed to describe the scattering process (Ishimaru 1978, 1997).

Wave scattering in clouds and precipitation can be described by three physical models: (i) single scattering, (ii) first-order multiple scattering, and (iii) multiple scattering. The three scattering models are sketched in Figure 4.1.

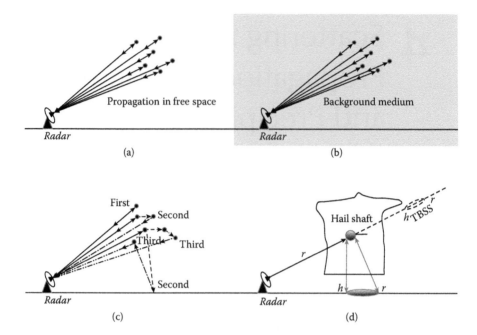

FIGURE 4.1   Three wave scattering models: (a) single scattering; (b) first-order multiple scattering; (c) multiple scattering, including the first, second, and third scattering paths indicted by solid, dashed, and dash-dotted lines, respectively; (d) conceptual sketch of three-body scattering signatures as an example of multiple scattering.

In the case of single scattering, the transmitted/incident wave is scattered once before arriving at the receiver, and the wave propagates in a deterministic medium such as the background medium of air (without particles and/or turbulence) before and after scattering. The first-order multiple scattering model also assumes the wave is scattered once before it is received, but the effects on wave propagation from the transmitter to the particle and that from the particle to the receiver are taken into account by using the wave number for the effective medium that includes the scatterers, rather than the wave number of the cloud-free medium. This way, attenuation, phase delay, and depolarization are included. The gray background in Figure 4.1b indicates the effective medium that has included the scatterers' effects on wave propagation.

For the multiple scattering model, the waves are not only scattered once, but two, three, or more times before reaching the receiver. The first-, second-, and third-order scatterings are indicated by the solid, dashed, and dash-dotted lines, respectively, in Figure 4.1c. One typical example of multiple scattering that occurs in weather radar observations is the *three-body scattering signature* (TBSS), in which the wave is first scattered by hydrometeors to the ground, then reflected back to the hydrometeors, and finally scattered to the radar antenna. When hailstones are present and cause strong scattering, TBSS occurs and yields an artifact that appears downrange from a high radar reflectivity core in a thunderstorm, due to the extra path (time delay) bounced to and from the ground (Zrnić et al. 2010b). This type of artifact needs to be recognized and removed to avoid misinterpretation of weather radar data (Mahale et al. 2014). Radar polarimetry has advantages for identifying this artifact with multiparameter measurements, which are shown in Chapter 6 with PRD-based classification.

In general, the single scattering model is applied to the case of a wave in a sparse medium (where the fractional volume occupied by the scatterers is typically less than 1% [Ishimaru 1978, 1997], a category into which clouds and precipitation fall) that is optically thin ($\int_0^r n\sigma_t(\ell)d\ell \ll 1$, which is an integral attenuation discussed in Section 4.3) when the propagation effects of attenuation and phase shift are negligible. This may be used for long wavelength (S-band or longer wavelength for light precipitation) radar data. The first-order multiple scattering model also deals with wave scattering and propagation in sparse media, but the media can be optically thick ($\int_0^r n\sigma_t(\ell)d\ell \sim$ or > 1) because it takes into account the propagation effects. This applies to most weather radar data, especially those short wavelength weather radars at the C- and X-band frequencies. The multiple scattering model is a rigorous approach that takes into account the mutual interaction between scatterers and applies to wave scattering and propagation in dense (>1% fractional volume) media such as vegetation or snowpack, as well as the cases of TBSS and cloud (W-band) radar observations (Battaglia et al. 2014). In the following sections, we focus on the single scattering and first-order multiple scattering models as well as their applications in weather radar polarimetry.

4.2 SINGLE SCATTERING MODEL

As mentioned earlier, in the single scattering model, the incident wave is scattered once by each particle before being received by the radar antenna. For simplicity, we start with a scalar wave (ignoring the polarization at the moment) to describe the concept and then extend it to a vector wave to address polarimetry. Because we are interested in wave scattering in both the backward and forward directions, to have generality, we use a bistatic scattering configuration for the following discussions.

4.2.1 COHERENT ADDITION APPROXIMATION

As illustrated in Figure 4.2, a plane wave E_i propagates in the $\hat{k}_i$ direction and is incident on a collection of particles. It is scattered once by each particle, and then the scattered waves are received. Let the incident wave be a unit plane wave, with its wave field represented by

$$E_i = E_0 e^{-j\vec{k}_i \bullet \vec{r}}. \tag{4.1}$$

In the coordinate system of Figure 4.2, the scattered wave field by the lth particle is then expressed by

$$E_{sl} = \frac{e^{-jk|\vec{r}_0 - \vec{r}_l|}}{|\vec{r}_0 - \vec{r}_l|} E_0 s_l\left(\hat{k}_s, \hat{k}_i\right) e^{-j\vec{k}_i \bullet \vec{r}_l}. \tag{4.2}$$

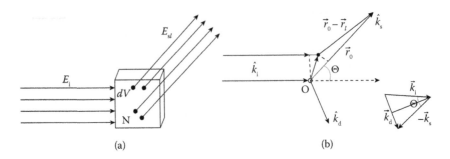

FIGURE 4.2 Single scattering model: (a) wave scattering by a collection of randomly distributed particles and (b) coordinate system of wave scattering by one of the particles.

In the far-field approximation $r_0 \gg r_l$, we have $|\vec{r}_0 - \vec{r}_l| \approx r_0 - \hat{k}_s \cdot \vec{r}_l$. Hence, Equation 4.2 becomes

$$E_{sl} \approx \frac{e^{-jkr_0}}{r_0} E_0 e^{-j(\vec{k}_i - \vec{k}_s)\cdot\vec{n}} s_l\left(\hat{k}_s, \hat{k}_i\right) = \frac{e^{-jkr_0}}{r_0} E_0 e^{-j\vec{k}_d\cdot\vec{n}} s_l\left(\hat{k}_s, \hat{k}_i\right), \qquad (4.3)$$

where $\vec{k}_d = \vec{k}_i - \vec{k}_s = 2k \sin(\Theta/2)\hat{k}_d$ and $\vec{k}_d = 2k\hat{k}_i$ for the backscattering direction, and $\vec{k}_d \cdot \vec{n}$ is the extra phase compared with the phase through the origin, which is the phase difference caused by the projection of mean velocity $\vec{n}$ on the incident and scattering directions, indicated by the segments in Figure 4.2b. The total scattered wave field is then the summation of the scattered wave fields by individual particles in the volume, specifically:

$$E_s = \sum_{l=1}^{N} E_{sl} \approx \frac{e^{-jkr_0}}{r_0} E_0 \sum_{l=1}^{N} s_l\left(\hat{k}_s, \hat{k}_i\right) e^{-j\vec{k}_d\cdot\vec{n}}. \qquad (4.4)$$

This is also called the *coherent addition approach* because the scattered wave fields are added coherently with their relative phase $\vec{k}_d \cdot \vec{n}$, is taken into account. In general, the total scattered wave field represented by Equation 4.4 is a random variable due to the random changes in the hydrometeors' positions and orientations. To describe a random variable, its statistical moments and probability density function (PDF) are studied.

4.2.2 MEAN WAVE FIELD

The first moment is the mean wave field, which is also called the *coherent field*,

$$\langle E_s \rangle = \sum_{l=1}^{N} \langle E_{sl} \rangle \approx \frac{e^{-jkr_0}}{r_0} E_0 \sum_{l=1}^{N} \left\langle s_l\left(\hat{k}_s, \hat{k}_i\right) e^{-j\vec{k}_d\cdot\vec{n}} \right\rangle \rightarrow \begin{cases} 0 & \left(\hat{k}_s \neq \hat{k}_i\right) \\ \dfrac{e^{-jkr_0}}{r_0} E_0 \displaystyle\sum_{l=1}^{N} \left\langle s_l\left(\hat{k}_s, \hat{k}_i\right) \right\rangle & \left(\hat{k}_s = \hat{k}_i\right), \end{cases}$$

$$(4.5)$$

where the angle brackets $\langle\cdots\rangle$ denote the ensemble average, yielding the expected value of the random wave field. Equation 4.5 shows that the coherent wave field exists in the forward direction only, whereas the mean scattered wave fields cancel in all other directions because of the random phase. This means that forward scattering contributes to the attenuation and phase shift of wave propagation in clouds and precipitation (which will be studied in Section 4.3), and the backscattering and bistatic scattering causes radar echoes in monostatic and bistatic radar, respectively. This is why the mean wave field is rarely studied and ignored most of the time in the weather radar community and why weather radars are also called *incoherent scatter radars* (because the coherent component disappears). However, it is important to understand the condition for the independent scattering approximation, which is that the random phase $\vec{k}_d \cdot \vec{r}_l$ is uniformly distributed over $[0, 2\pi]$. This can be understood further by looking at the second moments of the random wave field: wave intensity and correlation function.

4.2.3 Wave Intensity and Independent Scattering

Because the mean wave field is zero except for that in the forward direction, and the power flux density (S) is proportional to the wave intensity $(I = |E|^2)$ with $S = \dfrac{I}{2\eta_0}$ shown in Equation 3.43, we now examine the scattered wave intensity. From Equation 4.4, we have

$$I_s = |E_s|^2 = \sum_{l=1}^{N} E_{sl}^* \sum_{m=1}^{N} E_{sm} = \sum_{l=1}^{N} |E_{sl}|^2 + \sum_{l=1}^{N}\sum_{m\neq l}^{N} E_{sl}^* E_{sm}. \tag{4.6}$$

The mean scattered wave intensity is then

$$\langle I_s \rangle = \langle |E_s|^2 \rangle = \sum_{l=1}^{N} \langle |E_{sl}|^2 \rangle + \sum_{l=1}^{N}\sum_{m\neq l}^{N} \langle E_{sl}^* E_{sm} \rangle = I_{inc} + I_c, \tag{4.7}$$

where the first term, $I_{inc} = \sum_{l=1}^{N} \langle |E_{sl}|^2 \rangle = \sum_{l=1}^{N} \langle I_{sl} \rangle$, is called the *incoherent wave intensity*, which is simply a summation of the mean scattered wave intensities from each individual particle, specifically:

$$I_{inc} = \frac{|E_0|^2}{r_0^2} \sum_{l=1}^{N} \langle |s_l(\hat{k}_s, \hat{k}_i)|^2 \rangle = \frac{I_0}{r_0^2} N_t \langle |s_l(\hat{k}_s, \hat{k}_i)|^2 \rangle \equiv \frac{I_0}{r_0^2} \langle n|s_l(\hat{k}_s, \hat{k}_i)|^2 \rangle. \tag{4.8}$$

This shows that, in the case of independent scattering, the scattering cross sections are added together to yield the radar reflectivity, which is given in Equation 4.54 in Section 4.2.6.

The second term represents the correlation effect on the mean wave intensity. This is also called *coherent scattering*, which has been well studied in the EM and remote sensing communities (Ishimaru 1978, 1997; Tsang et al. 1985, 1995; Zhang et al. 1996), but has only recently received attention in the field of radar meteorology (Jameson and Kostinski 2010). Using Equation 4.4, we have

$$I_c = \sum_{l=1}^{N} \sum_{m \neq l}^{N} \left\langle E_{sl}^* E_{sm} \right\rangle$$

$$= \frac{1}{r_0^2} \sum_{l=1}^{N} \sum_{m \neq l}^{N} \left\langle s_l^* \left(\hat{k}_s, \hat{k}_i \right) s_m \left(\hat{k}_s, \hat{k}_i \right) e^{j \vec{k}_d \bullet (\vec{r}_l - \vec{r}_m)} \right\rangle \xrightarrow{\text{std} \left[\vec{k}_d \bullet \ (\vec{r}_l - \vec{r}_m) \right] \geq \pi} 0,$$

(4.9)

where the phase term $\vec{k}_d \bullet (\vec{r}_l - \vec{r}_m)$ represents the phase difference of the wave paths for the two particles (Figure 4.3). It is evident that this term, which is due to the correlation between particles' motion, disappears when the phase is uniformly distributed or its standard deviation is larger than 2π: $\text{std}\left[\vec{k}_d \bullet (\vec{r}_l - \vec{r}_m) \right] > 2\pi$. This would require that the particles be distributed uniformly or in clusters and that their relative motion or cluster size be larger than half a wavelength for a monostatic radar. This condition is mostly satisfied in radar measurements of precipitation because the raindrop number concentration is typically on the order of $N_t \sim 1000$ m^{-3} and the mean free distance among drops is on the order of 10 cm, which is on the order of a wavelength. In this case, the total mean wave intensity is equal to the incoherent wave intensity; hence, it is called *independent scattering*. Coherent scattering can happen when an individual scatterer or a few scatterers moving in a synchronized way dominate the radar echo. Examples include aircraft or ground clutter, which are discussed in the next chapter. Coherent scattering may also occur in wave scattering in turbulence, namely, due to random fluctuations of the refraction index. This is also called *Bragg scattering*. This topic is beyond the scope of this book—interested readers are referred to Tatarskii (1971), Ishimaru (1978, 1997), Gossard et al. (1998), and Zhang et al. (1990).

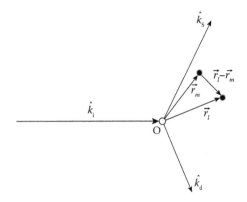

FIGURE 4.3 Geometric sketch of the coupling effect on mean wave intensity.

4.2.4 TIME-CORRELATED SCATTERING

In Section 4.2.3, we discussed mean wave intensity (which led to the definition of radar variables that is discussed in Section 4.2.6), in the case of independent scattering approximation. However, we did not mention how scattered waves received at different times are correlated. Time-correlated scattering is important in weather radar because (i) it allows us to obtain dynamic (motion) information of hydrometeors, and (ii) it determines the number of independent samples that is directly related to the accuracy of radar measurements. To study time correlation, we look at the autocorrelation function (ACF) of the scattered wave field:

$$R(\Delta t) = \left\langle E_s^*(t + \Delta t)E_s(t)\right\rangle. \tag{4.10}$$

Substituting Equation 4.4 into Equation 4.10 and using the independent scattering approximation, we have

$$R(\Delta t) = \frac{1}{r_0^2}\sum_{l=1}^{N}\left\langle s_l^*\left(\hat{k}_s, \hat{k}_i\right)s_l\left(\hat{k}_s, \hat{k}_i\right)e^{j\vec{k}_d\bullet[\vec{n}(t+\Delta t)-\vec{n}(t)]}\right\rangle. \tag{4.11}$$

The particle's random position can be expressed by $\vec{n}(t) = \vec{n}_0 + \vec{v}t$ with an initial position $\vec{n}_0$ and displacement by a random motion with a velocity $\vec{v}$. Hence, the time-correlated reflectivity becomes

$$R(\Delta t) = \frac{1}{r_0^2}\left\langle n\left|s_l\left(\hat{k}_s, \hat{k}_i\right)\right|^2\right\rangle\left\langle e^{j\vec{k}_d\bullet\vec{v}\Delta t}\right\rangle, \tag{4.12}$$

where $\left\langle n\left|s_l\left(\hat{k}_s, \hat{k}_i\right)\right|^2\right\rangle = \int\left|s\left(\hat{k}_s, \hat{k}_i; D\right)\right|^2 N(D)dD$. Consider the particle velocity as the sum of its mean ($\vec{v}_0$) and random fluctuation ($\vec{v}_1$), $\vec{v} = \vec{v}_0 + \vec{v}_1$, project $\vec{v}$ onto the radial direction $\hat{k}_d$, and assume that the radial component of the random fluctuation $v_{1r} = \vec{v}_{1r} \bullet \hat{k}_d$ follows Gaussian distribution, namely:

$$p(v_{1r}) = \frac{1}{\sqrt{2\pi}\sigma_v}\exp\left(-\frac{v_{1r}^2}{2\sigma_v^2}\right). \tag{4.13}$$

Using Equation 4.13 in Equation 4.12 and noting $\left\langle e^{jk_dv_1\Delta t}\right\rangle = \int e^{jk_dv_1\Delta t}p(v_1)dv_1$, we obtain

$$R(\Delta t) = \frac{1}{r_0^2}\left\langle n\left|s_l\left(\hat{k}_s, \hat{k}_i\right)\right|^2\right\rangle e^{-k_d^2\sigma_v^2\Delta t^2/2}e^{jk_dv_r\Delta t}, \tag{4.14}$$

where $v_r = \vec{v}_0 \cdot \hat{k}_d$ is the Doppler velocity, the projection of mean velocity $\vec{v}_0$ on the radial direction, and σ_v is the standard deviation of the velocity fluctuation, also called *spectrum width*. Both can be related to the autocorrelation as follows:

$$v_r = \angle\, R(\Delta t)/(k_d \Delta t) \tag{4.15}$$

$$\sigma_v = \left\{ 2\ln\left[|R(0)|/|R(\Delta t)|\right] \right\}^{1/2} \Big/ (k_d \Delta t). \tag{4.16}$$

Hence, Equations 4.15 and 4.16 constitute the independent Doppler weather radar foundation from the correlation function. The PDF is discussed next. Polarimetric radar variables are then defined in Section 4.2.6—those who are interested in understanding radar variables can go to Section 4.2.6 directly.

4.2.5 PDF OF SCATTERED WAVE FIELDS

We have discussed the first and second moments of the scattered wave field (which is treated as a random variable) and now know some of its statistical properties. A more complete description for a random scattered wave field, however, is its PDF (Ishimaru 1978, 1997; Tsang et al. 2000), which is discussed in this section.

4.2.5.1 Single-Polarization Wave Field

We now write the complex scattered wave field (Equation 4.4) in either amplitude (A) and phase (ϕ) form or real (X) and imaginary (Y) parts, as

$$E_s = Ae^{-j\phi} = X - jY, \tag{4.17}$$

where $X = A\cos\phi$ and $X = A\sin\phi$. The real and imaginary parts refer to the in-phase and quadrature radar signal (i/q data) because they are orthogonal to each other and the phases differ by $\pi/2$ (a quarter of the 2π period). As shown in Equation 4.4, the total scattered wave field E_s is the sum of the many particles' scattered fields, each of which can be treated as random variables. If there is no dominant random variable, the central limit theorem states that the probability distribution of a sum of N independent random variables approaches the normal distribution when $N \to \infty$, regardless of the distribution of the random variables (Papoulis 1991, 214).

From the central limit theorem, it can be assumed that the total scattered wave field E_s follows the normal (or Gaussian) distribution. Therefore, the X and Y are also normally distributed. Considering that the field amplitude A is related to the scattering amplitudes while the phase ϕ is determined by the random positions of the particles, as shown in Equation 4.4, it is reasonable to assume that the random amplitude A and phase ϕ are independent and the phase ϕ is uniformly distributed, as expressed in their PDFs:

$$p(A, \phi) = p(A)\, p(\phi) \tag{4.18}$$

$$p(\phi) = \frac{1}{2\pi} \quad (-\pi < \phi < \pi) \tag{4.19}$$

Using Equations 4.18 and 4.19, we can obtain the expected value of X and Y

$$\langle X \rangle = \langle A\cos\phi \rangle = \langle A \rangle \langle \cos\phi \rangle = 0 \tag{4.20}$$

$$\langle Y \rangle = \langle A\sin\phi \rangle = \langle A \rangle \langle \sin\phi \rangle = 0 \tag{4.21}$$

$$\langle XY \rangle = \langle A^2 \sin\phi\cos\phi \rangle = \langle A^2 \rangle \frac{1}{2} \langle \sin 2\phi \rangle = 0 \tag{4.22}$$

$$\langle X^2 \rangle = \langle A^2 \cos^2\phi \rangle = \langle A^2 \rangle \frac{1}{2} \langle 1 + \cos 2\phi \rangle = \frac{1}{2} \langle A^2 \rangle \equiv \sigma^2 = \langle Y^2 \rangle \tag{4.23}$$

$$\langle I \rangle = \langle A^2 \rangle = \langle X^2 \rangle + \langle Y^2 \rangle = 2\sigma^2. \tag{4.24}$$

Therefore, the joint PDF of the Gaussian distributed (X, Y) is

$$p(X, Y) = \frac{1}{2\pi\sigma^2} \exp\left(-\frac{X^2 + Y^2}{2\sigma^2}\right). \tag{4.25}$$

Transforming the differential areas by using $dXdY = AdAd\phi$, we get $p(X, Y)dXdY = p(X, Y) AdAd\phi = p(A,\phi)dAd\phi$, and from Equation 4.18 we have

$$P(A) = p(X, Y)A/p(\phi) = \frac{A}{\sigma^2} \exp\left(-\frac{A^2}{2\sigma^2}\right). \tag{4.26}$$

This shows that the random amplitude A follows the Rayleigh distribution. The PDFs $p(X)$, $p(Y)$, $p(A)$, and $p(\phi)$ are verified with Monte Carlo simulation (see Appendix 4A), shown in Figure 4.4. Also plotted are the theoretical results of Equations 4.18, 4.19, 4.25, and 4.26. It is evident that the simulation and theoretical results agree well with each other.

Further analysis of the random wave field can be obtained from calculating its statistical moments. The nth moment of the amplitude is

$$\langle A^n \rangle = \int_0^\infty A^n p(A)dA = \int_0^\infty \frac{A^{n+1}}{\sigma^2} \exp\left(-\frac{A^2}{2\sigma^2}\right)dA = \left(\sqrt{2}\sigma\right)^n \Gamma\left(\frac{n}{2}+1\right). \tag{4.27}$$

When $n = 1$ is used in Equation 4.27, it gives the mean value of the amplitude

$$\langle A \rangle = \left(\sqrt{2}\sigma\right)^1 \Gamma\left(\frac{3}{2}\right) = \sqrt{\frac{\pi}{2}}\sigma. \tag{4.28}$$

Hence, the variance of the random amplitude is

$$\sigma_A^2 = \langle A^2 \rangle - \langle A \rangle^2 = \left(2 - \frac{\pi}{2}\right)\sigma^2. \tag{4.29}$$

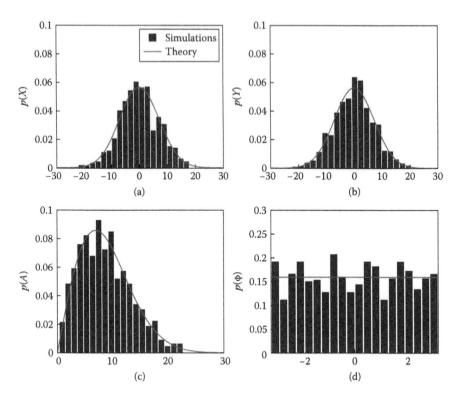

FIGURE 4.4 Statistics of single polarization scattered wave: (a) real part (X), (b) imaginary part (Y), (c) amplitude (A), and (d) phase (φ, radian). The simulation results agree well with the theory.

For the random intensity $I = A^2$, we have $p(I)dI = p(A)dA$ (Papoulis 1991). Using Equation 4.26, we obtain the PDF of the intensity

$$p(I) = p(A)\bigg/ \frac{dI}{dA} = \frac{1}{2\sigma^2}\exp\left(-\frac{I}{2\sigma^2}\right),$$ (4.30)

which is the exponential distribution. The nth moments of the intensity are then

$$\langle I^n \rangle = \int_0^\infty I^n p(I)dI = \int_0^\infty I^n \frac{1}{2\sigma^2}\exp\left(-\frac{I}{2\sigma^2}\right)dI = \left(2\sigma^2\right)^n n! = \langle I \rangle^n n!$$ (4.31)

While the mean intensity is known already from Equation 4.24 to be $\langle I \rangle = 2\sigma^2$, the variance of the random intensity can be found as follows:

$$\sigma_I^2 = \langle I^2 \rangle - \langle I \rangle^2 = 4\sigma^4 = \langle I \rangle^2.$$ (4.32)

Equation 4.32 shows that standard deviation of the intensity is equal to the mean intensity for the independent scattering from hydrometeors, which means that if there is only one independent sample, the power/intensity estimate has a relative error of 100%. That is why weather radars use a sequence of pulses to average for radar moment estimation.

4.2.5.2 Dual-Polarization Wave Fields

In the case of radar polarimetry, the scattered wave field can have arbitrary polarization, which requires two field components to be represented, as discussed in Section 3.1.4. Let E_h and E_v be the horizontal polarization and the vertical polarization fields, respectively. Like the expression in Equation 4.17, we can write the field components in either amplitude and phase form or in real and imaginary parts as follows:

$$E_h = A_h e^{-j\phi_h} = X_h - jY_h \tag{4.33}$$

$$E_v = A_v e^{-j\phi_v} = X_v - jY_v. \tag{4.34}$$

Therefore, X_h, Y_h, X_v, and Y_v are random variables, and they also follow a joint Gaussian distribution because of the central limit theorem. For the sake of simple expression, the four random variables are combined to form a vector as in Tsang et al. (2000):

$$\mathbf{Z} = [X_h, Y_h, X_v, Y_v], \tag{4.35}$$

and the covariance matrix $\mathbf{B}$ has 4×4 elements with each element expressed by

$$B_{ij} = \langle Z_i Z_j \rangle. \tag{4.36}$$

The joint Gaussian distribution can then be written as follows (Papoulis 1991):

$$p(X_h, Y_h, X_v, Y_v) = \frac{1}{\sqrt{(2\pi)^4 |\mathbf{B}|}} \exp\left(-\frac{1}{2}\mathbf{Z}\mathbf{B}^{-1}\mathbf{Z}^t\right), \tag{4.37}$$

where $|\mathbf{B}|$ is the determinant, the superscript -1 represents the inverse, and t means *transpose*.

From Equation 4.20 through 4.24, we have

$$\langle X_h \rangle = \langle Y_h \rangle = \langle X_v \rangle = \langle Y_v \rangle = 0 \tag{4.38}$$

$$\langle X_h Y_h \rangle = \langle X_v Y_v \rangle = 0 \tag{4.39}$$

$$\left\langle X_h^2 \right\rangle = \left\langle Y_h^2 \right\rangle = \frac{1}{2}\left\langle A_h^2 \right\rangle = \frac{1}{2}\langle I_h \rangle \tag{4.40}$$

$$\left\langle X_v^2 \right\rangle = \left\langle Y_v^2 \right\rangle = \frac{1}{2}\left\langle A_v^2 \right\rangle = \frac{1}{2}\langle I_v \rangle. \tag{4.41}$$

Noting the definition of the third and fourth Stokes parameters,

$$\langle U \rangle = 2\,\mathrm{Re}\left\langle E_h^* E_v \right\rangle = 2\langle X_h X_v \rangle + 2\langle Y_h Y_v \rangle, \tag{4.42}$$

$$\langle V \rangle = 2\,\mathrm{Im}\left\langle E_h^* E_v \right\rangle = -2\langle X_h Y_v \rangle + 2\langle X_v Y_h \rangle, \tag{4.43}$$

we obtain

$$\langle X_h X_v \rangle = \langle Y_h Y_v \rangle = \frac{1}{4}\langle U \rangle \tag{4.44}$$

$$-\langle X_h Y_v \rangle = \langle X_v Y_h \rangle = \frac{1}{4}\langle V \rangle. \tag{4.45}$$

Using Equations 4.36 and 4.38 through 4.45, we obtain the covariance matrix

$$\mathbf{B} = \begin{bmatrix} \dfrac{1}{2}\langle I_h \rangle & 0 & \dfrac{1}{4}\langle U \rangle & -\dfrac{1}{4}\langle V \rangle \\[2mm] 0 & \dfrac{1}{2}\langle I_h \rangle & \dfrac{1}{4}\langle V \rangle & \dfrac{1}{4}\langle U \rangle \\[2mm] \dfrac{1}{4}\langle U \rangle & \dfrac{1}{4}\langle V \rangle & \dfrac{1}{2}\langle I_v \rangle & 0 \\[2mm] -\dfrac{1}{4}\langle V \rangle & \dfrac{1}{4}\langle U \rangle & 0 & \dfrac{1}{2}\langle I_v \rangle \end{bmatrix}. \tag{4.46}$$

Once the covariance matrix is known, its determinant and inverse can be determined as follows:

$$|\mathbf{B}| = \left(\frac{1}{4}\langle I_h \rangle\langle I_v \rangle\right)^2 \left(1 - \rho_{hv}^2\right)^2, \tag{4.47}$$

where $\rho_{hv} = \sqrt{\dfrac{\langle U \rangle^2 + \langle V \rangle^2}{4\langle I_h \rangle\langle I_v \rangle}} = \dfrac{\left|\langle E_h^* E_v \rangle\right|}{\sqrt{\langle I_h \rangle\langle I_v \rangle}}$ is the correlation coefficient, and

$$\mathbf{B}^{-1} = \frac{4}{\langle I_h \rangle\langle I_h \rangle\left(1 - \rho_{hv}^2\right)} \begin{bmatrix} \dfrac{1}{2}\langle I_h \rangle & 0 & -\dfrac{1}{4}\langle U \rangle & \dfrac{1}{4}\langle V \rangle \\[2mm] 0 & \dfrac{1}{2}\langle I_h \rangle & -\dfrac{1}{4}\langle V \rangle & -\dfrac{1}{4}\langle U \rangle \\[2mm] -\dfrac{1}{4}\langle U \rangle & -\dfrac{1}{4}\langle V \rangle & \dfrac{1}{2}\langle I_v \rangle & 0 \\[2mm] \dfrac{1}{4}\langle V \rangle & -\dfrac{1}{4}\langle U \rangle & 0 & \dfrac{1}{2}\langle I_v \rangle \end{bmatrix}. \tag{4.48}$$

Substituting Equations 4.47 and 4.48 in Equation 4.37, we obtain the joint PDF of the X_h, Y_h, X_v, and Y_v:

$$p(X_h, Y_h, X_v, Y_v) = \frac{1}{\langle I_h \rangle \langle I_h \rangle \pi^2 \left(1 - \rho_{hv}^2\right)} \exp\left\{ -\frac{2}{\langle I_h \rangle \langle I_h \rangle \left(1 - \rho_{hv}^2\right)} \left[\frac{1}{2} \langle I_v \rangle \left(X_h^2 + Y_h^2\right) \right.\right.$$

$$\left.\left. + \frac{1}{2} \langle I_h \rangle \left(X_v^2 + Y_v^2\right) - \frac{1}{2} \langle U \rangle \left(X_h X_v + Y_h Y_v\right) + \frac{1}{2} \langle V \rangle \left(X_v Y_h - X_h Y_v\right) \right] \right\}. \tag{4.49}$$

Using the relation for the functional distribution, we obtain the joint PDF for the amplitudes and phases (A_h, ϕ_h, A_v, ϕ_v) as follows:

$$p(A_h, \phi_h, A_v, \phi_v) = A_h A_v \, p(X_h, Y_h, X_v, Y_v)$$

$$= \frac{A_h A_v}{\langle I_h \rangle \langle I_h \rangle \pi^2 \left(1 - \rho_{hv}^2\right)} \exp\left\{ -\frac{2}{\langle I_h \rangle \langle I_h \rangle \left(1 - \rho_{hv}^2\right)} \left[\frac{1}{2} \langle I_v \rangle A_h^2 + \frac{1}{2} \langle I_h \rangle A_v^2 \right.\right.$$

$$\left.\left. - \frac{1}{2} \langle U \rangle A_h A_v \cos\left(\phi_h - \phi_v\right) - \frac{1}{2} \langle V \rangle A_h A_v \sin\left(\phi_h - \phi_v\right) \right] \right\}. \tag{4.50}$$

Performing the integrations over the random phases (ϕ_h, ϕ_v), we obtain the joint PDF for the amplitudes (A_h, A_v) as follows:

$$p(A_h, A_v) = \frac{4 A_h A_v}{\langle I_h \rangle \langle I_v \rangle \left(1 - \rho_{hv}^2\right)} J_0 \left[\frac{2 \rho_{hv} A_h A_v}{\sqrt{\langle I_h \rangle \langle I_v \rangle} \left(1 - \rho_{hv}^2\right)} \right] \exp\left[\frac{\langle I_v \rangle A_h^2 + \langle I_h \rangle A_v^2}{\langle I_h \rangle \langle I_v \rangle \left(1 - \rho_{hv}^2\right)} \right], \tag{4.51}$$

where $J_0(\cdots)$ is the modified Bessel function of zero order. This joint PDF (Equation 4.51) for the amplitude simulation has also been verified with the Monte Carlo simulation, shown in Figure 4.5. Figure 4.5a is the simulation result and Figure 4.5b is the theoretical result of the joint PDF in Equation 4.51. Again, the theory agrees well with the simulation results, indicating that the assumption of Gaussian random variables for the scattered wave fields is valid for weather radar signals from hydrometeors.

By contrast, integrations over (A_h, A_v) with Equation 4.50 yield the joint PDF for the phases (ϕ_h, ϕ_v), giving

$$p(\phi_h, \phi_v) = \frac{\left(1 - \rho_{hv}^2\right)}{4\pi^2 \left(1 - \rho_d^2\right)} \left\{ 1 + \frac{\rho_d A_h A_v}{\sqrt{1 - \rho_d^2}} \left[\frac{\pi}{2} + \tan^{-1}\left(\frac{\rho_d}{\sqrt{1 - \rho_d^2}} \right) \right] \right\}, \tag{4.52}$$

where $\rho_d = \rho_{hv} \cos(\phi_d - \phi_{d0})$, with the phase difference of $\phi_d = \phi_h - \phi_v$ and the mean phase difference of $\phi_{d0} = \tan^{-1}\left(\langle V \rangle / \langle U \rangle\right)$. The fluctuation of the phase difference is associated with the random positions of the scatterers, whereas the mean phase difference is the sum of the mean scattering phase difference (δ) and the differential

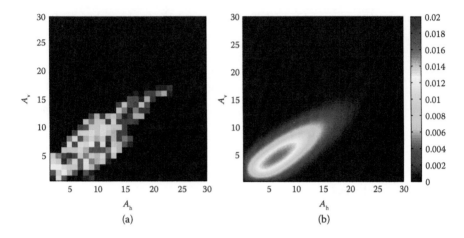

FIGURE 4.5 Statistics of dual-polarization scattered wave fields: (a) a simulated joint PDF of the scattered wave amplitudes $p(A_h, A_v)$; (b) a theoretical joint PDF of the scattered wave amplitudes $p(A_h, A_v)$.

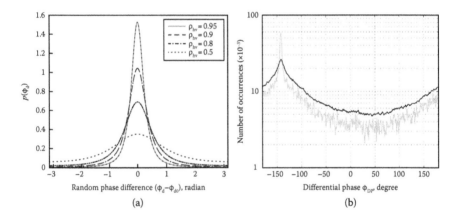

FIGURE 4.6 Dependence of the phase distribution on signal correlation. (a) Theoretical calculation from Equation 4.53 and (b) KOUN radar measurements. (From Zrnić, D. S., et al., 2006. *Journal of Atmospheric and Oceanic Technology*, 23, 381–394.)

propagation phase (ϕ_{dp}) between the H- and V-polarizations. Note that the joint distribution depends on only the phase difference. The PDF of the phase difference is then obtained by the integration of their sum, yielding

$$p(\phi_d) = \frac{\left(1 - \rho_{hv}^2\right)}{2\pi\left(1 - \rho_d^2\right)} \left\{ 1 + \frac{\rho_d A_h A_v}{\sqrt{1 - \rho_d^2}} \left[\frac{\pi}{2} + \tan^{-1}\left(\frac{\rho_d}{\sqrt{1 - \rho_d^2}} \right) \right] \right\}. \tag{4.53}$$

As shown in Figure 4.6a, the phase distribution depends on the correlation between the two orthogonal field components: the higher the correlation, the narrower the distribution. Normally, weather signals have a high correlation between the two channels, but clutter signals do not. This property can be used to separate clutter echoes from weather echoes, as shown in Figure 4.6b (Zrnić et al. 2006).

4.2.6 POLARIMETRIC RADAR VARIABLES

The theory and statistical properties of wave scattering were discussed in the two preceding subsections. It has been shown that independent scattering approximation is valid for radar observations of precipitation, in which the radar-received wave power flux density is proportional to the mean scattered wave intensity, and the scattered wave intensity is the sum of the scattered wave intensity by each individual particle (as shown in Equation 4.8), represented by radar reflectivity η, the backscattering cross section per unit volume. Taking size distribution into account, radar reflectivity is written as follows:

$$\eta = 4\pi \left\langle n\sigma_d \left(-\hat{k}_i, \hat{k}_i\right)\right\rangle = 4\pi \left\langle n \left|s\left(-\hat{k}_i, \hat{k}_i\right)\right|^2\right\rangle = 4\pi \left\langle n|s(\pi)|^2\right\rangle, \qquad (4.54)$$

where the angular bracket denotes the ensemble average over all particle size and canting angle distributions.

In the case of Rayleigh scattering of spherical particles, substituting Equation 3.62 into Equation 4.54 yields

$$\eta \equiv \frac{\pi^5 |K_w|^2}{\lambda^4} Z, \qquad (4.55)$$

where the dielectric constant factor of water is $K_w = \dfrac{\varepsilon_r - 1}{\varepsilon_r + 2}$. The reason that K_w is used is because hydrometeor species cannot be predetermined in a radar calibration. In the case of rain, the reflectivity factor is

$$Z = \left\langle nD^6\right\rangle = \int_0^{D_{max}} D^6 N(D) dD \equiv M_6. \qquad (4.56)$$

Z is also simply called *reflectivity* and is the sixth moment of a rain DSD. Z is the radar reflectivity η normalized with the radar wavelength and water dielectric constant factor K_w, and it depends on hydrometeor physics (i.e., DSD) only. Hence, the reflectivity factor Z [mm^6 m^{-3}] (instead of η) is usually provided by weather radars and used in radar meteorology.

When Rayleigh scattering approximation does not apply, the concept of the reflectivity factor is still used by introducing the effective reflectivity factor Z_e, defined following from Equation 4.55 and using Equation 4.54:

$$Z_e \equiv \frac{\lambda^4}{\pi^5 |K_w|^2} \eta = \frac{4\lambda^4}{\pi^4 |K_w|^2} \left\langle n|s(\pi)|^2\right\rangle. \qquad (4.57)$$

Expanding the definition of the reflectivity (Equation 4.57) for weather radar polarimetry, taking particle size distribution $N(D)$ and random orientation into account, and noting that

$$\left\langle ns_{pq}^{*}(\pi)s_{p'q'}(\pi)\right\rangle \equiv \int \left[s_{pq}^{*}(\pi; D, \vartheta, \varphi)s_{p'q'}(\pi; D, \vartheta, \varphi)\right] N(D)\,dD p(\vartheta, \varphi)\,d\vartheta\,d\varphi, \quad (4.58)$$

where $p(\vartheta, \varphi)$ is the PDF of the canting angles, we obtain the polarimetric radar reflectivity factors as follows:

$$Z_{hh} = \frac{4\lambda^4}{\pi^4 |K_w|^2} \left\langle n|s_{hh}(\pi)|^2 \right\rangle \quad (4.59)$$

for horizontal polarization,

$$Z_{vv} = \frac{4\lambda^4}{\pi^4 |K_w|^2} \left\langle n|s_{vv}(\pi)|^2 \right\rangle \quad (4.60)$$

for vertical polarization, and

$$Z_{hv} = \frac{4\lambda^4}{\pi^4 |K_w|^2} \left\langle n|s_{hv}(\pi)|^2 \right\rangle \quad (4.61)$$

for cross-polarization. Note that the reflectivity factors (Z_{pq}) are in units of [mm^6 m^{-3}], and they are usually represented in decibels as $Z_{PQ} = 10\log(Z_{pq})$ in [dBZ]. The polarization subscript is given in uppercase for the radar variables in the dB scale and in lowercase for the linear scale.

Taking ratios between the reflectivity factors and converting them to the dB scale, we have the definition of differential reflectivity

$$Z_{DR} = 10\log\left(\frac{Z_{hh}}{Z_{vv}}\right) = 10\log\left(\frac{\left\langle n|s_{hh}(\pi)|^2 \right\rangle}{\left\langle n|s_{vv}(\pi)|^2 \right\rangle}\right) \quad (4.62)$$

and the linear depolarization ratios for horizontally polarized transmission

$$LDR_H = 10\log\left(\frac{Z_{vh}}{Z_{hh}}\right) = 10\log\left(\frac{\left\langle n|s_{vh}(\pi)|^2 \right\rangle}{\left\langle n|s_{hh}(\pi)|^2 \right\rangle}\right) \quad (4.63)$$

and

$$LDR_V = 10\log\left(\frac{Z_{hv}}{Z_{vv}}\right) = 10\log\left(\frac{\left\langle n|s_{hv}(\pi)|^2 \right\rangle}{\left\langle n|s_{vv}(\pi)|^2 \right\rangle}\right) \quad (4.64)$$

for vertically polarized transmission. The differential reflectivity represents the difference in wave scattering by hydrometeors between horizontally polarized and vertically polarized waves, which is related to the shape and orientation of the particles. The linear depolarization ratios represent the effects of the particles' non-sphericity and canting in producing cross-polarized waves.

Similarly, the correlation coefficient magnitudes and phase can be defined. The correlation coefficients are

$$\rho_{hv} = \frac{\left|\left\langle ns_{hh}^*(\pi)s_{vv}(\pi)\right\rangle\right|}{\left(\left\langle n\left|s_{hh}(\pi)\right|^2\right\rangle\left\langle n\left|s_{vv}(\pi)\right|^2\right\rangle\right)^{1/2}} \tag{4.65}$$

$$\rho_{h} = \frac{\left|\left\langle ns_{hh}^*(\pi)s_{vh}(\pi)\right\rangle\right|}{\left(\left\langle n\left|s_{hh}(\pi)\right|^2\right\rangle\left\langle n\left|s_{vh}(\pi)\right|^2\right\rangle\right)^{1/2}} \tag{4.66}$$

$$\rho_{v} = \frac{\left|\left\langle ns_{vv}^*(\pi)s_{hv}(\pi)\right\rangle\right|}{\left(\left\langle n\left|s_{vv}(\pi)\right|^2\right\rangle\left\langle n\left|s_{hv}(\pi)\right|^2\right\rangle\right)^{1/2}} \tag{4.67}$$

for co-polar correlation, co-to-cross correlation for H-pol transmission, and co-to-cross correlation for V-pol transmission, respectively. The correlation coefficients represent the similarity of wave scattering between polarizations.

Corresponding to the three correlation coefficients above, there are three scattering differential phases:

$$\phi_{sd} = \angle\left\langle ns_{hh}^*(\pi)s_{vv}(\pi)\right\rangle \equiv \delta_d \tag{4.68}$$

$$\phi_{sh} = \angle\left\langle ns_{hh}^*(\pi)s_{vh}(\pi)\right\rangle \equiv \delta_h \tag{4.69}$$

$$\phi_{sv} = \angle\left\langle ns_{vv}^*(\pi)s_{hv}(\pi)\right\rangle \equiv \delta_v. \tag{4.70}$$

It is evident that 9 of the 12 polarimetric radar variables defined in Equations 4.59 through 4.70 are independent. These nine independent variables are associated with the covariance matrix for the full-polarization wave scattering (Zrnić 1991). Based on the reciprocity theorem (Tsang et al. 1985), the cross-polarization terms of the scattering matrix (Equation 3.129) are equal: $s_{hv} = s_{vh}$. The three complex random scattering amplitudes, s_{hh}, s_{vv}, and s_{hv}, constitute a scattering vector $\mathbf{s} = [s_{hh}, s_{vv}, s_{hv}]$ (Borgeaud et al. 1987; Jameson 1985). The covariance matrix is then defined as follows:

$$\mathbf{C} = \left\langle ns^+s\right\rangle = \begin{bmatrix} \left\langle n\left|s_{hh}\right|^2\right\rangle & \left\langle ns_{hh}^*s_{vv}\right\rangle & \left\langle ns_{hh}^*s_{hv}\right\rangle \\ \left\langle ns_{vv}^*s_{hh}\right\rangle & \left\langle n\left|s_{vv}\right|^2\right\rangle & \left\langle ns_{vv}^*s_{hv}\right\rangle \\ \left\langle ns_{hv}^*s_{hh}\right\rangle & \left\langle ns_{hv}^*s_{vv}\right\rangle & \left\langle n\left|s_{hv}\right|^2\right\rangle \end{bmatrix}, \tag{4.71}$$

noting that s^+ means *conjugate transpose* or *Hermitian transpose of* $\mathbf{s}$.

As in Equations 4.59 through 4.70, the 3×3 scattering covariance (Equation 4.71) also provides nine independent pieces of information for full polarimetry: three from the diagonal power terms, six from the three upper off diagonal correlation terms with each having magnitude and phase information. The matrix is symmetrical; the lower-left part produces no more information.

In the case of simultaneous dual-polarization measurements, s_{hv} is zero and only two elements of the scattering matrix are measured (s_{hh} and s_{vv}). Hence, only the upper-left 2×2 correlation functions of the 3×3 scattering covariance matrix can be obtained, producing four independent pieces of information.

In the covariance matrix (Equation 4.71), the diagonal terms are the ACFs of the co-polar and cross-polar scattering wave fields, representing power return and Doppler information. The off-diagonal terms are the cross-correlation functions between polarizations, which represent the correlativity and relative phases between different polarized wave fields. To calculate the correlation functions and the polarimetric radar variables, we use the scattering matrix (Equation 3.129) and decompose the average over the size distribution and the average over the canting angle distribution to obtain

$$\left\langle n|s_{hh}|^2\right\rangle = \left\langle n\left(As_a^* + Bs_b^*\right)\left(As_a + Bs_b\right)\right\rangle$$

$$= \left\langle A^2\right\rangle\left\langle n|s_a|^2\right\rangle + \left\langle B^2\right\rangle\left\langle n|s_b|^2\right\rangle + 2\left\langle AB\right\rangle\mathrm{Re}\left\langle ns_a^*s_b\right\rangle \tag{4.72}$$

$$\left\langle n|s_{vv}|^2\right\rangle = \left\langle D^2\right\rangle\left\langle n|s_a|^2\right\rangle + \left\langle C^2\right\rangle\left\langle n|s_b|^2\right\rangle + 2\left\langle CD\right\rangle\mathrm{Re}\left\langle ns_a^*s_b\right\rangle \tag{4.73}$$

$$\left\langle n|s_{hv}|^2\right\rangle = \left\langle n|s_{vh}|^2\right\rangle = \left\langle BC\right\rangle\left\langle n|s_a - s_b|^2\right\rangle \tag{4.74}$$

$$\left\langle ns_{hh}^*s_{vv}\right\rangle = \left\langle AD\right\rangle\left\langle n|s_a|^2\right\rangle + \left\langle BC\right\rangle\left\langle n|s_b|^2\right\rangle + \left\langle AC\right\rangle\left\langle ns_a^*s_b\right\rangle + \left\langle BD\right\rangle\left\langle ns_a s_b^*\right\rangle \tag{4.75}$$

$$\left\langle ns_{hh}^*s_{hv}\right\rangle = \left\langle A\sqrt{BC}\right\rangle\left\langle n|s_a|^2\right\rangle - \left\langle B\sqrt{BC}\right\rangle\left\langle n|s_b|^2\right\rangle + \left\langle A\sqrt{BC}\right\rangle\left\langle ns_a^*s_b\right\rangle - \left\langle B\sqrt{BC}\right\rangle\left\langle ns_a s_b^*\right\rangle \tag{4.76}$$

$$\left\langle ns_{vv}^*s_{hv}\right\rangle = \left\langle D\sqrt{BC}\right\rangle\left\langle n|s_a|^2\right\rangle - \left\langle C\sqrt{BC}\right\rangle\left\langle n|s_b|^2\right\rangle + \left\langle D\sqrt{BC}\right\rangle\left\langle ns_a^*s_b\right\rangle - \left\langle C\sqrt{BC}\right\rangle\left\langle ns_a s_b^*\right\rangle, \tag{4.77}$$

where the angular brackets $\langle\cdots\rangle$ denote the ensemble average. As in Equation 4.58, the average over particle size distribution can be calculated as

$$\left\langle nf(s_a, s_b)\right\rangle = \int f(s_a, s_b)N(D)dD \tag{4.78}$$

where $f(s_a, s_b)$ represents the function of s_a and s_b, and $N(D)$ is the particle size distribution (PSD), as described in Chapter 2. The backscattering amplitude s_a and s_b were provided in Chapter 3.

Similarly, the average over particle canting angles is represented by

$$\left\langle g\left(A, B, C, D\right)\right\rangle = \int g\left(A, B, C, D\right)p\left(\vartheta, \varphi\right)d\vartheta\, d\varphi. \tag{4.79}$$

To obtain the average over canting angles, we assume that the canting angles ϑ and φ are independent $p(\vartheta, \varphi) = p(\vartheta)p(\varphi)$ and Gaussian-distributed with PDFs of

$$p(\vartheta) = \frac{1}{\sqrt{2\pi}\sigma_\vartheta} \exp\left(-\frac{(\vartheta - \bar{\vartheta})^2}{2\sigma_\vartheta^2}\right) \tag{4.80}$$

$$p(\varphi) = \frac{1}{\sqrt{2\pi}\sigma_\varphi} \exp\left(-\frac{(\varphi - \bar{\varphi})^2}{2\sigma_\varphi^2}\right), \tag{4.81}$$

where $(\bar{\vartheta}, \bar{\varphi})$ are their mean and $(\sigma_\vartheta, \sigma_\varphi)$ are the standard deviations. After performing integrations (see Appendix 4B), we obtain the statistics of the angle-dependent factors in Equations 4B.12 through 4B.25.

Substitution of Equations 4B.12 through 4B.25 into Equations 4.72 through 4.77 gives the scattering covariance elements. Once the PSD and the canting angle statistics are specified, the polarimetric radar variables are ready to calculate from Equations 4.59 through 4.70 using the scattering amplitudes s_a and s_b described in Chapter 3. In a special case of particles canting in the polarization plane, $\vartheta = \sigma_\vartheta = 0$, with a zero mean, $\bar{\varphi} = 0$, and a small standard deviation, $\sigma_\varphi \ll 1$, we obtain the mean differential scattering cross sections and the correlation function as

$$\left\langle n|s_{hh}|^2 \right\rangle \approx \left(1 - 2\sigma_\varphi^2\right)\left\langle n|s_a|^2 \right\rangle + 2\sigma_\varphi^2 \operatorname{Re}\left\langle n s_a^* s_b \right\rangle \tag{4.82}$$

$$\left\langle n|s_{vv}|^2 \right\rangle \approx \left(1 - 2\sigma_\varphi^2\right)\left\langle n|s_b|^2 \right\rangle + 2\sigma_\varphi^2 \operatorname{Re}\left\langle n s_a^* s_b \right\rangle \tag{4.83}$$

$$\left\langle n|s_{hv}|^2 \right\rangle = \left\langle n|s_{vh}|^2 \right\rangle \approx \sigma_\varphi^2 \left\langle n|s_a - s_b|^2 \right\rangle \tag{4.84}$$

$$\left\langle n s_{hh}^* s_{vv} \right\rangle \approx \sigma_\varphi^2 \left\langle n|s_a|^2 \right\rangle + \sigma_\varphi^2 \left\langle n|s_b|^2 \right\rangle + \left(1 - 2\sigma_\varphi^2\right)\left\langle n s_a^* s_b \right\rangle. \tag{4.85}$$

From Equations 4.82 and 4.83, we can see that the difference between the two mean differential scattering cross sections $\left(\left\langle n|s_{hh}|^2 \right\rangle, \left\langle n|s_{vv}|^2 \right\rangle\right)$ decreases as the standard deviation of the canting angle increases, yielding a smaller Z_{DR} in Equation 4.62. This is expected because the particles are effectively more spherical if random orientation is taken into account. For cross-polarization scattering, however, $\left\langle n|s_{hv}|^2 \right\rangle$ and LDR in Equations 4.63 and 4.64 increase as the canting angle increases. This also makes a sense because canted particles produce wave scattering with polarization in their axes of symmetry that is not in the direction of the H or V reference polarizations, resulting in cross-polar scattering. Random orientation and size distribution of particles also cause the reduction of the co-polar correlation coefficient ρ_{hv}, especially for non-Rayleigh scattering when the scattering phase difference becomes important. This can be shown through the last term of Equation 4.85 as follows:

$$\left\langle n s_a^* s_b \right\rangle = \left\langle n|s_a|e^{j\delta_a}|s_b|e^{-j\delta_b} \right\rangle \approx \left\langle n|s_a||s_b| \right\rangle\left\langle e^{j(\delta_a - \delta_b)} \right\rangle = \left\langle n|s_a||s_b| \right\rangle e^{-\sigma_\delta^2/2} e^{j\delta}, \tag{4.86}$$

where the mean scattering phase difference is $\delta = \langle \delta_a - \delta_b \rangle$ which can bias the propagation differential phase ϕ_{dp} estimate, and the standard deviation is $\sigma_\delta = SD(\delta_a - \delta_b)$. It is clear that the co-polar correlation coefficient is reduced by a factor of $e^{-\sigma_\delta^2/2}$ due to random scattering phase difference in the case of melting snow, hail, and biological scatters, as well as ground clutter.

Note that all of the above defined polarimetric radar variables are intrinsic values without consideration of the wave propagation effects. The propagation effects are discussed next, and the propagation-included radar variables are discussed in Section 4.4.

4.3   COHERENT WAVE PROPAGATION

In the single scattering approximation discussed in the last section, it was assumed that a wave propagates in free space or in deterministic background media, and the effects of the presence of particles on wave propagation were ignored. This assumption is not valid in the case of the first-order multiple scattering and multiple scattering models where a particle's effects on wave propagation need to be taken into account (Ishimaru 1978, 1997; Twersky 1964). We will study coherent (mean) wave propagation with first-order multiple scattering and its effects on weather radar variables and measurements.

4.3.1   CONCEPT OF EFFECTIVE MEDIUM

Both the incident wave and scattered wave propagate in an atmosphere containing hydrometeor particles. The particles affect wave propagation through absorption and scattering. As discussed in Section 4.2.2, scattering contributes to the mean wave field only in the forward scattering direction. The propagation effects include attenuation, differential attenuation, phase shift, and differential phase, as well as depolarization. Effectively, these propagation effects can be represented by that of an effective medium characterized by an effective complex wave number or refraction index, as illustrated in Figure 4.7.

4.3.2   SCALAR WAVE PROPAGATION

To formulate coherent wave propagation in an atmosphere filled with hydrometeor particles, we divide the medium into multiple slabs, as shown in Figure 4.8.

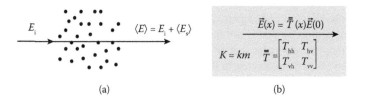

(a)   (b)

FIGURE 4.7 The first-order multiple scattering for wave propagation: (a) wave propagation in a medium of randomly distributed particles and (b) effective medium for wave propagation.

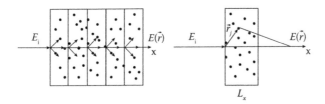

FIGURE 4.8 Conceptual diagram of the first-order multiple scattering and configuration of a plane wave incident on the volume of particles in a slab.

First, consider a unit plane wave incident on the volume of a slab, $V = L_x \times L_y \times L_z$. Hence, the total wave field in a given location is the sum of the incident field and the scattered wave field, expressed by

$$E = E_i + E_s = E_0 e^{-jkx} + \sum_{l=1}^{N} \frac{e^{-jk|\vec{r}-\vec{r}_l|}}{|\vec{r}-\vec{r}_l|} E_0 s_l \left(\hat{k}_s, \hat{k}_i\right) e^{-jkx_l}. \tag{4.87}$$

Taking the ensemble average of Equation 4.87 and using the PDF of $p(\vec{r}_l) = 1/V$, we obtain the coherent wave field as follows:

$$\langle E \rangle = E_0 e^{-jkx} + E_0 \sum_{l=1}^{N} \left\langle \frac{e^{-jk|\vec{r}-\vec{r}_l|}}{|\vec{r}-\vec{r}_l|} s_l \left(\hat{k}_s, \hat{k}_i\right) e^{-jkx_l} \right\rangle$$

$$= E_0 e^{-jkx} + E_0 \sum_{l=1}^{N} \int \frac{e^{-jk|\vec{r}-\vec{r}_l|}}{|\vec{r}-\vec{r}_l|} s_l \left(\hat{k}_s, \hat{k}_i\right) e^{-jkx_l} \frac{d\vec{r}_l}{V}$$

$$\approx E_0 e^{-jkx} + E_0 n \int_0^{L_x} \frac{dx_l}{(x-x_l)} \int_{-L_y/2}^{L_y/2} dy_l \int_{-L_z/2}^{L_z/2} dz_l \tag{4.88}$$

$$\times \exp\left\{-jk\left[(x-x_l) + \frac{y_l^2 + z_l^2}{2(x-x_l)}\right]\right\} s_l \left(\hat{k}_s, \hat{k}_i\right) e^{-jkx_l}$$

$$= E_0 e^{-jkx} \left[1 - \frac{2\pi j}{k} n s_l \left(\hat{k}_i, \hat{k}_i\right) L_x\right]$$

$$\approx E_0 e^{-jkx} \exp\left[-\frac{2\pi j}{k} n s_l \left(\hat{k}_i, \hat{k}_i\right) L_x\right].$$

Collecting the contributions from all the slabs and noting the drop size and canting angle distributions, we can express the coherent field by the following:

$$\langle E(x) \rangle = E_0 e^{-j\int_0^x K d\ell} \tag{4.89}$$

with

$$K = k + \frac{2\pi}{k} \langle ns(0) \rangle = km \tag{4.90}$$

as the effective propagation constant. The effects of the presence of the particles on wave propagation are an increase in the wave number and a decrease in wavelength. Hence, it attenuates wave intensity and slows down wave propagation. This leads to the definition of specific attenuation and specific phase of coherent wave propagation in a medium. *Specific attenuation (A)* is defined as "the attenuation that a wave experiences during its propagation in a unit distance (typically one kilometer), expressed as follows:

$$I_c(x) = \left|\langle E(x)\rangle\right|^2 = I_0 e^{-2\int_0^x \mathrm{Im}(K)\,d\ell} = I_0 e^{-2\mathrm{Im}(K)x} \tag{4.91}$$

$$A = 10\log\left(\frac{I_0}{I_c(1\mathrm{km})}\right) = 8.686 \times \mathrm{Im}(K) = 8.686\lambda \times \mathrm{Im}\langle ns(0)\rangle\left[\mathrm{dB\ km}^{-1}\right]. \tag{4.92}$$

From the optical theorem (Equation 3.51), the specific attenuation can be related to the extinction cross section as $A = 4.343\langle n\sigma_t\rangle$. This makes sense because attenuation is caused by the total cross section contributed by absorption and scattering. It is also common to use the extinction coefficient $\alpha = \langle n\sigma_t\rangle = 2\,\mathrm{Im}(K) = 0.23\,A\,[\mathrm{km}^{-1}]$ to represent the signal power loss per unit distance.

Similarly, the specific phase is the phase change of the wave propagation over a unit distance, which is

$$K' = \mathrm{Re}(K) = k + \lambda \times \mathrm{Re}\langle ns(0)\rangle. \tag{4.93}$$

Note that the result of the effective wave number K expressed by Equation 4.90 is consistent with the result of the effective dielectric constant for a mixture (discussed in Chapter 2). This can be reconciled as follows.

Clouds and precipitation are sparse media with a fractional volume of <0.001%, satisfying the sparse medium condition of 1%. Using the Maxwell-Garnet mixing formula (Equations 2.65 and 2.66) and treating air as the background and hydrometeor particles as inclusion, we have

$$\varepsilon_e = \varepsilon_1 \frac{1+2f_v y}{1-f_v y}; \quad y = \frac{\varepsilon_2 - \varepsilon_1}{\varepsilon_2 + 2\varepsilon_1} = \frac{\varepsilon_r - 1}{\varepsilon_r + 2} \tag{4.94}$$

$$\varepsilon_e \approx (1+3f_v y) = 1 + n\frac{4\pi a^3}{3}\frac{3(\varepsilon_r - 1)}{\varepsilon_r + 2}. \tag{4.95}$$

Noting the Rayleigh scattering amplitude (Equation 3.62), Equation 4.95 becomes

$$\varepsilon_e \approx 1 + \frac{4\pi}{k^2}ns(0). \tag{4.96}$$

Hence, the effective refraction index and wave number are

$$m = \sqrt{\varepsilon_e} \approx 1 + \frac{2\pi}{k^2}ns(0) \tag{4.97}$$

and

$$K = km \approx k\left(1 + \frac{2\pi}{k^2} ns(0)\right). \tag{4.98}$$

This is the same as that of Equation 4.90 when the particles are of the same size (monodispersion). Note that Equation 4.98 was obtained based on the assumption of a sparse medium and Rayleigh scattering. However, Equation 4.90 applies to non-Rayleigh scattering and can be extended to those media that consist of nonspherical particles.

4.3.3 POLARIZED WAVE PROPAGATION

In the previous subsection, the effective propagation constant was derived based on the assumption of scalar wave propagation in an isotropic medium composed of spherical particles. Clouds and precipitation are not necessarily isotropic when non-spherical hydrometeor particles are present with orientation. Hence, the properties of polarized wave propagation need to be characterized. Coherent wave propagation was thoroughly studied during the 1980s when satellite communication started to become popular (Oguchi 1983; Olsen 1982).

4.3.3.1 No Canting Angle

First, consider polarized wave propagation in a medium consisting of nonspherical particles with canting. The major and minor axes align with the wave polarization directions. In this case, there is no depolarization (defined as one polarization coupled to another polarization). Hence, the wave field in each polarization can be treated as a scalar wave propagation like those formulated in the last subsection,

$\langle E(x) \rangle \equiv E_c(x) = E_c(0)e^{-j\int_0^x K d\ell}$, but with a different propagation constant K:

$$K_a = k + \frac{2\pi}{k}\langle ns_a(0)\rangle \tag{4.99}$$

$$K_b = k + \frac{2\pi}{k}\langle ns_b(0)\rangle. \tag{4.100}$$

Therefore, coherent wave propagation can be expressed by the transmission matrix $\overline{T}$ as follows:

$$\begin{bmatrix} E_h(x) \\ E_v(x) \end{bmatrix} = \begin{bmatrix} e^{-j\int_0^x K_a d\ell} & 0 \\ 0 & e^{-j\int_0^x K_b d\ell} \end{bmatrix} \begin{bmatrix} E_h(0) \\ E_v(0) \end{bmatrix} \equiv \begin{bmatrix} T_{hh} & T_{hv} \\ T_{vh} & T_{vv} \end{bmatrix} \begin{bmatrix} E_h(0) \\ E_v(0) \end{bmatrix} \tag{4.101}$$

This would yield different specific attenuations and phase shifts for different polarizations.

4.3.3.2 Random Orientation with Zero Means

In practice, each particle's major and minor axes may not align with the reference H- and V-polarizations, but their mean canting angles are zero. In this case, there is still no depolarization because of the statistical symmetry of the configuration. Equations 4.99 through 4.101 can be extended to take the statistical orientation into account, as expressed by the following:

$$
\begin{bmatrix} E_h(x) \\ E_v(x) \end{bmatrix} = \begin{bmatrix} e^{-j\int_0^x K_h d\ell} & 0 \\ 0 & e^{-j\int_0^x K_v d\ell} \end{bmatrix} \begin{bmatrix} E_h(0) \\ E_v(0) \end{bmatrix}, \tag{4.102}
$$

where the effective propagation constants for the H- and V-polarizations, using the forward-scattering amplitude (Equation 3.129 through 3.133), are as follows:

$$
K_h = k + \frac{2\pi}{k}\langle ns_{hh}(0)\rangle
$$
$$
= k + \frac{2\pi}{k}\langle n[s_a(\cos^2\varphi + \sin^2\vartheta\sin^2\varphi) + s_b\cos^2\vartheta\sin^2\varphi]\rangle \tag{4.103}
$$

$$
K_v = k + \frac{2\pi}{k}\langle ns_{vv}(0)\rangle
$$
$$
= k + \frac{2\pi}{k}\langle n[(s_a(\sin^2\varphi + \sin^2\vartheta\cos^2\varphi) + s_b\cos^2\vartheta\cos^2\varphi]\rangle. \tag{4.104}
$$

The difference of the effective propagation constants between the H- and V-polarizations is

$$
K_h - K_v = \frac{2\pi}{k}\left\langle n\begin{bmatrix} s_a((\cos^2\varphi - \sin^2\varphi) + \sin^2\vartheta(\sin^2\varphi - \cos^2\varphi)) \\ + s_b\cos^2\vartheta(\sin^2\varphi - \cos^2\varphi) \end{bmatrix}\right\rangle
$$
$$
= \frac{2\pi}{k}\langle n[s_a(\cos 2\varphi - \sin^2\vartheta\cos 2\varphi) - s_b\cos^2\vartheta\cos 2\varphi]\rangle
$$
$$
= \frac{2\pi}{k}\langle n[s_a(1 - \sin^2\vartheta)\cos 2\varphi - s_b\cos^2\vartheta\cos 2\varphi]\rangle \tag{4.105}
$$
$$
= \frac{2\pi}{k}\langle n(s_a - s_b)\cos^2\vartheta\cos 2\varphi\rangle
$$
$$
= \frac{2\pi}{k}\langle n(s_a - s_b)\rangle\langle\cos^2\vartheta\cos 2\varphi\rangle
$$
$$
\approx \frac{2\pi}{k}\langle n(s_a - s_b)\rangle\frac{1}{2}\left(1 + e^{-2\sigma_\vartheta^2}\right)e^{-2\sigma_\varphi^2}.
$$

Equation 4.105 gives the expression of the difference between the effective propagation constants. As expected, $\Delta K = K_h - K_v$ is proportional to the ensemble average of the difference of the forward-scattering amplitudes $\langle n(s_a - s_b) \rangle$, and the effect of random orientation is to reduce ΔK, as $\frac{1}{2}\left(1 + e^{-2\sigma_\theta^2}\right)e^{-2\sigma_\phi^2}$ is always less than unity.

The specific differential phase K_{DP}, defined as the real part of the propagation constant difference in degrees per kilometer [degree/km], is

$$K_{DP} = \frac{180}{\pi} \text{Re}\left[K_h - K_v\right]$$

$$\approx \frac{180\lambda}{\pi} \int \text{Re}\left[s_a(0, D) - s_b(0, D)\right] N(D)dD\frac{1}{2}\left(1 + e^{-2\sigma_\theta^2}\right)e^{-2\sigma_\phi^2}.$$

$$(4.106)$$

Similarly, the specific differential attenuation (A_{DP}, dB) is the difference of the specific attenuations between the H- and V-polarizations expressed in Equations 4.91 and 4.92, as given by the following:

$$A_{DP} = A_h - A_v$$

$$= 8.686\,\text{Im}\left[K_h - K_v\right]$$

$$= 8.686\lambda \int \text{Im}\left[s_a(0, D) - s_b(0, D)\right]N(D)dD\frac{1}{2}\left(1 + e^{-2\sigma_\theta^2}\right)e^{-2\sigma_\phi^2}.$$

$$(4.107)$$

4.3.3.3 General Formulation for Coherent Wave Propagation

In general, the mean canting angles may not be zero as shown in Figure 4.9, and the polarization state would change during propagation, as formulated and documented by Oguchi (1975, 1983). In addition to the co-polar components that have

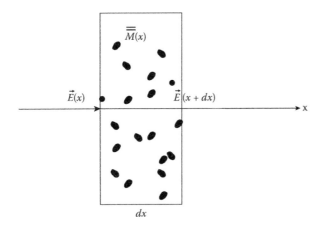

FIGURE 4.9 Configuration of a plane wave incident on the volume of particles in a slab.

been examined, there would be cross-polarization components due to scattering, which is called *depolarization in propagation*. Hence, the transmission matrix is not diagonal anymore. In this case, the coherent wave equation (Equation 4.102) $\vec{E}(x) = \bar{\bar{T}}(x)\vec{E}(0)$ is rewritten in a differential form that includes the effects of depolarization, as follows:

$$\frac{d\vec{E}}{dx} = \bar{\bar{M}}\vec{E}, \tag{4.108}$$

where $\bar{\bar{M}}$ is the medium characteristic matrix, and its elements are

$$M_{hh} = -jk - j\frac{2\pi}{k}\langle ns_{hh}(0)\rangle = -jk - j\frac{2\pi}{k}\langle n\left[s_a\left(\cos^2\varphi + \sin^2\vartheta\sin^2\varphi\right) + s_b\cos^2\vartheta\sin^2\varphi\right]\rangle$$

$$M_{hv} = -j\frac{2\pi}{k}\langle ns_{hv}(0)\rangle = -j\frac{2\pi}{k}\langle n\left[(s_a - s_b)\cos^2\vartheta\sin\varphi\cos\varphi)\right]\rangle$$

$$M_{vh} = -j\frac{2\pi}{k}\langle ns_{vh}(0)\rangle = -j\frac{2\pi}{k}\langle n\left[(s_a - s_b)\cos^2\vartheta\sin\varphi\cos\varphi)\right]\rangle$$

$$M_{vv} = -jk - j\frac{2\pi}{k}\langle ns_{vv}(0)\rangle = -jk - j\frac{2\pi}{k}\langle n\left[(s_a\left(\sin^2\varphi + \sin^2\vartheta\cos^2\varphi\right) + s_b\cos^2\vartheta\cos^2\varphi\right]\rangle.$$
$$\tag{4.109}$$

The solution of Equation 4.108 can be expressed by the eigenvalues (λ_1, λ_2) and eigenvectors represented by ϕ as follows:

$$\begin{bmatrix} E_h(x) \\ E_v(x) \end{bmatrix} = \begin{bmatrix} e^{\lambda_1 x}\cos^2\phi + e^{\lambda_2 x}\sin^2\phi & \left(e^{\lambda_1 x} - e^{\lambda_2 x}\right)\sin\phi\cos\phi \\ \left(e^{\lambda_1 x} - e^{\lambda_2 x}\right)\sin\phi\cos\phi & e^{\lambda_2 x}\cos^2\phi + e^{\lambda_1 x}\sin^2\phi \end{bmatrix} \begin{bmatrix} E_h(0) \\ E_v(0) \end{bmatrix} \tag{4.110}$$

where the eigenvalues are the solution of the determinant

$$\begin{vmatrix} M_{hh} - \lambda & M_{hv} \\ M_{vh} & M_{vv} - \lambda \end{vmatrix} = 0 \tag{4.111}$$

with

$$\lambda_{1,2} = \frac{M_{hh} + M_{vv}}{2} \pm \sqrt{\frac{(M_{hh} - M_{vv})^2}{4} + M_{hv}M_{vh}} \tag{4.112}$$

and

$$\tan 2\phi = \frac{2M_{vh}}{M_{hh} - M_{vv}}. \tag{4.113}$$

Assume the canting angle statistics are Gaussian distributed: $\vartheta : N\left(\overline{\vartheta}, \sigma_{\vartheta}\right)$ and $\varphi : N\left(\overline{\varphi}, \sigma_{\varphi}\right)$. We have

$$M_{hv} = M_{vh} = -j\frac{2\pi}{k}\langle n(s_a - s_b)\rangle \frac{1}{4}\left(1 + e^{-2\sigma_\vartheta} \cos 2\overline{\vartheta}\right)e^{-2\sigma_\varphi} \sin 2\overline{\varphi} \qquad (4.114)$$

and

$$M_{hh} - M_{vv} = -j\frac{2\pi}{k}\langle n(s_a - s_b)\rangle \frac{1}{2}\left(1 + e^{-2\sigma_\vartheta} \cos 2\overline{\vartheta}\right)e^{-2\sigma_\varphi} \cos 2\overline{\varphi}. \qquad (4.115)$$

Using Equations 4.114 and 4.115 in Equations 4.112 and 4.113, we have

$$\lambda_1 - \lambda_2 = -j\frac{2\pi}{k}\langle n(s_a - s_b)\rangle \frac{1}{2}\left(1 + e^{-2\sigma_\vartheta} \cos 2\overline{\vartheta}\right)e^{-2\sigma_\varphi} \equiv -j(K_h - K_v) \quad (4.116)$$

and

$$\tan 2\phi = \tan 2\overline{\varphi} \longrightarrow \phi = \overline{\varphi}, \qquad (4.117)$$

respectively. Hence, we have the specific differential phase

$$K_{DP} = \frac{180}{\pi}\mathrm{Re}\left[K_h - K_v\right]$$

$$\approx \frac{180\lambda}{\pi}\int \mathrm{Re}\left[s_a(0, D) - s_b(0, D)\right]N(D)dD\frac{1}{2}\left(1 + e^{-2\sigma_\vartheta^2} \cos 2\overline{\vartheta}\right)e^{-2\sigma_\varphi^2} \qquad (4.118)$$

and then the specific differential attenuation as

$$A_{DP} = 8.686\,\mathrm{Im}\left[K_h - K_v\right]$$

$$= 8.686\lambda\int \mathrm{Im}\left[s_a(0, D) - s_b(0, D)\right]N(D)dD\frac{1}{2}\left(1 + e^{-2\sigma_\vartheta^2} \cos 2\overline{\vartheta}\right)e^{-2\sigma_\varphi^2}. \qquad (4.119)$$

Equations 4.118 and 4.119 constitute the formulations for calculating the specific differential phase and the specific differential attenuation from particle DSDs and orientation angle statistics.

4.4 PROPAGATION-INCLUDED SCATTERING

Thus far, we have described both wave scattering and propagation in a medium consisting of random particles, but we have addressed them separately. In weather measurements, however, the radar-transmitted EM wave propagates in clouds/precipitation consisting of hydrometeors, is scattered by hydrometeor particles, and then propagates back to the radar. Radar measurements contain both the scattering and propagation effects, and it is important to know the formulation for these effects.

4.4.1 Transmission-Included Scattering Matrix

In Section 4.3, wave scattering by a particle was represented by the scattering matrix, whereas the incident and scattered waves were assumed to propagate in the background medium with the wave number (denoted by k) without consideration of the propagation effects from the existence of other particles.

To take into account these propagation effects, we use the transmission matrix $\overline{\overline{T}}$ to replace the propagation term e^{-jkr}. Therefore, the received wave field is expressed by the incident wave field, which in turn is expressed by the transmitted field as

$$\vec{E}_r = \frac{1}{r}\overline{\overline{T}}\,\overline{\overline{S}}\vec{E}_i \quad \text{and} \quad \vec{E}_i = \overline{\overline{T}}\vec{E}_t. \tag{4.120}$$

Hence, in the case of backscattering, the received wave field is expressed by the transmitted wave field

$$
\begin{aligned}
\begin{bmatrix} E_{rh} \\ E_{rv} \end{bmatrix} &= \frac{1}{r}\begin{bmatrix} T_{hh} & T_{hv} \\ T_{vh} & T_{vv} \end{bmatrix}\begin{bmatrix} s_{hh}(\pi) & s_{hv}(\pi) \\ s_{vh}(\pi) & s_{vv}(\pi) \end{bmatrix}\begin{bmatrix} E_{ih} \\ E_{iv} \end{bmatrix} \\[2mm]
&= \frac{1}{r}\begin{bmatrix} T_{hh} & T_{hv} \\ T_{vh} & T_{vv} \end{bmatrix}\begin{bmatrix} s_{hh}(\pi) & s_{hv}(\pi) \\ s_{vh}(\pi) & s_{vv}(\pi) \end{bmatrix}\begin{bmatrix} T_{hh} & T_{hv} \\ T_{vh} & T_{vv} \end{bmatrix}\begin{bmatrix} E_{th} \\ E_{tv} \end{bmatrix} \\[2mm]
&= \frac{1}{r}\begin{bmatrix} T_{hh}^2 s_{hh} + 2T_{hh}T_{hv}s_{vh} + T_{hv}^2 s_{vv} & T_{hh}T_{hv}s_{hh} + (T_{hv}^2 + T_{hh}T_{vv})s_{vh} + T_{hv}T_{vv}s_{vv} \\ T_{hh}T_{hv}s_{hh} + (T_{hv}^2 + T_{hh}T_{vv})s_{vh} + T_{hv}T_{vv}s_{vv} & T_{hv}^2 s_{hh} + 2T_{hv}T_{vv}s_{vh} + T_{vv}^2 s_{vv} \end{bmatrix} \\[2mm]
&\quad \times \begin{bmatrix} E_{th} \\ E_{tv} \end{bmatrix}
\end{aligned}
$$

$$\tag{4.121}$$

which can be written as

$$\begin{bmatrix} E_{rh} \\ E_{rv} \end{bmatrix} = \frac{1}{r}\begin{bmatrix} s'_{hh}(\pi) & s'_{hv}(\pi) \\ s'_{vh}(\pi) & s'_{vv}(\pi) \end{bmatrix}\begin{bmatrix} E_{th} \\ E_{tv} \end{bmatrix}. \tag{4.122}$$

$\overline{\overline{S}}'$ is the transmission-included scattering matrix, which includes both the propagation and scattering effects. This is also called *first-order multiple scattering*. In general, the transmission-included scattering matrix contains both co-polar (diagonal: s'_{hh} and s'_{vv}) and cross-polar (off-diagonal: s'_{hv} and s'_{vh}) terms that are contributed by propagation and scattering; the propagation and scattering effects are generally coupled, as shown in Equation 4.121.

Two extreme cases are discussed as follows:

1. **Transmission-induced cross-polarization**

Consider the intrinsic scattering matrix $\bar{\bar{S}}$ with only diagonal terms, with no cross-polarization component in the scattering ($s_{hv} = s_{vh} = 0$). Wave scattering by rain, for example, would satisfy this condition. Hence, Equation 4.121 becomes

$$\begin{bmatrix} E_{rh} \\ E_{rv} \end{bmatrix} = \frac{1}{r} \begin{bmatrix} T_{hh}^2 s_{hh} + T_{hv}^2 s_{vv} & T_{hh}T_{hv}s_{hh} + T_{hv}T_{vv}s_{vv} \\ T_{hh}T_{hv}s_{hh} + T_{hv}T_{vv}s_{vv} & T_{hv}^2 s_{hh} + T_{vv}^2 s_{vv} \end{bmatrix} \begin{bmatrix} E_{th} \\ E_{tv} \end{bmatrix}. \quad (4.123)$$

In this case, co-polar bias and cross-polar components are caused by the depolarization during propagation (T_{hv}/T_{vh}). These effects occur on the all the measurements beyond the gate that has propagation depolarization. Depolarization-caused Z_{DR} bias has been observed in clouds with the reorientation of ice crystals after lightning (Ryzhkov and Zrnić 2007) and has also been studied for simultaneous transmission mode by Hubbert et al. (2010a, 2010b).

Figure 4.10 shows polarimetric radar measurements of Z_H(upper) and Z_{DR}(lower) made by the NCAR S-Pol with both alternate and simultaneous modes. While there is no stripe feature in the Z_H and Z_{DR} fields for the alternate transmission mode and in the reflectivity field, there are radial stripes in the Z_{DR} image (lower-right) that occur after the 45-km range ring (above bright band at ~30 km). These stripes are likely caused by a nonzero mean canting angle of ice crystals.

2. **Scattering-induced cross-polarization**

Here, we consider the transmission matrix $\bar{\bar{T}}$ with only diagonal terms and no propagation depolarization, so that $T_{hv} = T_{vh} = 0$. Using these properties in Equation 4.121, we obtain

$$\begin{bmatrix} E_{rh} \\ E_{rv} \end{bmatrix} = \frac{1}{r} \begin{bmatrix} T_{hh}^2 s_{hh} & T_{hh}T_{vv}s_{hv} \\ T_{hh}T_{vv}s_{vh} & T_{vv}^2 s_{vv} \end{bmatrix} \begin{bmatrix} E_{th} \\ E_{tv} \end{bmatrix}$$

$$= \frac{1}{r} \begin{bmatrix} s_{hh}e^{-2j\int_0^r K_h d\ell} & s_{hv}e^{-j\int_0^r (K_h + K_v)d\ell} \\ s_{vh}e^{-j\int_0^r (K_h + K_v)d\ell} & s_{vv}e^{-2j\int_0^r K_v d\ell} \end{bmatrix} \begin{bmatrix} E_{th} \\ E_{tv} \end{bmatrix}. \quad (4.124)$$

In this special case, the propagation and scattering terms are decoupled in the transmission-included scattering matrix. Each of the s_{pq} terms represents the co-polar and cross-polar scattering, and each of their corresponding phasor terms represents the phase change for H-pol transmission and H-pol reception, V-pol transmission and H-pol reception, H-pol transmission and V-pol reception, and V-pol transmission and V-pol reception.

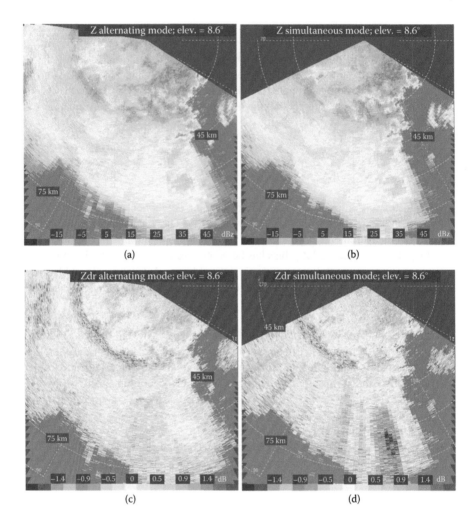

FIGURE 4.10 Propagation effects on polarimetric radar measurements made with NCAR S-Pol. (a, c) ATSR (alternate transmission and simultaneous reception) mode and (b, d) STSR (simultaneous transmission and simultaneous reception) mode. (From Hubbert, J., et al., 2010b. *Journal of Atmospheric and Oceanic Technology*, 27, 1599–1607.)

4.4.2 PROPAGATION-INCLUDED RADAR VARIABLES

In the section 4.2.6, intrinsic polarimetric radar variables were defined based on the single scattering approximation, which is only valid when propagation effects can be ignored. In the first-order multiple scattering approximation, wave scattering was characterized by the transmission-included scattering matrix, where wave propagation was represented by the transmission matrix. Following the definition of the intrinsic polarimetric radar variables (Equations 4.59 through 4.70), the propagation-included polarimetric radar variables are defined based on the transmission-included scattering matrix, including reflectivity factors:

$$Z'_{hh} = \frac{4\lambda^4}{\pi^4 \mid K_w \mid^2} \left\langle n \mid s'_{hh} \mid^2 \right\rangle \tag{4.125}$$

$$Z'_{vv} = \frac{4\lambda^4}{\pi^4 \mid K_w \mid^2} \left\langle n \mid s'_{vv} \mid^2 \right\rangle \tag{4.126}$$

$$Z'_{hv} = \frac{4\lambda^4}{\pi^4 \mid K_w \mid^2} \left\langle n \mid s'_{hv} \mid^2 \right\rangle, \tag{4.127}$$

differential reflectivity,

$$Z'_{DR} = 10\log\frac{Z'_{hh}}{Z'_{vv}} = 10\log\frac{\left\langle n \mid s'_{hh} \mid^2 \right\rangle}{\left\langle n \mid s'_{vv} \mid^2 \right\rangle} \tag{4.128}$$

the linear depolarization ratio,

$$LDR'_H = 10\log\frac{Z'_{vh}}{Z'_{hh}} = 10\log\frac{\left\langle n \mid s'_{vh} \mid^2 \right\rangle}{\left\langle n \mid s'_{hh} \mid^2 \right\rangle} \tag{4.129}$$

$$LDR'_V = 10\log\frac{Z'_{hv}}{Z'_{vv}} = 10\log\frac{\left\langle n \mid s'_{hv} \mid^2 \right\rangle}{\left\langle n \mid s'_{vv} \mid^2 \right\rangle} \tag{4.130}$$

co-polar correlation coefficient,

$$\rho'_{hv} = \frac{\left\langle n s'^*_{hh} s'_{vv} \right\rangle}{\sqrt{\left\langle n \mid s'_{hh} \mid^2 \right\rangle \left\langle n \mid s'_{vv} \mid^2 \right\rangle}} \tag{4.131}$$

and the co-/cross-polar correlation coefficient

$$\rho'_h = \frac{\left\langle n s'^*_{hh} s'_{vh} \right\rangle}{\sqrt{\left\langle n \mid s'_{hh} \mid^2 \right\rangle \left\langle n \mid s'_{vh} \mid^2 \right\rangle}} \tag{4.132}$$

$$\rho'_v = \frac{\left\langle n s'^*_{vv} s'_{hv} \right\rangle}{\sqrt{\left\langle n \mid s'_{vv} \mid^2 \right\rangle \left\langle n \mid s'_{hv} \mid^2 \right\rangle}}. \tag{4.133}$$

Thus, Equations 4.125 through 4.133 constitute definitions of propagation-included polarimetric radar variables. In general, they depend on the forward scattering properties of hydrometeors along the path to and from the radar resolution volume and the scattering properties of hydrometeors within the volume.

If we ignore the propagation depolarization, the transmission-included scattering matrix is represented by Equation 4.124. Substitution of Equation 4.124 into Equations 4.125 through 4.133 yields

$$Z'_{hh} = \frac{4\lambda^4}{\pi^4 \, | \, K_w \, |^2} \left\langle n|s_{hh}|^2 \right\rangle e^{-4\int_0^r \mathrm{Im}(K_h)d\ell} = Z_{hh} e^{-4\int_0^r \mathrm{Im}(K_h)d\ell} \tag{4.134}$$

$$Z'_{vv} = \frac{4\lambda^4}{\pi^4 \, | \, K_w \, |^2} \left\langle n|s_{vv}|^2 \right\rangle e^{-4\int_0^r \mathrm{Im}(K_v)d\ell} = Z_{vv} e^{-4\int_0^r \mathrm{Im}(K_v)d\ell} \tag{4.135}$$

$$Z'_{hv} = \frac{4\lambda^4}{\pi^4 \, | \, K_w \, |^2} \left\langle n|s_{hv}|^2 \right\rangle e^{-2\int_0^r \mathrm{Im}(K_h+K_v)d\ell} = Z_{hv} e^{-2\int_0^r \mathrm{Im}(K_h+K_v)d\ell}. \tag{4.136}$$

Rewriting Equations 4.134 through 4.136 in dBZ values and noting $A_{h,v} = 20\log(e) \times \mathrm{Im}(K_{h,v}) = 8.686 \times \mathrm{Im}(K_{h,v})$ for specific attenuation, we obtain

$$Z'_H = Z_H - 2\int_0^r A_H(\ell)d\ell \equiv Z_H - PIA_H \tag{4.137}$$

$$Z'_V = Z_V - 2\int_0^r A_V(\ell)d\ell \equiv Z_V - PIA_V \tag{4.138}$$

$$Z'_{HV} = Z_{HV} - \int_0^r \left[A_H(\ell) + A_V(\ell)\right]d\ell \equiv Z_{HV} - \frac{1}{2}(PIA_H + PIA_V), \tag{4.139}$$

where $PIA \equiv 2\int_0^r A(\ell)d\ell$ stands for the two-way path-integrated attenuation (PIA). The negative sign in PIA terms indicates wave propagation energy loss between the radar location and the targeted medium, which needs to be taken into account or compensated for in PRD analysis. The compensation of the PIA is called *attenuation correction* and is discussed in Chapter 6.

Similarly, we have the differential reflectivity (Z_{DR}) in dB as follows:

$$Z'_{DR} = Z_{DR} - 2\int_0^r A_{DP}(\ell)d\ell \equiv Z_{DR} - PIA_{DP}, \tag{4.140}$$

where $PIA_{DP} \equiv 2\int_0^r A_{DP}(\ell)d\ell$ is the two-way path-integrated differential attenuation. Because the attenuation of the H-polarized wave is generally larger than that of the V-polarized wave, PIA_{DP} is typically positive and causes a negative bias to the differential reflectivity. This bias is shown in Equation 4.140 and needs to be taken into account. Taking into account the attenuation and differential attenuation is especially important in analyzing X- and C-band polarimetric weather radar measurements, where attenuation cannot be ignored most of the time.

For the linear depolarization ratio, wave polarization is the same for the transmission path—the polarization is different only in the receiving path, which causes a difference in attenuation and hence biased LDR. From Equations 4.124, 4.129, and 4.130 and the PIA_{DP} above, we have

$$LDR'_H = LDR_H + \frac{1}{2} PIA_{DP} \tag{4.141}$$

$$LDR'_V = LDR_V - \frac{1}{2} PIA_{DP}. \tag{4.142}$$

Next, let us discuss the correlation coefficients. Using the transmission-included scattering amplitudes of Equation 4.124 in the definition of the propagation-included co-polar correlation coefficient, we have

$$\tilde{\rho}'_{hv} = \frac{\left\langle ns^*_{hh}s_{vv} \right\rangle}{\sqrt{\left\langle n|s_{hh}|^2 \right\rangle \left\langle n|s_{vv}|^2 \right\rangle}} e^{2j\int_0^r \text{Re}(K_h - K_v)d\ell} = \tilde{\rho}_{hv} e^{j\Phi_{dp}}. \tag{4.143}$$

Writing the correlation coefficients in terms of magnitudes and phases, that is, $\tilde{\rho}'_{hv} = \rho_{hv} e^{j\Phi_{dp}}$ and $\tilde{\rho}_{hv} = \rho_{hv} e^{j\delta_{hv}}$, we obtain the result that the magnitude of the correlation coefficient with propagation effects, ρ'_{hv}, is equal to that without propagation effects, ρ_{hv}. However, the correlation phases are related by the following:

$$\Phi_{DP} = \delta + \phi_{DP} = \frac{180}{\pi}(\delta_d + \phi_{dp}) \tag{4.144}$$

and

$$\phi_{DP} = 2\int_0^r K_{DP}(\ell)d\ell. \tag{4.145}$$

The equations show that the total differential phase is the sum of the differential scattering phase (δ) and the differential propagation phase (ϕ_{DP}). The differential propagation phase [also called the *differential phase* (ϕ_{DP})] is the integral of the specific differential phase over the propagation path, given in Equation 4.145. Therefore, the specific differential phase is expressed as the derivative of the differential phase by the following:

$$K_{DP} = \frac{1}{2} \frac{d\phi_{DP}}{dr}. \tag{4.146}$$

Note that ϕ_{DP} in Equation 4.146 is the differential propagation phase, whereas what a polarimetric radar measures is the estimates of the total differential phase ($\Phi_{DP} = \delta + \phi_{DP}$). Hence, Equation 4.146 cannot be used to estimate the specific differential phase from the measured differential phase when the differential scattering phase is present as in the case of hail, melting snow, and biological scatterers.

APPENDIX 4A: MONTE CARLO SIMULATION

In Section 4.2, we studied wave scattering by randomly distributed particles by understanding the analytical theory of wave statistics of moments and PDFs. That is, we only used deterministic values or functions to characterize the random scattering problem. In this subsection, we introduce the Monte Carlo method.

Unlike analytical methods, the Monte Carlo method is a numerical method that simulates the randomly scattered wave field, which allows for the study of wave statistics. In general, the Monte Carlo method includes

1. Specifying the problem and domain, including radar constants and scattering amplitudes
2. Generating random numbers for particle position and motion
3. Performing calculations of the scattered wave field by each particle
4. Calculating the total scattered wave fields
5. Performing statistical analysis of the simulated wave fields

As in the provided code, we use 100 randomly moving scatterers to model hydrometeor particles and simulate wave scattering at S-band frequency ($f = 3$ GHz). Their initial positions are within a $10 \times 10 \times 10$ (m³) volume, as shown in Figure 4A.1. Their mean velocities are 10 (m/s) in each of the three (x, y, z) directions. The standard deviation of the component motion is 1.0 (m/s). The wave is incident in the direction of the x-axis. By ignoring the spherical wave factor in Equation 4.4, the bistatic scattered wave fields are calculated at every millisecond, which corresponds to the pulse repetition time. Sample time-series data are shown in Figure 4A.2 with the real part, imaginary part, amplitude, and phase plotted as a function of time.

The time-series data can be used to obtain wave statistics and to verify the theoretical results in Equations 4.18, 4.19, 4.25, and 4.26. This simulation and analysis have also been done for dual-polarization and the joint PDF of the amplitudes and phases.

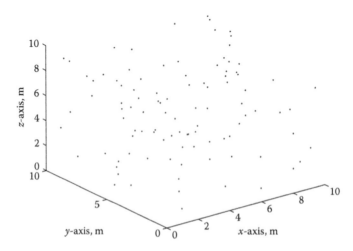

FIGURE 4A.1   Generation of randomly distributed particles to simulate hydrometeor scattering.

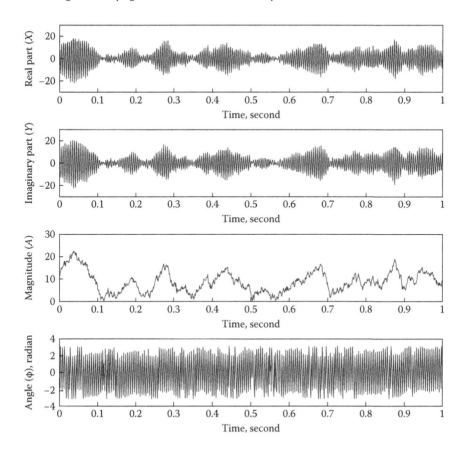

FIGURE 4A.2 Simulated time-series data.

APPENDIX 4B: STATISTICS OF RANDOM ORIENTATION ANGLE TERMS

Assume that the canting angles υ and φ are independent and Gaussian-distributed with a PDF of

$$p(\vartheta) = \frac{1}{\sqrt{2\pi}\sigma_\vartheta} \exp\left(-\frac{(\vartheta - \bar{\vartheta})^2}{2\sigma_\vartheta^2}\right) \tag{4B.1}$$

$$p(\varphi) = \frac{1}{\sqrt{2\pi}\sigma_\varphi} \exp\left(-\frac{(\varphi - \bar{\varphi})^2}{2\sigma_\varphi^2}\right). \tag{4B.2}$$

where $\left(\bar{\vartheta}, \bar{\varphi}\right)$ are the mean and $(\sigma_\upsilon, \sigma_\varphi)$ are the standard deviations.

Using the integral formulas on page 515 of Gradshteyn and Ryzhik (1994)

$$\int \exp\left(-\frac{x^2}{2\sigma_x^2}\right) \cos \alpha x\, dx = \sqrt{2\pi}\sigma_x \exp\left(-\frac{\alpha^2\sigma_x^2}{2}\right) \tag{4B.3}$$

and

$$\int \exp\left(-\frac{x^2}{2\sigma_x^2}\right)\cos\alpha x \cos\beta x\, dx = \frac{1}{2}\sqrt{2\pi}\sigma_x\left(e^{-\frac{(\alpha-\beta)^2\sigma_x^2}{2}} - e^{-\frac{(\alpha+\beta)^2\sigma_x^2}{2}}\right), \quad (4B.4)$$

we obtain the ensemble average of function for a Gaussian random variable x as

$$\langle\cos\alpha x\rangle = \int \cos\alpha x\, p(x)dx = \cos\alpha\bar{x}\, e^{-\frac{\alpha^2\sigma_x^2}{2}} \tag{4B.5}$$

$$\langle\cos^2 x\rangle = \frac{1}{2}\langle 1+\cos 2x\rangle = \frac{1}{2}\left(1+\cos 2\bar{x}\, e^{-2\sigma_x^2}\right) \tag{4B.6}$$

$$\langle\cos^4 x\rangle = \frac{1}{4}\langle(1+\cos 2x)^2\rangle = \frac{1}{8}\left(3+4\cos 2\bar{x}\, e^{-2\sigma_x^2} + \cos 4\bar{x}\, e^{-8\sigma_x^2}\right) \tag{4B.7}$$

$$\langle\sin^2 x\rangle = \langle 1-\cos^2 x\rangle = \frac{1}{2}\left(1-\cos 2\bar{x}\, e^{-2\sigma_x^2}\right) \tag{4B.8}$$

$$\begin{aligned}
\langle\sin^4 x\rangle &= \langle(1-\cos^2 x)^2\rangle \\
&= \langle 1-2\cos^2 x + \cos^4 x\rangle \\
&= 1-\left(1+\cos 2\bar{x}\, e^{-2\sigma_x^2}\right) + \frac{1}{8}\left(3+4\cos 2\bar{x}\, e^{-2\sigma_x^2} + \cos 4\bar{x}\, e^{-8\sigma_x^2}\right) \\
&= \frac{1}{8}\left(3-4\cos 2\bar{x}\, e^{-2\sigma_x^2} + \cos 4\bar{x}\, e^{-8\sigma_x^2}\right)
\end{aligned} \tag{4B.9}$$

$$\langle\sin^2 x\cos^2 x\rangle = \langle\cos^2 x - \cos^4 x\rangle = \frac{1}{8}\left(1-\cos 4\bar{x}\, e^{-8\sigma_x^2}\right) \tag{4B.10}$$

$$\begin{aligned}
\langle\sin x\cos x\rangle &= \frac{1}{2}\langle\sin 2x\rangle = \frac{1}{2}\langle\sin 2(\bar{x}+x')\rangle \\
&= \frac{1}{2}\langle(\sin 2\bar{x}\cos 2x' + \cos 2\bar{x}\sin 2x')\rangle \\
&= \frac{1}{2}\sin 2\bar{x}\langle\cos 2x'\rangle \\
&= \frac{1}{2}\sin 2\bar{x}\, e^{-2\sigma_x^2}.
\end{aligned} \tag{4B.11}$$

After performing the integrations using Equations 4B.5 through 4B.11, we obtain the statistics of the angle-dependent factors as follows:

$$\langle A^2 \rangle = \left\langle \left(\cos^2 \varphi + \sin^2 \vartheta \sin^2 \varphi \right)^2 \right\rangle$$

$$= \left\langle \cos^4 \varphi + \sin^4 \vartheta \sin^4 \varphi + 2 \sin^2 \vartheta \sin^2 \varphi \cos^2 \varphi \right\rangle$$

$$= \left\langle \cos^4 \varphi + \sin^4 \vartheta \sin^4 \varphi + \frac{1}{2} \sin^2 \vartheta \sin^2 2\varphi \right\rangle$$

$$= \frac{1}{8} \left(3 + 4 \cos 2\bar{\varphi} e^{-2\sigma_\varphi^2} + \cos 4\bar{\varphi} e^{-8\sigma_\varphi^2} \right) \tag{4B.12}$$

$$+ \frac{1}{64} \left(3 - 4 \cos 2\bar{\vartheta} e^{-2\sigma_\vartheta^2} + \cos 4\bar{\vartheta} e^{-8\sigma_\vartheta^2} \right) \left(3 - 4 \cos 2\bar{\varphi} e^{-2\sigma_\varphi^2} + \cos 4\bar{\varphi} e^{-8\sigma_\varphi^2} \right)$$

$$+ \frac{1}{8} \left(1 - \cos 2\bar{\vartheta} e^{-2\sigma_\vartheta^2} \right) \left(1 - \cos 4\bar{\varphi} e^{-8\sigma_\varphi^2} \right)$$

$$\langle D^2 \rangle = \left\langle \left(\sin^2 \varphi + \sin^2 \vartheta \cos^2 \varphi \right)^2 \right\rangle$$

$$= \left\langle \sin^4 \varphi + \sin^4 \vartheta \cos^4 \varphi + 2 \sin^2 \vartheta \sin^2 \varphi \cos^2 \varphi \right\rangle$$

$$= \left\langle \sin^4 \varphi + \sin^4 \vartheta \cos^4 \varphi + \frac{1}{2} \sin^2 \vartheta \sin^2 2\varphi \right\rangle$$

$$= \frac{1}{8} \left(3 - 4 \cos 2\bar{\varphi} e^{-2\sigma_\varphi^2} + \cos 4\bar{\varphi} e^{-8\sigma_\varphi^2} \right) \tag{4B.13}$$

$$+ \frac{1}{64} \left(3 - 4 \cos 2\bar{\vartheta} e^{-2\sigma_\vartheta^2} + \cos 4\bar{\vartheta} e^{-8\sigma_\vartheta^2} \right) \left(3 + 4 \cos 2\bar{\varphi} e^{-2\sigma_\varphi^2} + \cos 4\bar{\varphi} e^{-8\sigma_\varphi^2} \right)$$

$$+ \frac{1}{8} \left(1 - \cos 2\bar{\vartheta} e^{-2\sigma_\vartheta^2} \right) \left(1 - \cos 4\bar{\varphi} e^{-8\sigma_\varphi^2} \right).$$

Following the same procedure as in deriving Equation 4B.12, we have

$$\langle C^2 \rangle = \left\langle \cos^4 \vartheta \cos^4 \varphi \right\rangle$$

$$= \frac{1}{64} \left(3 + 4 \cos 2\bar{\vartheta} e^{-2\sigma_\vartheta^2} + \cos 4\bar{\vartheta} e^{-8\sigma_\vartheta^2} \right) \left(3 + 4 \cos 2\bar{\varphi} e^{-2\sigma_\varphi^2} + \cos 4\bar{\varphi} e^{-8\sigma_\varphi^2} \right) \tag{4B.14}$$

$$\langle B^2 \rangle = \left\langle \cos^4 \vartheta \sin^4 \varphi \right\rangle$$

$$= \frac{1}{64} \left(3 + 4 \cos 2\bar{\vartheta} e^{-2\sigma_\vartheta^2} + \cos 4\bar{\vartheta} e^{-8\sigma_\vartheta^2} \right) \left(3 - 4 \cos 2\bar{\varphi} e^{-2\sigma_\varphi^2} + \cos 4\bar{\varphi} e^{-8\sigma_\varphi^2} \right) \tag{4B.15}$$

$$\langle BC \rangle = \langle \cos^4 \vartheta \sin^2 \varphi \cos^2 \varphi \rangle$$

$$= \frac{1}{64}\left(3 + 4\cos 2\overline{\vartheta}e^{-2\sigma_\vartheta^2} + \cos 4\overline{\vartheta}e^{-8\sigma_\vartheta^2}\right)\left(1 - \cos 4\overline{\varphi}e^{-8\sigma_\varphi^2}\right) \tag{4B.16}$$

$$\langle AB \rangle = \langle \left(\cos^2 \varphi + \sin^2 \vartheta \sin^2 \varphi\right)\cos^2 \vartheta \sin^2 \varphi \rangle$$

$$= \frac{1}{16}\left(1 + \cos 2\overline{\vartheta}e^{-2\sigma_\vartheta^2}\right)\left(1 - \cos 4\overline{\varphi}e^{-8\sigma_\varphi^2}\right) \tag{4B.17}$$

$$+ \frac{1}{64}\left(1 - \cos 4\overline{\vartheta}e^{-8\sigma_\vartheta^2}\right)\left(3 - 4\cos 2\overline{\varphi}e^{-2\sigma_\varphi^2} + \cos 4\overline{\varphi}e^{-8\sigma_\varphi^2}\right)$$

$$\langle CD \rangle = \langle \left(\sin^2 \varphi + \sin^2 \vartheta \cos^2 \varphi\right)\cos^2 \vartheta \cos^2 \varphi^2 \rangle$$

$$= \frac{1}{16}\left(1 + \cos 2\overline{\vartheta}e^{-2\sigma_\vartheta^2}\right)\left(1 - \cos 4\overline{\varphi}e^{-8\sigma_\varphi^2}\right) \tag{4B.18}$$

$$+ \frac{1}{64}\left(1 - \cos 4\overline{\vartheta}e^{-8\sigma_\vartheta^2}\right)\left(3 + 4\cos 2\overline{\varphi}e^{-2\sigma_\varphi^2} + \cos 4\overline{\varphi}e^{-8\sigma_\varphi^2}\right)$$

$$\langle AC \rangle = \langle \left(\cos^2 \varphi + \sin^2 \vartheta \sin^2 \varphi\right)\cos^2 \vartheta \cos^2 \varphi \rangle$$

$$= \frac{1}{16}\left(1 + \cos 2\overline{\vartheta}e^{-2\sigma_\vartheta^2}\right)\left(3 + 4\cos 2\overline{\varphi}e^{-2\sigma_\varphi^2} + \cos 4\overline{\varphi}e^{-8\sigma_\varphi^2}\right) \tag{4B.19}$$

$$+ \frac{1}{64}\left(1 - \cos 4\overline{\vartheta}e^{-8\sigma_\vartheta^2}\right)\left(1 - \cos 4\overline{\varphi}e^{-8\sigma_\varphi^2}\right)$$

$$\langle BD \rangle = \langle \left(\sin^2 \varphi + \sin^2 \vartheta \cos^2 \varphi\right)\cos^2 \vartheta \sin^2 \varphi \rangle$$

$$= \frac{1}{16}\left(1 + \cos 2\overline{\vartheta}e^{-2\sigma_\vartheta^2}\right)\left(3 - 4\cos 2\overline{\varphi}e^{-2\sigma_\varphi^2} + \cos 4\overline{\varphi}e^{-8\sigma_\varphi^2}\right) \tag{4B.20}$$

$$+ \frac{1}{64}\left(1 - \cos 4\overline{\vartheta}e^{-8\sigma_\vartheta^2}\right)\left(1 - \cos 4\overline{\varphi}e^{-8\sigma_\varphi^2}\right)$$

$$\langle AD \rangle = \langle \left(\cos^2 \varphi + \sin^2 \vartheta \sin^2 \varphi\right)\left(\sin^2 \varphi + \sin^2 \vartheta \cos^2 \varphi\right) \rangle$$

$$= \left\langle \begin{array}{l} \cos^2 \varphi \sin^2 \varphi + \cos^4 \varphi \sin^2 \vartheta \\ + \sin^4 \varphi \sin^2 \vartheta + \sin^2 \varphi \cos^2 \varphi \sin^4 \vartheta \end{array} \right\rangle$$

$$= \frac{1}{8}\left(3 + \cos 4\overline{\varphi}e^{-8\sigma_\varphi^2}\right)\left(1 - \cos 2\overline{\vartheta}e^{-2\sigma_\vartheta^2}\right) \tag{4B.21}$$

$$+ \frac{1}{64}\left(1 - \cos 4\overline{\varphi}e^{-8\sigma_\varphi^2}\right)\left(11 - 4\cos 2\overline{\vartheta}e^{-2\sigma_\vartheta^2} + \cos 4\overline{\vartheta}e^{-8\sigma_\vartheta^2}\right)$$

$$\left\langle A\sqrt{BC}\right\rangle = \left\langle \left(\cos^2\varphi + \sin^2\vartheta\sin^2\varphi\right)\sqrt{\cos^4\vartheta\sin^2\varphi\cos^2\varphi}\right\rangle$$

$$= \left\langle \frac{1}{2}\left(\cos^2\varphi\cos^2\vartheta\sin 2\varphi + \cos^2\vartheta\sin 2\varphi\sin^2\vartheta\sin^2\varphi\right)\right\rangle$$

$$= \left\langle \frac{1}{4}\cos^2\vartheta\left(\sin 2\varphi + \cos 2\varphi\sin 2\varphi\right) + \frac{1}{16}\sin^2 2\vartheta\left(\sin 2\varphi - \cos 2\varphi\sin 2\varphi\right)\right\rangle$$

$$= \frac{1}{16}\left(2\sin 2\bar{\varphi}e^{-2\sigma_\varphi^2} + \sin 4\bar{\varphi}e^{-8\sigma_\varphi^2}\right)\left(1 + \cos 2\bar{\vartheta}e^{-2\sigma_\vartheta^2}\right)$$

$$+ \frac{1}{64}\left(2\sin 2\bar{\varphi}e^{-2\sigma_\varphi^2} - \sin 4\bar{\varphi}e^{-8\sigma_\varphi^2}\right)\left(1 - \cos 4\bar{\vartheta}e^{-8\sigma_\vartheta^2}\right) \qquad \text{(4B.22)}$$

$$\left\langle B\sqrt{BC}\right\rangle = \left\langle \cos^2\vartheta\sin^2\varphi\sqrt{\cos^4\vartheta\sin^2\varphi\cos^2\varphi}\right\rangle$$

$$= \left\langle \cos^4\vartheta\sin^2\varphi\sin\varphi\cos\varphi\right\rangle$$

$$= \frac{1}{4}\left\langle \cos^4\vartheta\left(1 - \cos 2\varphi\right)\sin 2\varphi\right\rangle$$

$$= \frac{1}{4}\left\langle \cos^4\vartheta\left(\sin 2\varphi - \sin 2\varphi\cos 2\varphi\right)\right\rangle$$

$$= \frac{1}{64}\left(2\sin 2\bar{\varphi}e^{-2\sigma_\varphi^2} - \sin 4\bar{\varphi}e^{-8\sigma_\varphi^2}\right)\left(3 + 4\cos 2\bar{\vartheta}e^{-2\sigma_\vartheta^2} + \cos 4\bar{\vartheta}e^{-8\sigma_\vartheta^2}\right)$$

$$\text{(4B.23)}$$

$$\left\langle C\sqrt{BC}\right\rangle = \left\langle \cos^2\vartheta\cos^2\varphi\sqrt{\cos^4\vartheta\sin^2\varphi\cos^2\varphi}\right\rangle$$

$$= \left\langle \cos^4\vartheta\cos^2\varphi\sin\varphi\cos\varphi\right\rangle$$

$$= \frac{1}{4}\left\langle \cos^4\vartheta\left(1 + \cos 2\varphi\right)\sin 2\varphi\right\rangle$$

$$= \frac{1}{4}\left\langle \cos^4\vartheta\left(\sin 2\varphi + \sin 2\varphi\cos 2\varphi\right)\right\rangle$$

$$= \frac{1}{64}\left(2\sin 2\bar{\varphi}e^{-2\sigma_\varphi^2} + \sin 4\bar{\varphi}e^{-8\sigma_\varphi^2}\right)\left(3 + 4\cos 2\bar{\vartheta}e^{-2\sigma_\vartheta^2} + \cos 4\bar{\vartheta}e^{-8\sigma_\vartheta^2}\right)$$

$$\text{(4B.24)}$$

$$\left\langle D\sqrt{BC}\right\rangle = \left\langle \left(\sin^2\varphi + \sin^2\vartheta\cos^2\varphi\right)\sqrt{\cos^4\vartheta\sin^2\varphi\cos^2\varphi}\right\rangle$$

$$= \left\langle \frac{1}{2}\left(\sin^2\varphi\cos^2\vartheta\sin 2\varphi + \cos^2\vartheta\sin 2\varphi\sin^2\vartheta\cos^2\varphi\right)\right\rangle$$

$$= \left\langle \frac{1}{4}\cos^2\vartheta\left(\sin 2\varphi - \cos 2\varphi\sin 2\varphi\right) + \frac{1}{16}\sin^2 2\vartheta\left(\sin 2\varphi + \cos 2\varphi\sin 2\varphi\right)\right\rangle$$

$$= \frac{1}{16}\left(2\sin 2\overline{\varphi}e^{-2\sigma_\varphi^2} - \sin 4\overline{\varphi}e^{-8\sigma_\varphi^2}\right)\left(1 + \cos 2\overline{\vartheta}e^{-2\sigma_\vartheta^2}\right)$$

$$+ \frac{1}{64}\left(2\sin 2\overline{\varphi}e^{-2\sigma_\varphi^2} + \sin 4\overline{\varphi}e^{-8\sigma_\varphi^2}\right)\left(1 - \cos 4\overline{\vartheta}e^{-8\sigma_\vartheta^2}\right). \tag{4B.25}$$

Problems

4.1 Give definitions for the first-order scattering model, the first-order multiple scattering model, and the multiple scattering model. Also explain the conditions under which they apply.

4.2 Show that the mean scattering wave intensity (Equation 4.7) is equal to the independent scattering wave intensity for all directions other than the forward direction. Discuss when the independent scattering approximation is not valid.

4.3 The total backscattered wave field from many randomly distributed particles can be represented by

$$E_b = Ae^{-j\phi} = X - jY$$

where A is the amplitude, ϕ is the phase, and X and Y are the quadrature components. From the central limit theorem, X and Y are normally (Gaussian) distributed and independent, each with a mean of zero and a standard deviation of σ.

a. Find the PDF for the amplitude A and that for the intensity $I_b = A^2$. Give the names for their distributions.

b. Find the expected values $\langle A\rangle$ and $\langle I_b\rangle$ in terms of σ and the relation between $\langle I_b\rangle$ and $\langle A\rangle$.

c. Use and modify the provided Monte Carlo code to verify the PDFs for X, Y, A, ϕ, and I_b.

d. Use the simulations to estimate $\langle X\rangle$, $\langle Y\rangle$, $\langle\phi\rangle$, $\langle A\rangle$, and $\langle I_b\rangle$ and verify the relations you found in Problem 4.3b.

e. Estimate the radial wind v_r and turbulence σ_v.

f. Assuming $|s_{hh}| = 2\,|s_{vv}|$ and $SD(\sigma_h) = SD(\sigma_v) = 5°$, extend the code to simulate dual-polarization wave fields. Estimate Z_{DR} and ρ_{hv} and compare your estimation with the theoretical results.

4.4 Show that the representation of a sparse random medium by the effective propagation constant (Equation 4.90) is equivalent to that by the effective dielectric constant derived from the Maxwell-Garnet mixing formula when the conditions for Rayleigh scattering approximation apply.

4.5 Using the given scattering amplitudes in the forward direction and those in the backward direction of raindrops calculated using the T-matrix method and ignoring canting angle effects, calculate and plot the polarimetric radar variables Z_H, Z_{DR}, A_H, A_{DP}, and K_{DP} at S-band ($f = 2.8$ GHz).

a. Use the Marshall–Palmer DSD model, $N(D) = 8000\exp(-\Lambda D)$ [# m^{-3} mm^{-1}], with $\Lambda = 4.1R^{-0.21}$ [mm^{-1}] and $|K|^2$ factor at a temperature of 10°. Calculate and plot the polarimetric radar variables as functions of rainfall rate changing from 1 mm/hr to 100 mm/hr.

b. Repeat Problem 4.5a with the real rain DSD data provided in Problem 2.4. Use the velocity formula, $v(D) = -0.1021 + 4.932D - 0.9551D^2 + 0.07934D^3 - 0.002362D^4$, to calculate the rainfall rate and plot the polarimetric radar variables and rainfall rate versus time. The starting time for the DSD data is 0700 UTC and the time interval is 1 min. Also plot Z_H and Z_{DR} as functions of rainfall rate and compare the results with those obtained in Problem 4.5a.

c. Plot the KOUN radar measurements of Z_H and Z_{DR} provided in "koun.dat" versus time, along with those you calculated in Problem 4.5b. Compare the radar measurements with DSD-based calculations and discuss the results.

5 Radar Measurements and Improvement of Data Quality

A weather radar transmits and receives EM wave signals from which radar variables are estimated. We can use these estimates to obtain moment data for weather applications. Whereas in previous chapters we discussed the definitions of polarimetric radar variables and the information they contain about the physical characteristics of hydrometeors, it is also important to know the practical issues that arise in radar measurements and in estimating moment data in order to effectively utilize these data. These issues include sample errors in moment estimates, noise effects in polarimetric measurements, clutter contamination, and mitigation. This chapter provides the fundamentals of the radar equation and signal processing to estimate polarimetric radar variables. It also discusses recent advances in signal processing and clutter detection and filtering that allow us to obtain high quality polarimetric radar data.

5.1 THE POLARIMETRIC WEATHER RADAR SYSTEM AND EQUATION

5.1.1 FUNDAMENTALS AND THE POLARIMETRIC RADAR EQUATION

Radar is a remote sensing system that consists of an antenna, transmitter, receiver, signal processer, display, and control units. The fundamentals of weather radar were well documented by Doviak and Zrnić (1984, 1993, 2006) and Bringi and Chandrasekar (2001). Figure 5.1 shows a sketch of the components of the WSR-88D (Doviak et al. 2000). Whereas the yellow-highlighted portion of the figure shows the original radar data acquisition (RDA) of its single polarization system, the red portion shows the additional second receiver for the dual-polarization upgrade. Note that the receiver location shown in the figure is for the prototype KOUN WSR-88D radar with polarimetric capability. The operational WSR-88Ds use antenna-mounted receivers.

A radar-transmitted wave field can be expressed by the following:

$$\vec{E}_t(\theta,\phi) = A(\theta,\phi)\frac{e^{-jkr}}{r}\hat{e}_t,\tag{5.1}$$

where $A(\theta,\phi)$ is the amplitude and $\hat{e}$ is the polarization. The transmitted power flux at a range of r is then

$$\vec{S}(\theta,\phi) = \frac{1}{2\eta_0}\left|A(\theta,\phi)\right|^2\frac{\hat{r}}{r^2}.\tag{5.2}$$

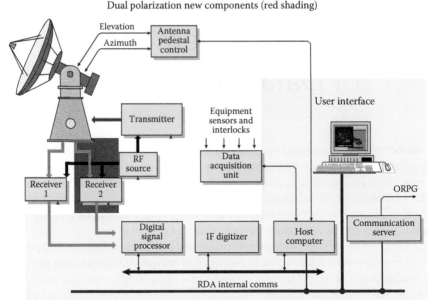

Open RDA new components (yellow shading)

Dual polarization new components (red shading)

FIGURE 5.1 A block diagram of the polarimetric WSR-88D radar system. (From Doviak, R. J., et al., 2000. *Journal of Atmospheric and Oceanic Technology*, 17, 257–278.)

Hence, the total transmitted power is the integral of the power flux over the $4\pi r^2$ closed area, expressed as

$$P_t = \oint S(\theta,\phi)\,da = \frac{1}{2\eta_0}\oint_{4\pi}\left|A(\theta,\phi)\right|^2 d\Omega. \tag{5.3}$$

The *gain* is defined as the ratio of the actual power density (Equation 5.2) over the power density, calculated as if the total power is uniformly distributed in all directions $P_t/(4\pi r^2)$, giving

$$G(\theta,\phi) = \frac{\left|A(\theta,\phi)\right|^2/2\eta_0}{P_t/4\pi}. \tag{5.4}$$

Solving the amplitude $A(\theta, \phi)$ from Equation 5.4 and substituting it into Equation 5.1, we obtain

$$\vec{E}_t = \sqrt{\frac{P_t\eta_0}{2\pi}}\sqrt{G_t}\,\frac{e^{-jkr}}{r}\hat{e}_t. \tag{5.5}$$

This transmitted wave field propagates in the atmosphere, is scattered by hydrometeors, and then propagates back to the radar antenna, as expressed by Equations 4.121 and 4.122. The received power is a product of the receiving power flux density and the effective antenna area $A_r(\theta, \phi)$, which is related to the receiving antenna gain $G_r(\theta, \phi)$ by

$$A_r(\theta, \phi) = \frac{\lambda^2}{4\pi} G_r(\theta, \phi). \tag{5.6}$$

Hence, the effective antenna size vector is

$$\vec{D}_p = \frac{\lambda}{\sqrt{4\pi}} \sqrt{G_R} \hat{e}_{rp}. \tag{5.7}$$

In the case of radar polarimetry, the pattern function becomes the pattern matrix

$$\sqrt{G} \rightarrow \overline{F} = \begin{bmatrix} F_{hh} & F_{hv} \\ F_{vh} & F_{vv} \end{bmatrix} \tag{5.8}$$

with its elements expressed by

$$F_{pq}(\theta, \phi) = \sqrt{G_{pq}(0)} g_{pq}(\theta, \phi), \tag{5.9}$$

where the peak power gain is $G_{pq}(0)$ and $g_{pq}(\theta, \phi)$ is the normalized pattern function. The diagonal terms of (F_{hh}, F_{vv}) and the off-diagonal terms of (F_{hv}, F_{vh}) represent the co-polar and cross-polar radiation functions, respectively. It is important to maintain polarization purity in order to have accurate polarimetric radar measurements of weather. Figure 5.2 shows the normalized co-polar and cross-polar radiation patterns for a center-fed parabolic reflector antenna that is used on WSR-88D radars (Fradin 1961). Note that there is a null at the beam axis for the cross-polar pattern and the four equal lobes have an alternate phase that cause cancellation of the strongest cross-polar component (Zrnić et al. 2010a). Hence, high-quality polarimetric measurements can be achieved.

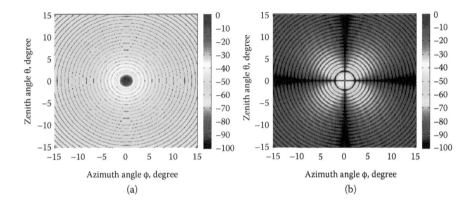

FIGURE 5.2 (a) Co-polar and (b) cross-polar radiation patterns of the WSR-88D antenna.

From Equations 5.7 through 5.9, we have the received complex voltage in p-polarization for the q-polarized wave transmission:

$$V_{pq} = \frac{1}{\sqrt{2\eta_0}} \vec{D}_p \cdot \sum_l \vec{E}_{rl} = \frac{\lambda}{\sqrt{8\pi\eta_0}} \sqrt{\frac{P_t\eta_0}{2\pi}} \sum_l \frac{1}{r^2} \hat{e}_{rp} \overline{\overline{F}} \, \overline{\overline{S_l'}} \, \overline{\overline{F}} \hat{e}_{tq}, \qquad (5.10)$$

that is

$$\begin{bmatrix} V_{hh} & V_{hv} \\ V_{vh} & V_{vv} \end{bmatrix} = \frac{\lambda}{4\pi r^2} \sqrt{P_t} \sum_{l=1}^{N} \begin{bmatrix} F_{hh} & F_{hv} \\ F_{vh} & F_{vv} \end{bmatrix} \begin{bmatrix} s_{hh}' & s_{hv}' \\ s_{vh}' & s_{vv}' \end{bmatrix} \begin{bmatrix} F_{hh} & F_{hv} \\ F_{vh} & F_{vv} \end{bmatrix}. \qquad (5.11)$$

Following the procedure for obtaining the weather radar equation (Equation 4.25) of Doviak and Zrnic (2006), we obtain the received power (P_{rp}) in p-polarization for a transmitted power (P_{tq}) in q-polarization as follows:

$$P_{rp} = \left\langle |V_{pq}|^2 \right\rangle = \frac{\lambda^2 G^2}{(4\pi)^2 r^4} \int dv \left\langle n|s_{pq}'|^2 \right\rangle P_{tq} = P_{tq} \frac{\lambda^2 G^2 \pi \theta_1^2 c\tau}{(4\pi)^3 r^2 16\ln 2} \eta_{pq}. \qquad (5.12)$$

where $\eta_{pq} = 4\pi \left\langle n|s_{pq}'|^2 \right\rangle$ is the radar reflectivity for q-pol transmission and p-pol reception; θ_1 is the one-way 3 dB beamwidth, and τ is the pulse width. Equation 5.12 serves as the polarimetric weather radar equation. There are different ways to measure the complex voltages with polarimetric radar—these are discussed in Section 5.1.2.

5.1.2 POLARIZATION MODES OF RADAR OPERATION

In weather radar practice, the complex voltages are measured by one of the following approaches:

1. *Alternate transmission and alternate reception (ATAR) mode.* In this mode, only one transmitter and one receiver are needed. By changing the transmitted wave polarization from $\hat{e}_{th} = [1\ 0]'$ to $\hat{e}_{tv} = [0\ 1]'$ and the reception polarization from $\hat{e}_{rh} = [1\ 0]'$ to $\hat{e}_{rv} = [0\ 1]'$, the radar receiver can obtain the complex voltages of V_{hh}, V_{vh}, V_{hv}, and V_{vv}.

2. *Alternate transmission and simultaneous reception (ATSR) mode.* In this mode, one transmitter and two receivers are needed. The NCAR S-pol radar is capable of operating in this mode. First an H-polarization pulse is transmitted so that both H- and V-pol signals are received, which allows the receivers to obtain V_{hh} and V_{vh}; then a V-polarization pulse is transmitted to obtain V_{hv} and V_{vv}. For alternate transmission, a high-power switch is required, which is expensive and costly to replace.

3. *Simultaneous transmission and simultaneous reception (STSR) mode.* Instead of alternating the transmitted wave polarization, the transmitted power is split into the two (H and V) channels for transmission, and

the return signals are received by the two (H and V) receivers to obtain the V_{hh} and V_{vv} co-polar signals simultaneously. The cross-polar signals are not measured and are therefore assumed to not exist, which may not always be true. This is the mode currently adopted by most polarimetric weather radars, including the dual-polarization WSR-88D radars. The main advantage of the STSR mode is that there is no need for the expensive high-power switch that is required for the ATSR mode. Furthermore, the STSR mode is easy to implement and its signal processing and data analysis are compatible with that of the single polarization radar system. The disadvantage is that there is a stringent requirement of 40 dB polarization isolation in comparison with 20 dB isolation for the ATSR mode. This is to ensure that the coupling through hardware is controlled. Otherwise, the co-polar signals would be biased by the antenna cross-polar leakage. This is because cross-pol coupling occurs twice in both transmission and reception in the STSR mode, whereas the cross-pol coupling occurs only once at reception in the ATSR mode and there is no cross-pol coupling in the transmission.

Nevertheless, the STSR mode is currently the most popular approach in the weather radar community; it is the focus of the rest of this chapter. Once the complex voltages V_h (which stands for V_{hh}) and V_v (which stands for V_{vv}) are measured, the polarimetric radar variables can be obtained from estimates of the correlation functions.

5.2 REGULAR ESTIMATION OF POLARIMETRIC RADAR VARIABLES

The ACF and cross-correlation function (CCF) (Doviak and Zrnić 2006) are now extended for polarimetric weather radar signals (complex voltages: V_h and V_v). As shown in Equation 5.10, the complex voltages are proportional to the received wave fields scattered from particles, and the signal statistics are the same as the scattered wave statistics except for the difference in the proportional constant. As in Equation 4.14, the correlation functions of weather radar signals follow the Gaussian form, consistent with the work of Janssen and Van Der Spek (1985) and Doviak and Zrnić (2006, 125). It is now extended to include the ACF for the horizontally (H) and vertically (V) polarized signals as well as their CCF.

Let time lag $\tau = nT_s$ with $n = 0,1,2 \ldots N$ as the lag number and T_s as the pulse repetition time (PRT). Note that $k_d = 2k = \dfrac{4\pi}{\lambda}$ for backscattering measurements with a monostatic radar. The time correlation term in Equation 4.14 is then

$$\rho(nT_s) = \exp\left[-2k^2\sigma_v^2(nT_s)^2\right] = \exp\left[-\frac{(nT_s)^2}{2\tau_c^2}\right] \tag{5.13}$$

with the correlation time of $\tau_c = \dfrac{\lambda}{4\pi\sigma_v}$, where λ is the wavelength and σ_v is the spectrum width.

Note that the phase wraps around every 2π, yielding a radial velocity that is limited within the Nyquist velocity $[-v_N, v_N]$, with range $v_N = \dfrac{\lambda}{4T_s}$. Hence, the phase term in Equation 4.14 becomes

$$\exp(2jkv_r nT) = \exp\left(\frac{j\pi n v_r}{v_N}\right). \tag{5.14}$$

Therefore, extending Equation 4.14 for polarimetric radar signals with Equations 5.13 and 5.14, we obtain the expected ACF as

$$R_{h,v}(nT_s) \equiv \left\langle V_{h,v}^*(t+nT_s)V_{h,v}(t)\right\rangle = S_{h,v}\rho(nT_s)\exp\left(\frac{j\pi n v_r}{v_N}\right) + \mathbb{N}_{h,v}\delta_m, \tag{5.15}$$

and the expected CCF is given by

$$C_{hv}(nT_s) \equiv \left\langle V_h^*(t+nT_s)V_v(t)\right\rangle = \sqrt{S_h S_v}\,\rho_{hv}\rho(nT_s)\exp\left(\frac{j\pi n v_r}{v_N} + j\phi_{dp}\right), \tag{5.16}$$

where the subscripts h, v, and hv mean that the parameters are calculated by using signals from the H-channel, V-channel, or both the H- and V-channels (i.e., for $C_{hv}(nT_s)$). For consistency with that in Doviak and Zrnic (2006), S is also denoted as the signal power $S \equiv P_S$. The co-polar correlation coefficient at lag 0 is ρ_{hv}, and ϕ_{dp} is the differential phase. If the beams are matched, the Doppler mean velocities v_r and the correlation times τ_c are almost the same for weather signals from the H- and V-channels (e.g., $v_{rh} = v_{rv} = v_r$) (Bringi and Chandrasekar 2001; Melnikov and Zrnić 2007; Sachidananda and Zrnić 1985, 1986).

Let $V_h(m)$ and $V_v(m)$ be the dual-polarization co-polar (HH and VV) radar signals (time-series data) for the horizontal and vertical polarization channels, respectively; let $V_{hv}(m)$ be the cross-polarization signal. The argument m denotes the mth sample/pulse. Hence, the correlation functions can be estimated from the time-series data $V_h(m)$ and $V_v(m)$. The ensemble average in Equations 5.15 and 5.16 is replaced by the time average. Hence the ACF and CCF estimates are expressed by

$$\hat{R}_{h,v}(n) = \frac{1}{M-n}\sum_{m=1}^{M-n} V_{h,v}^*(m+n)V_{h,v}(m) \tag{5.17}$$

$$\hat{C}_{hv}(n) = \frac{1}{M-n}\sum_{m=1}^{M-n} V_h^*(m+n)V_v(m), \tag{5.18}$$

where $\wedge$ denotes the estimated value. Once the correlation estimates are obtained, the polarimetric radar parameters or moment data are estimated. In this section, we discuss the regular estimation method in the absence of noise $(\mathbb{N}_{h,v} = 0)$, and in the next section we introduce a more advanced method to mitigate noise effects.

5.2.1 REFLECTIVITY ESTIMATION

Ignoring noise effects, reflectivity is estimated from a power estimate that contains an estimation (sampling) error. The total co-polar powers are estimated from the average of instantaneous powers, written as

$$\hat{P}_{h,v} = \hat{R}_{h,v}(0) = \frac{1}{M} \sum_{m=1}^{M} |V_{h,v}(m)|^2. \tag{5.19}$$

Taking the ensemble average of Equation 5.19 and assuming a stationary random process for the signals, we can easily obtain the expected value of the estimates as

$$\left\langle \hat{P}_{h,v} \right\rangle = \frac{1}{M} \sum_{m=1}^{M} \left\langle |V_{h,v}^{(s)}(m)|^2 \right\rangle = P_{sh,sv} \equiv S_{h,v}. \tag{5.20}$$

It is evident that the expected values of power estimates are the mean signal powers.

Because we use a time average with a finite number of pulses in Equation 5.19, the power estimates contain a sampling (fluctuation) error that needs to be quantified. From the theory of weather signal processing (Doviak and Zrnić 1993: Chapter 6), the variance of the power estimates is as follows:

$$\text{var}\left(\hat{P}_{h,v} \right) = \left\langle \hat{P}_{h,v}^2 \right\rangle - \left\langle \hat{P}_{h,v} \right\rangle^2 = \frac{S_{h,v}^2}{M_I} \tag{5.21}$$

and the standard deviation of the power estimates is

$$\text{SD}\left(\hat{P}_{h,v} \right) = \frac{S_{h,v}}{\sqrt{M_I}}. \tag{5.22}$$

where $M_I = \dfrac{T_d}{\sqrt{\pi \tau_c}}$ is the number of independent samples with the correlation time τ_c and dwell time T_d. For $\sigma_v = 1$ and 4 m/s, the correlation time $\tau_c = \dfrac{\lambda}{4\pi\sigma_v}$ is about 8.0 and 2.0 ms, respectively, at the S-band frequency of 3 GHz. Hence, the short PRT of 0.8 ms is used for Doppler mode on WSR-88D so that pulse-to-pulse signals are highly correlated and radial velocity can be accurately estimated. A long PRT of 3.2 ms is used for surveillance mode to avoid range ambiguity and maintain independence.

From Equation 5.22, the standard deviation of the reflectivity estimates (which are proportional to the power estimates) is expressed in decibel units as follows:

$$\text{SD}\left(\hat{Z}_{H,V} \right) \approx 10 \log\left[1 + 1/\sqrt{M_I} \right]. \tag{5.23}$$

Figure 5.3a shows a plot of the standard deviations of the reflectivity estimates at the S-band frequency of 3 GHz for various turbulences as a function of dwell time. The SD of 2.0 dB corresponds to the dwell time T_d of about 12 ms for $\sigma_v = 4$ m/s and about 43 ms for $\sigma_v = 1$ m/s. These dwell times determine the number of pulses needed for each estimate.

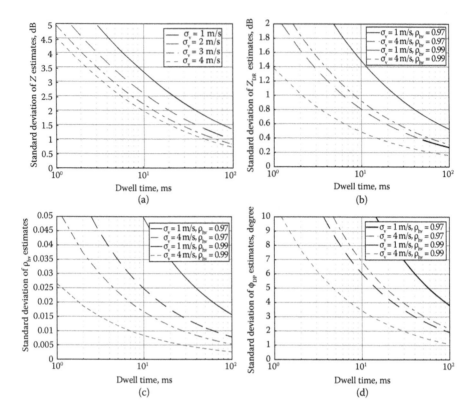

FIGURE 5.3   The standard deviations of the polarimetric radar variable estimates of (a) reflectivity, (b) differential reflectivity, (c) co-polar correlation coefficient, and (d) differential phase at S-band (3 GHz) frequency as a function of dwell time for various spectrum widths.

5.2.2   DIFFERENTIAL REFLECTIVITY

Differential reflectivity is estimated from the ratio of the power estimates for the horizontal and vertical polarization signals and is represented by

$$\hat{Z}_{DR} = 10\log\left(\frac{\hat{P}_h}{\hat{P}_v}\right). \tag{5.24}$$

Hence, the expected value of the differential reflectivity estimates is

$$\left\langle \hat{Z}_{DR} \right\rangle = 10\log\left(\frac{\left\langle \hat{P}_h \right\rangle}{\left\langle \hat{P}_v \right\rangle}\right) = 10\log\left(\frac{P_h}{P_v}\right) = Z_{DR}. \tag{5.25}$$

Because $Z_{DR} = Z_{DR}(P_h, P_v)$, the sampling/fluctuation error can be represented by

$$\delta Z_{DR} = \frac{\partial Z_{DR}}{\partial P_h}\delta P_h + \frac{\partial Z_{DR}}{\partial P_v}\delta P_v = \frac{10}{\ln 10}\left(\frac{\delta P_h}{P_h} - \frac{\delta P_v}{P_v}\right) \tag{5.26}$$

and

$$\operatorname{var}(\hat{Z}_{DR}) = \left\langle \left| \delta Z_{DR} \right|^2 \right\rangle$$

$$= \left(\frac{10}{\ln 10} \right)^2 \left[\frac{\operatorname{var}(\hat{P}_h)}{P_h^2} + \frac{\operatorname{var}(\hat{P}_v)}{P_v^2} - 2\frac{\operatorname{cov}(\hat{P}_h, \hat{P}_v)}{P_h P_v} \right] \qquad (5.27)$$

$$= \left(\frac{10}{\ln 10} \right)^2 \left[\frac{2}{M_I} \left(1 - \rho_{hv}^2 \right) \right].$$

Accurate estimation of polarimetric radar variables is crucial for accurate weather quantification. The accuracy of polarimetric variable estimates, however, depends highly on the system performance and the signal-to-noise ratio (SNR; SNR effects are discussed in the next section). The standard deviation of the differential reflectivity (Z_{DR}) estimates is the square root of Equation 5.27, namely

$$\operatorname{SD}(\hat{Z}_{DR}) \approx 4.343 \sqrt{\frac{2}{M_I} \left(1 - \rho_{hv}^2 \right)}. \qquad (5.28)$$

For $\rho_{hv} = 0.99$ and 0.97, the standard deviation of the relative differential reflectivity estimates is plotted as a function of dwell time for a spectrum width of 1.0 and 4.0 m/s, respectively, in Figure 5.3b. It can be seen that an SD of 0.4 dB corresponds to a dwell time of 60 ms for turbulence of 1 m/s. This would require 20 pulses for a long PRT of 3.0 ms.

5.2.3 CO-POLAR CORRELATION COEFFICIENT

The co-polar cross-correlation coefficient (ρ_{hv}) is a polarimetric parameter that plays an important role in determining system performance and in classifying radar echo types, which can be estimated by the following:

$$\hat{\rho}_{hv} = \frac{\left| \hat{C}_{hv}(0) \right|}{\sqrt{\hat{P}_h \hat{P}_v}}. \qquad (5.29)$$

Following the same procedure as for deriving the estimation error for differential reflectivity, the standard deviation of the correlation coefficient estimates can be expressed approximately by the following (Doviak and Zrnic 1993):

$$\operatorname{SD}(\hat{\rho}_{hv}) = \frac{1 - \rho_{hv}^2}{\sqrt{2M_I}}. \qquad (5.30)$$

The standard deviations of the co-polar correlation coefficient estimates for a spectrum width of $\sigma_v = 1.0$ and 4.0 m/s and $\rho_{hv} = 0.99$ and 0.97 are plotted in Figure 5.3c. It is evident that 20 long PRT pulses (60 ms) would achieve $\operatorname{SD}\left(\hat{\rho}_{hv} \right) < 0.007$.

5.2.4 DIFFERENTIAL PHASE

The differential phase (ϕ_{DP}) and its derived variable, the specific differential phase (K_{DP}), can be used for QPE without the bias caused by reflectivity measurement bias or error. The differential phase estimate and its standard deviation can be expressed approximately by

$$\hat{\phi}_{DP} = \frac{180}{\pi} \angle \hat{C}_{hv}(0) \tag{5.31}$$

and

$$SD(\hat{\phi}_{DP}) = \frac{180}{\pi \rho_{hv}} \sqrt{\frac{1 - \rho_{hv}^2}{2M_I}}. \tag{5.32}$$

The results are plotted in Figure 5.3d. The maximum requirement of 3.0° SD is satisfied with a dwell time longer than 60 ms (20 long PRT pulses). As in Equations 5.28, 5.30, and 5.32, the accuracy of polarimetric radar variable estimates depends on the intrinsic value of the co-polar correlation coefficient: the higher the correlation, the more accurate the polarimetric radar estimates. Because of the requirement for the high correlation ρ_{hv} between the dual-polarization signals, the requirement for radar hardware is also high: any mismatch between the H- and V-channels and signal instability would affect measurement accuracy.

5.2.5 RADIAL VELOCITY AND SPECTRUM WIDTH

From Equations 4.15 and 4.16, radial velocity and spectrum width can be estimated from the phase and magnitude of ACF estimates, expressed by

$$\hat{v}_r = \angle \hat{R}_{h,v}(T_s) \times \lambda / (4\pi T_s) \tag{5.33}$$

$$\hat{\sigma}_v = \left\{ 2 \ln \left[\left| \hat{R}_{h,v}(0) \right| / \left| \hat{R}_{h,v}(T_s) \right| \right] \right\}^{1/2} \times \lambda / (4\pi T_s). \tag{5.34}$$

The standard deviation of the radial velocity estimates and that of the spectrum width estimates were derived by Doviak and Zrnic (1993; Eq. 6.22), given as

$$SD(\hat{v}_r) = \frac{\lambda}{4\pi\rho(T_s)T_s\sqrt{2M}} \left[1 - \rho^2(T_s) \right]^{1/2} \tag{5.35}$$

$$SD(\hat{\sigma}_v) = \frac{\lambda^2}{16\pi^2\rho(T_s)\sigma_v T_s^2 \sqrt{2M}} \left[\left(1 - \rho^2(T_s) \right)^2 \right]^{1/2}. \tag{5.36}$$

The accuracy requirement of 1 m/s can be satisfied with a dwell time of 40 ms for S-band, which is less stringent than that for polarimetric radar variables.

5.3   MULTILAG CORRELATION ESTIMATORS

As noted in Section 5.2, signal noise has been ignored in the estimation of the polarimetric radar variables and their error analysis. The noise, however, causes bias and
extra errors in moment estimates and polarimetric radar data. Data quality degrades
when SNR decreases. This limits the usage of PRD to the high SNR region. For example, most hydrometeor classification algorithms and rain estimations use the Z_{DR} and ρ_{hv}
data only in the regions where the SNR is larger than 5 dB (Park et al. 2009).

Attempts have been made to correct the bias caused by the noise power. For
example, signal power is estimated by subtracting noise power $\hat{N}_{h,v}$ from the autocorrelation estimate as

$$\hat{P}^{(c)}_{sh,sv} = |\hat{R}_{h,v}(0)| - \hat{N}_{h,v}. \tag{5.37}$$

Then, the polarimetric radar variables Z_{DR} and ρ_{hv} are corrected accordingly.
Below is an alternate way of correcting the Z_{DR} and ρ_{hv} biases caused by noise:

$$\hat{Z}^{(c)}_{DR} = \hat{Z}_{DR} + 10 \cdot \log\left(\frac{1 + 1/SNR_v}{1 + 1/SNR_h}\right) \tag{5.38}$$

$$\hat{\rho}^{(c)}_{hv} = \hat{\rho}_{hv}\sqrt{(1 + 1/SNR_h)(1 + 1/SNR_v)}, \tag{5.39}$$

where SNR_h and SNR_v are the SNR for the H- and V-polarization channels,
respectively.

In practice, however, it is difficult to accurately estimate noise powers and correct
them in moment data because the noise power depends on the weather environment,
which varies. That is, the estimation and correction of noise power need to be done
depending on the beam direction (Ivić 2014; Ivić et al. 2013), which is being implemented on the WSR-88D.

There have been other attempts to improve the quality of PRD by reducing the
effects of noise. For example, an LDR estimator introduced by Hubbert et al. (2003)
is effective in low-SNR regions. It uses the cross-to-cross covariances as opposed to
just the autocovariance of the cross-polar time series to calculate cross-polar power. In
addition, *one-lag estimators* were introduced to estimate Z_{DR} and ρ_{hv} from lag 1 correlations (Melnikov and Zrnić 2007). The one-lag estimators avoid the use of lag 0 data
to estimate the polarimetric parameters Z_{DR} and ρ_{hv}, and hence the one-lag estimates
of these two parameters are unbiased by noise. Furthermore, the standard deviations
of the one-lag estimators of Z_{DR} and ρ_{hv} are virtually the same as the conventional estimates when spectral width is less than 6 m/s and SNRs are larger than 5dB; at wider
spectral widths, the standard deviations of one-lag estimates are larger than those from
the conventional algorithm (Doviak and Zrnić 2006). To minimize the noise effects
on the estimation of polarimetric radar variables, multilag correlation estimators were
developed (Lei et al. 2012). This is based on the fact that correlation estimators that do
not use autocorrelation estimates at lag 0 are free from asymptotic noise bias. The multilag correlation estimator was originally developed for crossbeam wind measurement

using spaced antenna interferometry (Zhang and Doviak 2007; Zhang et al. 2004b) and was extended to estimate polarimetric radar variables.

5.3.1 Concept of Multilag Correlation Estimation

The idea of a multilag estimator is to use many available and informative lag correlation estimates to fit a Gaussian function and hence to obtain more accurate estimates of spectral moments and PRD at low SNRs than those obtained using conventional estimators. Depending on the correlation time (which is inversely proportional to spectrum width) of the weather signal, the number of sequential correlation lags, excluding zero lag, can be two, three, four, and so on. They are correspondingly called *two-lag, three-lag, four-lag,* and x-*lag estimators* (Lei 2009; Lei et al. 2009b).

To illustrate the idea, correlation estimates from spaced antenna radar signals and from simulated weather signals are shown in Figure 5.4. Figure 5.4a displays the result from NCAR's multiple antenna profiler radar (Zhang et al. 2004b), and it shows a sharp peak in the autocorrelation (but not cross correlation) at lag 0. Figure 5.4b shows a numerical simulation result for a weather signal generated using the spectrum method of Zrnić (1975). In this case, the frequency is 3 GHz, the number of pulses is 128, and the PRT is 0.001 s. Figure 5.4b shows simulated estimates and fitted Gaussian functions for an SNR of 3 dB and a spectrum width of 3 m/s. The raw ACF estimates are shown as solid dots connected with a dotted line. The fitted Gaussian functions for two-lag and four-lag estimators are shown as dashed and solid lines, respectively. Because the ACF at lag 0 is excluded from the fittings, the resulting fitted Gaussian functions provide better estimates of the spectral moments than those obtained using the conventional pulse pair processor (PPP) in which an independent measurement of noise power is used to estimate the signal power from the ACF datum at zero lag. The fitted Gaussian function is then used to calculate moments (i.e., power, velocity, spectrum width, etc.) and PRD. The detailed fitting procedure and radar moment estimators are described in the following subsections.

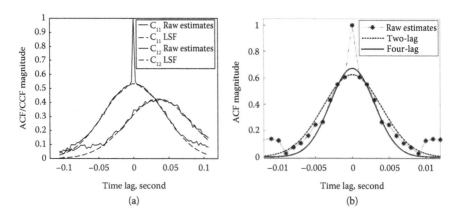

FIGURE 5.4 Concept of multilag correlation estimators for (a) multiple antenna profiler radar signals and from (b) simulated weather signals.

5.3.2 GENERAL EXPRESSIONS

Recall from the equations of the expected ACF and CCF in Equations 5.15 and 5.16 that we have

$$R_{h,v}(nT_s) = P_{sh,sv} \exp\left[-\frac{(nT_s)^2}{2\tau_c^2}\right] \exp\left(-\frac{j\pi n v_r}{v_N}\right) + \mathbb{N}_{h,v}\,\delta_n, \tag{5.40}$$

$$C_{hv}(nT_s) = \sqrt{P_{sh}P_{sv}}\,\rho_{hv} \exp\left[-\frac{(nT_s)^2}{2\tau_c^2}\right] \exp\left(-\frac{j\pi n v_r}{v_N} + j\phi_{dp}\right). \tag{5.41}$$

Taking the natural logarithm of the magnitude of both sides of Equation 5.40, the expected ACF is rewritten as

$$y_n = \ln\left(\left|R_{h,v}(nT_s)\right|\right) = an^2 T_s^2 + b, \tag{5.42}$$

for $n = 1, 2, 3, \ldots$, where $-\dfrac{1}{2\tau_c^2} = a$, $\ln(S_{h,v}) = b$. Then the estimated y_n is given by $\ln\left(\left|\hat{R}_{h,v}(nT_s)\right|\right) = \hat{y}_n$, and $\wedge$ denotes the estimated value. Using both the expected and estimated ACF to form the merit function $J(a, b)$, we obtain

$$J(a,b) = \sum_{n=1}^{N}(n^2 aT_s^2 + b - \hat{y}_n)^2 = a^2 T_s^4 \sum_{n=1}^{N} n^4 + 2abT_s^2 \sum_{n=1}^{N} n^2$$

$$- 2aT_s^2 \sum_{n=1}^{N} n^2 \hat{y}_n + Nb^2 - 2b\sum_{n=1}^{N} \hat{y}_n + \sum_{n=1}^{N} \hat{y}_n^2, \tag{5.43}$$

where N is the number of lags used and $N \geq 2$ (e.g., $N = 3$ indicates that data at lags 1, 2, and 3 are used). Let $\hat{a}$ (or $\hat{b}$) be the optimal estimate of a (or b) obtained from the fitted Gaussian correlation function using N lags, which leads to the merit function (Equation 5.43) reaching its minimum value when

$$\left| \begin{array}{l} \dfrac{\partial J(\hat{a},\hat{b})}{\partial \hat{a}} = 0 \\[3mm] \dfrac{\partial J(\hat{a},\hat{b})}{\partial \hat{b}} = 0. \end{array} \right.$$

These two equations are solved jointly for $\hat{a}$ and $\hat{b}$, and after manipulation the solution can be expressed in the reduced form

$$\hat{a} = \frac{30\sum_{n=1}^{N}\left[6n^2 - (N+1)(2N+1)\right]\hat{y}_n}{T_s^2 N(N-1)(N+1)(2N+1)(8N+11)}, \tag{5.44}$$

$$\hat{b} = \frac{6 \sum_{n=1}^{N} \left(3N^2 + 3N - 1 - 5n^2\right) \hat{y}_n}{N(N-1)(8N+11)},$$ (5.45)

where $N \geq 2$.

From the above equations and the definitions of $\hat{a}$, $\hat{b}$, and $\hat{y}_n$, the general expression for multilag power and spectrum width for H- or V-channel signals can be written as

$$\hat{P}_{sh,sv}^{(N)} \equiv \hat{S}_{h,v}^{(N)} = \left|\hat{R}_{h,v}^{(N)}(0)\right| = \exp\left\{ \frac{6 \sum_{n=1}^{N} \left[\left(3N^2 + 3N - 1 - 5n^2\right) \ln\left(\left|\hat{R}_{h,v}(nT_s)\right|\right)\right]}{N(N-1)(8N+11)} \right\},$$ (5.46)

$$\hat{\sigma}_{h,v}^{(N)} = \frac{\lambda\sqrt{-2\hat{a}}}{4\pi} = \frac{\lambda\sqrt{2}}{4\pi} \sqrt{-\frac{30 \sum_{n=1}^{N} \left\{\left[6n^2 - (N+1)(2N+1)\right]\ln\left(\left|\hat{R}_{h,v}(nT_s)\right|\right)\right\}}{T_s^2 N(N-1)(N+1)(2N+1)(8N+11)}},$$ (5.47)

where the superscript N indicates an N-lag fitted estimate. From the signal power estimates in Equation 5.46, the differential reflectivity estimator is obtained as

$$\hat{Z}_{DR}^{(N)} = 10 \cdot \log\left(\frac{\hat{P}_{sh}^{(N)}}{\hat{P}_{sv}^{(N)}}\right).$$ (5.48)

To obtain the co-polar correlation coefficient, the CCF is estimated by fitting the data to the Gaussian function in Equation 5.41, a similar procedure as that used to estimate the ACF from Equations 5.44 and 5.45. The only difference is that negative lags are used together with zero and positive lags, specifically, $n = -N, -(N-1), \ldots,$ $-1, 0, 1, \ldots, N$, because the CCF is not symmetric and lag 0 is not biased by noise. After applying the similar least-squares (LS) fit used to estimate the ACF, the result of the CCF fitting is

$$\left|\hat{C}_{hv}^{(N)}(0)\right| = \exp\left\{ \frac{3 \sum_{n=-N}^{N} \left[(3N^2 + 3N - 1 - 5n^2)\ln\left(\left|\hat{C}_{hv}(nT_s)\right|\right)\right]}{(2N-1)(2N+1)(2N+3)} \right\},$$ (5.49)

and the co-polar correlation coefficient is

$$\hat{\rho}_{hv}^{(N)} = \frac{\left|\hat{C}_{hv}^{(N)}(0)\right|}{\sqrt{\hat{P}_{sh}^{(N)}\hat{P}_{sv}^{(N)}}}.$$ (5.50)

The multilag method can also be used to obtain the Doppler velocity v_r and the differential phase ϕ_{DP} from the multilag data of the ACF and CCF angles. The mean Doppler velocity is estimated by fitting the unwrapped phase angles of ACFs at multiple lags to a linear line as

$$\hat{v}_{rh,rv}^{(N)} = -\frac{1}{N} \sum_{n=1}^{N} \left\{ \frac{v_N}{n\pi} \left[\angle \hat{R}_{h,v}(nT_s) + 2\pi q_n \right] \right\}, \tag{5.51}$$

where q_n is the integer used to unwrap the phase for lag n. Similarly, the differential phase is obtained by fitting the phase angles of the product of estimated CCFs, that is,

$$\hat{\phi}_{DP}^{(N)} = \frac{180}{2\pi(N+1)} \sum_{n=0}^{N} \angle \left[\hat{C}_{hv}(nT_s) \hat{C}_{hv}(-nT_s) \right]. \tag{5.52}$$

The products of the CCFs at positive and negative lags are used to cancel the Doppler phase shift (Sachidananda and Zrnić 1989).

Equations 5.46 through 5.52 contain the general expressions of power, spectrum width, differential reflectivity, co-polar cross-correlation coefficient, Doppler velocity, and differential phase estimates obtained using multilag estimators. For example, if $N = 4$, it means that lag 1, lag 2, lag 3, and lag 4 of the ACF data and lag 0, lag ±1, lag ±2, lag ±3, and lag ±4 of the CCF data are used to estimate the polarimetric parameters. The value of N chosen depends on the SNR, number of pulses, spectrum width, noise type, and so on. The phase-based estimations for Doppler velocity and differential phase are straightforward—they are the linear average of the estimates from the ACF/CCF angle at different lags. The following section provides the detailed multilag correlation estimators from the ACF/CCF magnitudes for $N = 2$, 3, and 4.

5.3.3 Specific Estimators

5.3.3.1 Two-Lag Estimator

The two-lag estimator uses lags 1 and 2 of the ACF and lags 0, ±1, and ±2 of the CCF to estimate the polarimetric parameters. Substitution of $N = 2$ into Equations 5.46 through 5.52 yields the formulas to estimate power, spectrum width, differential reflectivity, and correlation coefficient, given by

$$\hat{P}_{sh,sv}^{(2)} = \frac{\left| \hat{R}_{h,v}(T_s) \right|^{\frac{4}{3}}}{\left| \hat{R}_{h,v}(2T_s) \right|^{\frac{1}{3}}} \tag{5.53}$$

$$\hat{\sigma}_{h,v}^{(2)} = \frac{\lambda}{\sqrt{24}\pi T_s} \cdot \sqrt{\ln \left| \hat{R}_{h,v}(T_s) \right| - \ln \left| \hat{R}_{h,v}(2T_s) \right|} \tag{5.54}$$

$$\hat{Z}_{DR}^{(2)} = 10 \cdot \log \left(\frac{\left| \hat{R}_h(T_s) \right|^{\frac{4}{3}}}{\left| \hat{R}_h(2T_s) \right|^{\frac{1}{3}}} \cdot \frac{\left| \hat{R}_v(2T_s) \right|^{\frac{1}{3}}}{\left| \hat{R}_v(T_s) \right|^{\frac{4}{3}}} \right) \tag{5.55}$$

$$\hat{\rho}_{hv}^{(2)} = \left|\hat{C}_{hv}^{(2)}(0)\right| \cdot \frac{\left(\left|\hat{R}_h(2T_s)\right| \cdot \left|\hat{R}_v(2T_s)\right|\right)^{\frac{1}{6}}}{\left(\left|\hat{R}_h(T_s)\right| \cdot \left|\hat{R}_v(T_s)\right|\right)^{\frac{2}{3}}}, \tag{5.56}$$

where $\left|\hat{C}_{hv}^{(2)}(0)\right|$ is obtained by substituting $N = 2$ into Equation 5.49.

For spectrum width estimates, the two-lag estimator (i.e., using lag 1 and lag 2) has better performance, as will be shown in Figures 5.5 and 5.6, than the conventional estimator (i.e., using lag 0 and lag 1) if the SNR is low and spectrum width is narrow (Doviak and Zrnić 1984, 1993, 2006; Srivastava et al. 1979). However, for other parameters such as power and differential reflectivity, the improvement is marginal. In order to decrease the exponent and improve the statistical performance, more lags are needed (e.g., three or four lags), as shown next.

5.3.3.2 Three-Lag Estimator

The three-lag estimator uses lags 1, 2, and 3 of the ACF and lags 0, ±1, ±2, and ±3 of the CCF to estimate the radar parameters. This is the case for $N = 3$ in the general formulas (Equations 5.46 through 5.52). Substituting $N = 3$ into the general expressions, the formulas to estimate spectrum width, differential reflectivity, and correlation coefficient are given by

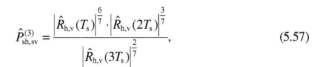

$$\hat{P}_{sh,sv}^{(3)} = \frac{\left|\hat{R}_{h,v}(T_s)\right|^{\frac{6}{7}} \cdot \left|\hat{R}_{h,v}(2T_s)\right|^{\frac{3}{7}}}{\left|\hat{R}_{h,v}(3T_s)\right|^{\frac{2}{7}}}, \tag{5.57}$$

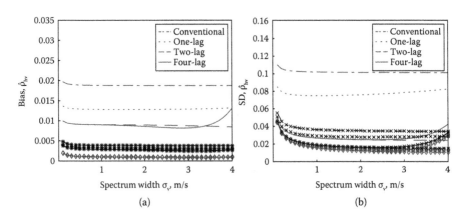

FIGURE 5.5 Bias and standard deviation of co-polar cross-correlation coefficient estimates resulting from (a) estimation error with noise correction and (b) standard deviation of the co-polar cross-correlation coefficient estimates for $\rho_{hv} = 0.97$, $Z_{DR} = 1$ dB, $M = 128$, $T_s = 0.001$ s, and $\lambda = 0.1$ m. The unmarked lines are for a signal-to-noise ratio (SNR) of 0 dB, the lines marked with × signs are for SNR = 5 dB, and those with diamond signs are for SNR = 10 dB.

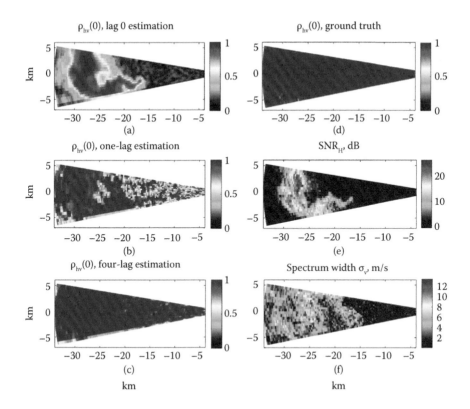

FIGURE 5.6 Comparison of co-polar correlation estimates made using the (a) conventional, (b) one-lag, and (c) four-lag estimators. Also plotted are the ground truth (d) ρ_{hv}, (e) SNR, and (f) σ_v.

$$\hat{\sigma}_{h,v}^{(3)} = \frac{\lambda}{28\pi T_s} \cdot \sqrt{11 \cdot \ln\left|\hat{R}_{h,v}(T_s)\right| + 2 \cdot \ln\left|\hat{R}_{h,v}(2T_s)\right| - 13 \cdot \ln\left|\hat{R}_{h,v}(3T_s)\right|}, \quad (5.58)$$

$$\hat{Z}_{DR}^{(3)} = 10 \cdot \log \frac{\left|\hat{R}_h(T_s)\right|^{\frac{6}{7}} \cdot \left|\hat{R}_h(2T_s)\right|^{\frac{3}{7}} \cdot \left|\hat{R}_v(3T_s)\right|^{\frac{2}{7}}}{\left|\hat{R}_h(3T_s)\right|^{\frac{2}{7}} \cdot \left|\hat{R}_v(T_s)\right|^{\frac{6}{7}} \cdot \left|\hat{R}_v(2T_s)\right|^{\frac{3}{7}}}, \quad (5.59)$$

$$\hat{\rho}_{hv}^{(3)} = \left|\hat{C}_{hv}^{(3)}(0)\right| \cdot \frac{\left[\left|\hat{R}_h(3T_s)\right| \cdot \left|\hat{R}_v(3T_s)\right|\right]^{\frac{1}{7}}}{\left[\left|\hat{R}_h(T_s)\right| \cdot \left|\hat{R}_v(T_s)\right|\right]^{\frac{3}{7}} \cdot \left[\left|\hat{R}_h(2T_s)\right| \cdot \left|\hat{R}_v(2T_s)\right|\right]^{\frac{3}{14}}}. \quad (5.60)$$

In Equation 5.59, the exponential factors are 6/7, 3/7, and 2/7, which are much smaller than the exponential factor of the two-lag estimator in Equation 5.55 (which can be as large as 4/3). We expect the statistical performance of the three-lag estimator to be better than the two-lag estimator and conventional estimator at a low SNR and narrow spectrum width.

5.3.3.3 Four-Lag Estimator

The four-lag estimators for the spectral moments and PRD are

$$
\hat{P}^{(4)}_{\text{sh,sv}} = \frac{\left|\hat{R}_{\text{h,v}}(T_{\text{s}})\right|^{\frac{54}{86}} \cdot \left|\hat{R}_{\text{h,v}}(2T_{\text{s}})\right|^{\frac{39}{86}} \cdot \left|\hat{R}_{\text{h,v}}(3T_{\text{s}})\right|^{\frac{14}{86}}}{\left|\hat{R}_{\text{h,v}}(4T_{\text{s}})\right|^{\frac{21}{86}}},
\tag{5.61}
$$

$$
\hat{\sigma}^{(4)}_{\text{h,v}} = \frac{\lambda}{4\sqrt{129}\pi T_{\text{s}}} \cdot \sqrt{\begin{array}{l} 13 \cdot \ln\left|\hat{R}_{\text{h,v}}(T_{\text{s}})\right| + 7 \cdot \ln\left|\hat{R}_{\text{h,v}}(2T_{\text{s}})\right| \\ -3 \cdot \ln\left|\hat{R}_{\text{h,v}}(3T_{\text{s}})\right| - 17 \cdot \ln\left|\hat{R}_{\text{h,v}}(4T_{\text{s}})\right| \end{array}},
\tag{5.62}
$$

$$
\hat{Z}^{(4)}_{\text{DR}} = 10 \cdot \log\left[\frac{\left|\hat{R}_{\text{h}}(T_{\text{s}})\right|^{\frac{54}{86}} \cdot \left|\hat{R}_{\text{h}}(2T_{\text{s}})\right|^{\frac{39}{86}} \cdot \left|\hat{R}_{\text{h}}(3T_{\text{s}})\right|^{\frac{14}{86}} \cdot \left|\hat{R}_{\text{v}}(4T_{\text{s}})\right|^{\frac{21}{86}}}{\left|\hat{R}_{\text{h}}(4T_{\text{s}})\right|^{\frac{21}{86}} \cdot \left|\hat{R}_{\text{v}}(T_{\text{s}})\right|^{\frac{54}{86}} \cdot \left|\hat{R}_{\text{v}}(2T_{\text{s}})\right|^{\frac{39}{86}} \cdot \left|\hat{R}_{\text{v}}(3T_{\text{s}})\right|^{\frac{14}{86}}}\right],
\tag{5.63}
$$

$$
\hat{\rho}^{(4)}_{\text{hv}} = \left|\hat{C}^{(4)}_{\text{hv}}(0)\right|
$$

$$
\cdot \frac{\left[\left|\hat{R}_{\text{h}}(4T_{\text{s}})\right| \cdot \left|\hat{R}_{\text{v}}(4T_{\text{s}})\right|\right]^{\frac{21}{172}}}{\left[\left|\hat{R}_{\text{h}}(T_{\text{s}})\right| \cdot \left|\hat{R}_{\text{v}}(T_{\text{s}})\right|\right]^{\frac{27}{86}} \cdot \left[\left|\hat{R}_{\text{h}}(2T_{\text{s}})\right| \cdot \left|\hat{R}_{\text{v}}(2T_{\text{s}})\right|\right]^{\frac{39}{172}} \cdot \left[\left|\hat{R}_{\text{h}}(3T_{\text{s}})\right| \cdot \left|\hat{R}_{\text{v}}(3T_{\text{s}})\right|\right]^{\frac{7}{86}}}.
\tag{5.64}
$$

In Equations 5.63 and 5.64, the exponents are smaller than the corresponding exponents for the two- and three-lag estimators. Therefore, we expect the performance of the four-lag estimator to be better at a low SNR and narrow spectrum width (i.e., at long correlation time). The higher order multilag (e.g., four-lag) estimators give better estimates at narrow spectrum widths. However, the maximum number of lags to be used is determined by the correlation time (which is inversely proportional to the spectrum width). If the lag is larger than the correlation time, using this lag will introduce more error because the relative error of ACFs and CCFs at larger lags is larger than that at small lags. There is a limit as to how many lags to use in multilag estimators; this limit depends on the number of pulses, the SNR, and the correlation time. For example, the absolute value of the ACF is used to weight the multilag poly pulse pair estimate for Doppler velocity (Lee 1978; May and Strauch 1989). The number of lags to be used can be determined according to measured radar parameters, typically no more than the correlation time (Cao et al. 2012b; Lei et al. 2012).

5.3.4 Performance of the Estimators

The performance of the multilag estimators is examined through perturbation analysis and compared with the performance of conventional estimators. Theoretical statistical biases and standard deviations of power, spectrum width, differential reflectivity estimates, and co-polar correlation coefficient estimates were calculated and verified by simulations by Lei et al. (2012). Here, only the general procedure for the error analysis and the results for the correlation coefficient are briefly described below.

To calculate the bias and standard deviation of multilag estimators, perturbation analysis is used (Zhang et al. 2004b). Terms up to the second order of the Taylor expansion are retained. For example, the following is the Taylor expansion of signal power in several variables, $|\hat{R}(T_s)|, |\hat{R}(2T_s)|,$ to $|\hat{R}(NT_s)|$:

$$\hat{P}_s^{(N)}\left[|\hat{R}(T_s)|,...,|\hat{R}(NT_s)|\right]$$

$$= \sum_{n_1=0}^{\infty}...\sum_{n_N=0}^{\infty}\left\{\begin{array}{c}\dfrac{\left[|\hat{R}(T_s)|-|R(T_s)|\right]^{n_1}...\left[|\hat{R}(NT_s)|-|R(NT_s)|\right]^{n_N}}{n_1!...n_N!}\\\left[\dfrac{\partial^{n_1+...+n_N}S^{(N)}}{\partial|\hat{R}(T_s)|^{n_1}...\partial|\hat{R}(NT_s)|^{n_N}}\right]\left[|R(T_s)|,...,|R(NT_s)|\right]\end{array}\right\}.$$ (5.65)

$R(T_s)$ is the ACF and is the signal power estimated by using N lags of the ACF; $n_1...n_N$ is the number of derivatives for each variable in the Taylor expansion. Similar expansions can also be applied to other radar parameters. If an N lag estimator is used, $P_{sh,sv}$ and σ_v both have N variables, but Z_{DR} has $2N$ variables and ρ_{hv} has $4N+1$ variables. The difference between the estimated and true $S^{(N)}$ is expressed by

$$\delta P_s^{(N)} = \hat{P}_s^{(N)}\left[|\hat{R}(T_s)|,...,|\hat{R}(NT_s)|\right] - P_s^{(N)}\left[|R(T_s)|,...,|R(NT_s)|\right]$$

$$= \sum_{n_1=0}^{\infty}...\sum_{n_N=0}^{\infty}\left\{\begin{array}{c}\dfrac{\left[|\hat{R}(T_s)|-|R(T_s)|\right]^{n_1}...\left[|\hat{R}(NT_s)|-|R(NT_s)|\right]^{n_N}}{n_1!...n_N!}\\\left[\dfrac{\partial^{n_1+...+n_N}S^{(N)}}{\partial|\hat{R}(T_s)|^{n_1}...\partial|\hat{R}(NT_s)|^{n_N}}\right]\left[|R(T_s)|,...,|R(NT_s)|\right]\end{array}\right\}$$

$$- S^{(N)}\left[|R(T_s)|,...,|R(NT_s)|\right],$$ (5.66)

$$= \sum_{n_1=0}^{\infty}...\sum_{n_N=0}^{\infty}\left\{\begin{array}{c}\dfrac{\left[|\hat{R}(T_s)|-|R(T_s)|\right]^{n_1}...\left[|\hat{R}(NT_s)|-|R(NT_s)|\right]^{n_N}}{n_1!...n_N!}\\\left[\dfrac{\partial^{n_1+...+n_N}P_s^{(N)}}{\partial|\hat{R}(T_s)|^{n_1}...\partial|\hat{R}(NT_s)|^{n_N}}\right]\left[|R(T_s)|,...,|R(NT_s)|\right]\end{array}\right\}$$

if $n_1, n_2, ..., n_N$, the number of derivatives of the Taylor expansion is not simultaneously equal to zero.

The statistical bias and variance of the signal power estimate $\hat{P}_s$ are defined as

$$\text{bias}(\hat{P}_s^{(N)}) = \langle \delta P_s^{(N)} \rangle, \tag{5.67}$$

$$\text{var}(\hat{P}_s^{(N)}) = \langle (\delta P_s^{(N)})^2 \rangle. \tag{5.68}$$

Substitution of Equation 5.66 into Equations 5.67 and 5.68 yields the expression for the bias

$$\text{bias}\left(\hat{P}_s^{(N)}\right) = \left\langle \sum_{n_1=0}^{\infty} \cdots \sum_{n_N=0}^{\infty} \left\{ \frac{\left[\left|\hat{R}(T_s)\right| - \left|R(T_s)\right|\right]^{n_1} \cdots \left[\left|\hat{R}(NT_s)\right| - \left|R(NT_s)\right|\right]^{n_N}}{n_1! \cdots n_N!} \cdot \left[\frac{\partial^{n_1 + \cdots + n_N} P_s^{(N)}}{\partial\left|\hat{R}(T_s)\right|^{n_1} \cdots \partial\left|\hat{R}(NT_s)\right|^{n_N}}\right] \left[\left|R(T_s)\right|, ..., \left|R(NT_s)\right|\right] \right\} \right\rangle, \tag{5.69}$$

and the expression for the variance

$$\text{var}\left(\hat{P}_s^{(N)}\right) = \left\langle \left\{ \sum_{n_1=0}^{\infty} \cdots \sum_{n_N=0}^{\infty} \left\{ \frac{\left[\left|\hat{R}(T_s)\right| - \left|R(T_s)\right|\right]^{n_1} \cdots \left[\left|\hat{R}(NT_s)\right| - \left|R(NT_s)\right|\right]^{n_N}}{n_1! \cdots n_N!} \cdot \left[\frac{\partial^{n_1 + \cdots + n_N} \hat{P}_s^{(N)}}{\partial\left|\hat{R}(T_s)\right|^{n_1} \cdots \partial\left|\hat{R}(NT_s)\right|^{n_N}}\right] \left[\left|R(T_s)\right|, ..., \left|R(NT_s)\right|\right] \right\} \right\}^2 \right\rangle. \tag{5.70}$$

In Equations 5.69 and 5.70, $n_1, n_2, ..., n_N$ are not simultaneously equal to zero. Following the same procedure used to obtain the bias and variance of the signal power estimates, the bias and variance of $\hat{\sigma}_v^{(N)}$, $\hat{Z}_{DR}^{(N)}$, and $\hat{\rho}_{hv}^{(N)}$ can be obtained (see detailed analysis in Lei et al. 2012).

The co-polar correlation coefficient $\hat{\rho}_{hv}^{(N)}$ is used as an example to demonstrate the performance of the multilag estimator for a PRT of $T_s = 0.001$s at S-band of $\lambda = 0.1$ m. The bias and standard deviation for four types of estimators—conventional, one-lag, two-lag, and four-lag estimators—are examined in Figure 5.5a and b. It is shown that the four-lag estimator produces significantly less biased estimates than do the other estimators at low SNR (e.g., <5 dB) and for spectrum widths less than about 3.5 m/s. However, if there is bias in the measured noise power as in Figure 5.5b, the improvement of multilag estimators over the conventional estimator is even larger. Furthermore, the $\text{SD}\left(\hat{\rho}_{hv}\right)$ for the four-lag estimator is significantly lower at a low SNR (Figure 5.5b). These theoretical results agree well with simulations except for small truncation errors due to the limitations of the perturbation method (figure 13 in Lei et al. 2012).

Nevertheless, the multilag estimator performs better than the conventional estimator, which tends to yield low ρ_{hv} if noise power is not corrected for.

Figure 5.6 shows the results of ρ_{hv} estimates with the multilag estimators using NWP simulation. The input of this radar simulation comes from the Advanced Regional Prediction System (ARPS) NWP model (Lei 2009; Lei et al. 2009a; Xue et al. 2000, 2001). The prognostic state variables include the three wind components, potential temperature, pressure, turbulent kinetic energy, mixing ratios for water vapor, rainwater, cloud water, cloud ice, snow, and hail. The ground truth ρ_{hv}, SNR, and spectrum width are calculated from the ARPS model and shown in the right column (Figure 5.6d through f). The model-generated ground truth correlation coefficient is larger than 0.96 in most places, even though the SNR is as small as 0 dB. Estimates made with the four-lag estimator are closer to the ground truth than those made with the one-lag and conventional estimators. For this simulation, the noise power was not calculated because there is not a standard procedure for noise correction for all radar parameters; thus the noise power was not subtracted to estimate the signal power needed in Equation 5.37 for the conventional estimate of ρ_{hv} (Figure 5.6a). The omission of noise power from the calculation of signal power causes significant bias when compared with the ground truth (Figure 5.6d); the bias is largest where the SNR (Figure 5.6e) is weakest. Figure 5.6c shows that $\hat{\rho}_{hv}$ estimation improves significantly when the four-lag estimator is used.

The multilag correlation estimators were also implemented by Cheong et al. (2013) on a real radar system (PX-1000), a low power X-band polarimetric radar. Figure 5.7 shows the reflectivity factor on the left and the co-polar correlation coefficient on the right. The top row is from the S-band KTLX WSR-88DP radar with high sensitivity. The conventional estimate is in the middle row, and the multilag result is in the bottom row. It is evident that the multilag estimators produced better polarimetric radar data than the conventional estimators. Whereas the KTLX data show high ρ_{hv} values except for the region of the melting layer, the conventional estimator produced low ρ_{hv} values at near range (short pulse) and on the storm's edges where the SNR was low. Had these data been used without correction, the storm's edges might have been interpreted as biological scatterers or mixed phase. With multilag estimators, the ρ_{hv} values were recovered at the edges and near range gates, and the result was more reasonable, correctly detecting the melting layer with low ρ_{hv} values.

5.4 CLUTTER DETECTION

Besides noise, ground clutter is another issue in weather radar measurement because it causes bias and error in weather radar data and hence affects rain estimation and microphysics studies. Ground clutter tends to be stationary, and the clutter contamination is conventionally reduced by applying a band-stop (notch) filter centered at zero (Groginsky and Glover 1980). This method, however, can cause bias in weather moment estimates for narrow-band zero-velocity weather signals, because the weather power components are suppressed by the filter. Therefore, it is best to first detect the locations of ground clutter and then apply an optimal estimator or filter on those contaminated signals in order to obtain high quality weather data without bias from the estimation or filtering.

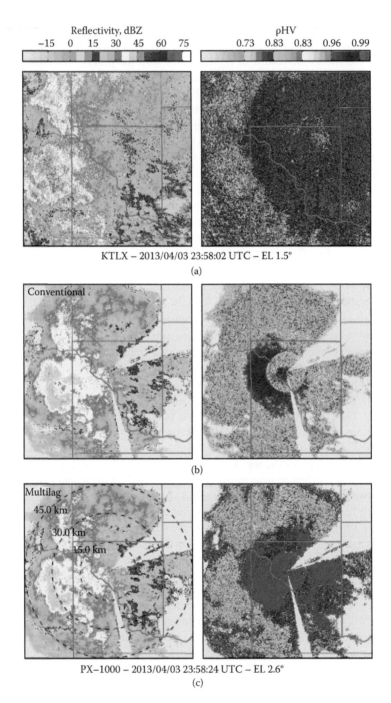

FIGURE 5.7 An example of multilag estimation of X-band PX-1000 data (c), compared with conventional estimation (b) and with KTLX WSR-88DP radar data (a). Reflectivity is on the left and the co-polar correlation coefficient is on the right.

5.4.1 Background of Clutter Detection

Traditionally, clutter-contaminated gates are identified by using a static clutter map determined from data collected in clear-air conditions (Meischner 2004). However, the clutter maps can change depending on weather conditions, for example, clutter contamination that appears only under anomalous propagation (AP) conditions. Thus, an adaptive ground clutter detection algorithm is needed to detect ground clutter under both normal propagation (NP) and AP conditions.

One such adaptive approach is the decision tree algorithm introduced by Lee et al. (1995), which makes a clutter/nonclutter decision using radial velocity, spectrum width, minimum detectable signal, one-lag and two-lag signal fluctuations, the vertical gradient of reflectivity, and a continuously updated clutter map. A radar echo classifier (REC) proposed by Kessinger et al. (2003) was deployed within the National Weather Service's WSR-88D ORPG (Open Radar Product Generator) Build 2—it classifies radar echoes, including ground scatterers seen via AP, precipitation, insects, and clutter associated with sea waves. Another approach introduced by Steiner and Smith (2002) uses the three-dimensional reflectivity structure to detect ground clutter observed under both NP and AP conditions. The clutter mitigation decision (CMD) algorithm introduced by Hubbert et al. (2009a, 2009b) and tested by Ice et al. (2009) combines three discriminants: clutter phase alignment (CPA), texture of reflectivity, and spatial variability of reflectivity field (SPIN). (The SPIN parameter indicates the number of fluctuations when a reflectivity increases [decreases] in a radial direction from one pixel to the next by more than 2 dBZ, using a fuzzy logic approach to determine the existence of clutter; it is currently used in the WSR-88D RDA subsystem.) More recently, Torres and Warde (2014) introduced the CLEAN-AP (clutter environment analysis using adaptive processing) filter, which automates the detection and mitigation of ground clutter contamination in both NP and AP conditions, using the characteristics of ground clutter. The CLEAN-AP filter is currently being tested for inclusion in the WSR-88D radar network.

Considering the difference in spectrum properties between clutter and weather signals, Li et al. (2013b) recently introduced a spectrum clutter identification (SCI) that included four discriminants: spectral power distribution, spectral phase fluctuation, power texture, and spectrum width texture for clutter detection.

All of these clutter detection methods include (i) finding discriminants, which are parameters that differ for clutter versus weather and (ii) empirically combining the discriminants to reach a decision on whether the signal comes from weather or clutter. In the following sections, we define discriminants based on the properties of polarimetric radar signals and introduce the simple Bayesian classifier (SBC) for clutter detection.

5.4.2 Definition and Property of Discriminants

5.4.2.1 Power Ratio Discriminant

As discussed in Section 4.2 of Chapter 4, in general, wave scattering from randomly distributed hydrometeors yields an incoherent wave field and intensity, but not a coherent (average) wave field and intensity, in all directions other than the forward direction.

Deterministic scatterers such as ground clutter, however, cause coherent intensity/power. Hence, the ratio between coherent power $\left(\left|\langle V\rangle\right|^2\right)$ and incoherent power $\left(\langle|V|^2\rangle - \left|\langle V\rangle\right|^2\right)$ can used to distinguish clutter from weather signals. Therefore, the power ratio (PR) can be defined from time series data with M samples:

$$\hat{pr} = \frac{\left|\langle V\rangle\right|^2}{\langle|V|^2\rangle - \left|\langle V\rangle\right|^2} \approx \frac{\left|\dfrac{1}{M}\displaystyle\sum_{m=1}^{M} V(m)\right|^2}{\dfrac{1}{M}\displaystyle\sum_{m=1}^{M}|V(m)|^2 - \left|\dfrac{1}{M}\displaystyle\sum_{m=1}^{M} V(m)\right|^2}. \tag{5.71}$$

Equation 5.71 is equivalent to the ratio of the spectral power of the zero Doppler (DC or coherent power) to the total spectral power of all other spectra (AC or incoherent power) in the frequency domain, but is computationally more efficient to calculate in the time domain. It is expected that the pr will be large for ground clutter and small for weather echoes. This property is similar to that of the CPA introduced in Hubbert et al. (2009b) as

$$CPA = \frac{\left|\displaystyle\sum_{m=1}^{M} V(m)\right|}{\displaystyle\sum_{m=1}^{M}|V(m)|}. \tag{5.72}$$

Whereas the CPA ranges from zero for weather to one for clutter, the pr in Equation 5.71 has a larger dynamic range from zero for weather to infinite for clutter. Furthermore, the coherent and incoherent power in Equation 5.71 are all directly estimated from the coherent and incoherent intensity defined in Equations 4.8 and 4.9. Hence, we choose to use the power ratio $PR = 10\log(pr)$ in dB for echo separation.

Three types of radar echoes are considered: (i) clutter (c), (ii) narrow-band zero-velocity weather ($w0$) (i.e., $|v_r| < 1$ m/s, $\sigma_v \leq 1$ m/s), and (iii) nonzero velocity weather (w) (i.e., $|v_r| > 1$ m/s, $\sigma_v > 1$ m/s). We separate $w0$ from w because it is more difficult to distinguish narrow-band zero-velocity weather echo from ground clutter than it is broadband nonzero velocity weather echo. Figure 5.8 shows the PDFs of PR for c, $w0$, and w for the data collected with OU-PRIME, a C-band polarimetric radar with a beamwidth of 0.5°. It is clear the PR for clutter is generally much larger than that for weather. There is very little overlap (<0.02 common area) in the PR between clutter and nonzero velocity weather. By contrast, there is substantial overlap (~0.35 common area) between clutter and narrow-band zero velocity weather, which require further measurement information to separate.

5.4.2.2 Dual-Polarization Discriminants

Because the single polarization power ratio is not sufficient to distinguish clutter from narrow-band zero-velocity weather, it is reasonable and desirable to use

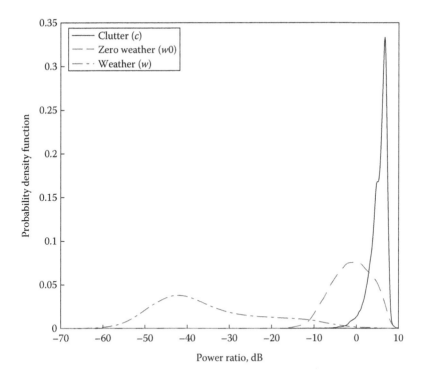

FIGURE 5.8 Probability density functions of the power ratio (PR, in dB) obtained from OU-PRIME data: (1) clutter, (2) narrow-band zero-velocity weather (i.e., $|v_r| < 1$ m/s, $\sigma_v \leq 1$ m/s), and (3) nonzero velocity weather (i.e., $|v_r| > 1$ m/s, $\sigma_v > 1$ m/s).

polarimetric radar signals to define discriminants for clutter detection. Applying the definition of the PR in Equation 5.71 to dual-polarization radar signals and including the polarimetric radar variables Z_{DR} and ρ_{hv}, we obtain four discriminants from the dual-polarization radar signals V_h and V_v as follows (Li et al. 2014):

$$\widehat{PR}_H = 10\log\left(\frac{\left|\frac{1}{M}\sum_{m=1}^{M}V_h(m)\right|^2}{\frac{1}{M}\sum_{m=1}^{M}|V_h(m)|^2 - \left|\frac{1}{M}\sum_{m=1}^{M}V_h(m)\right|^2}\right), \qquad (5.73)$$

$$\widehat{PR}_v = 10\log\left(\frac{\left|\frac{1}{M}\sum_{m=1}^{M}V_v(m)\right|^2}{\frac{1}{M}\sum_{m=1}^{M}|V_v(m)|^2 - \left|\frac{1}{M}\sum_{m=1}^{M}V_v(m)\right|^2}\right), \qquad (5.74)$$

$$
\hat{Z}_{DR} = 10 \log \left(\frac{\left| \dfrac{1}{M-l} \displaystyle\sum_{m=1}^{M-l} V_h^*(m+l)V_h(m) \right|}{\left| \dfrac{1}{M-l} \displaystyle\sum_{m=1}^{M-l} V_v^*(m+l)V_v(m) \right|} \right), \quad l = 0,1, \tag{5.75}
$$

$$
\hat{\rho}_{hv} = \frac{\left| \dfrac{1}{M-l} \displaystyle\sum_{m=1}^{M-l} V_h^*(m+l)V_v(m) \right| + \left| \dfrac{1}{M-l} \displaystyle\sum_{m=1}^{M-l} V_h(m)V_v^*(m+l) \right|}{2\sqrt{\left| \dfrac{1}{M-l} \displaystyle\sum_{m=1}^{M-l} V_h^*(m+l)V_h(m) \right| \left| \dfrac{1}{M-l} \displaystyle\sum_{m=1}^{M-l} V_v^*(m+l)V_v(m) \right|}}, \quad l = 0,1. \tag{5.76}
$$

PR_H and PR_V are the ratios between the coherent power and incoherent power of the horizontal and vertical channels, respectively. In Equations 5.73 and 5.76, $l = 0$ is used for a high SNR (>20 dB) and $l = 1$ for a low SNR (≤20 dB) to minimize noise effects.

Using a data set collected by OU-PRIME, the property of the discriminants can be quantified by calculating their PDFs for a given class of echo. The PDFs for the four discriminants $\widehat{PR}_H$, $\widehat{PR}_V$, $\hat{Z}_{DR}$, and $\hat{\rho}_{hv}$ are shown in Figure 5.9.

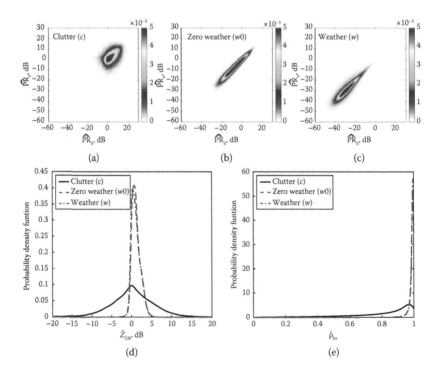

FIGURE 5.9 The joint probability density functions $p(PR_H, PR_V)$ for (a) clutter (c), (b) zero weather ($w0$), and (c) weather (w); the probability density functions of (d) $\hat{Z}_{DR}$ and (e) $\hat{\rho}_{hv}$ for c, $w0$, and w (From Li, Y., et al., 2014. *IEEE Transactions on Signal Processing*, 62, 597–606.)

It is evident from Figure 5.9a through c that there is almost no overlap between $p(PR_H, PR_V)$ for clutter (Figure 5.9a) and that for weather (Figure 5.9c), but it is more difficult to distinguish clutter from zero weather. In Figure 5.9d and e, the PDFs of $\hat{Z}_{DR}$ and $\hat{\rho}_{hv}$ for w and $w0$ are almost the same, but very different from that of clutter. $\hat{Z}_{DR}$ has a much larger spread for clutter than weather, but the middle values overlap. The PDFs, $p(\hat{\rho}_{hv})$, are almost equal to zero for weather when $\hat{\rho}_{hv}$ is less than 0.85 and the peak is located at $\hat{\rho}_{hv} = 0.995$; $\hat{\rho}_{hv}$ has a much larger spread for c and the peak of $p(\hat{\rho}_{hv}|c)$ is located at $\hat{\rho}_{hv} = 0.96$. The joint PDF for $\widehat{PR}_H$ and $\widehat{PR}_V$, and the PDFs for $\hat{Z}_{DR}$ and $\hat{\rho}_{hv}$ provide further information to distinguish clutter from weather.

5.4.2.3   Dual-Scan Discriminant

Considering that the correlation time of radar echoes from hydrometeors is typically much shorter than that from ground clutter, the scan-to-scan (dual-scan) correlation can be used to separate clutter from weather echoes (Li et al. 2013a). Typically, the correlation time of weather signals from a 10-cm wavelength radar is less than 10 ms (8 ms for $\sigma_v = 1.0$ m/s) for most weather phenomena (as discussed in Chapter 4), and it is even shorter for C- or X-band radar signals. These weather signals do not correlate from one 360° azimuthal scan to the next. Ground clutter is typically stationary and the clutter signals have a long correlation time, on the order of a minute. These clutter signals can be correlated from one scan to next, which is distinctly different from weather signals.

The dual-scan data are available for most operational radars at low elevations: one of the modes is the reflectivity (surveillance) mode and the other is the Doppler mode. The two mode signals can be jointly processed for clutter detection. In the volume coverage patterns of the WSR-88D, two sequential azimuthal scans with different PRTs are used at each of the two low elevation angles (i.e., 0.5° and 1.5°) to mitigate range-velocity ambiguities (Handbook 2006). The first azimuthal scan collects time-series data using a long PRT (e.g., $T_{s1} = 3.10$ ms) and the second scan at the same elevation angles uses a short PRT (e.g., $T_{s2} = 0.973$ ms). To find the scan-to-scan correlation, the short PRT data are non-uniformly down-sampled to match the time as closely as possible to the time series data collected with long PRT.

Once the two time series data sets are obtained with equal intervals and lengths, the cross-correlation coefficient between two scans ρ_{12} for $l = 0$ or 1 can be estimated as follows:

$$\hat{\rho}_{12} = \frac{\left|\dfrac{1}{M-l}\displaystyle\sum_{m=1}^{M-l} V_1^*(m+l)V_2(m)\right| + \left|\dfrac{1}{M-l}\displaystyle\sum_{m=1}^{M-l} V_1(m)V_2^*(m+l)\right|}{2\sqrt{\left|\dfrac{1}{M-l}\displaystyle\sum_{m=1}^{M-l} V_1^*(m+l)V_1(m)\right|\left|\dfrac{1}{M-l}\displaystyle\sum_{m=1}^{M-l} V_2^*(m+l)V_2(m)\right|}}, \quad l = 0,1 \qquad (5.77)$$

where M is the number of samples in the dwell time; V_1 is the time series of voltage samples from the first scan; and V_2 is the down-sampled time series of voltage samples from the second scan from the same volume but collected at a higher pulse repetition frequency (PRF).

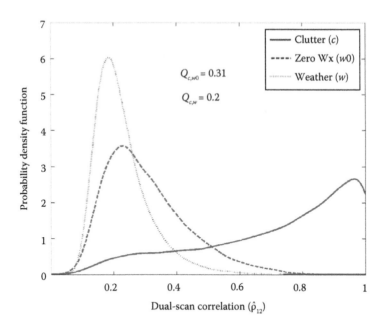

FIGURE 5.10 Probability density functions of cross-correlation coefficients $\hat{\rho}_{12}$ between two scan signals for clutter, zero weather, and weather, respectively. (From Li, Y., et al., 2013b. *IEEE Transactions on Geoscience and Remote Sensing*, 51, 2373–2387.)

Because the correlation time for ground clutter is typically much longer than that for weather signals, ρ_{12} is expected to be much larger for ground clutter than for either weather signal (w or $w0$). Figure 5.10 shows the PDFs of $\hat{\rho}_{12}$ for clutter, zero weather, and weather signals. As expected, clutter has, in general, a much larger ρ_{12} than that of weather, and the overlap area between weather and clutter in the PDF plot is less than 31%. Hence, the scan-to-scan correlation provides useful information for detecting clutter from weather echoes.

5.4.2.4 Dual-Pol Dual-Scan Discriminants (DPDS)

As shown earlier, both dual-pol and dual-scan discriminants provide useful information to separate ground clutter from weather signals. Hence, it is reasonable to combine the dual-pol and dual-scan discriminants for clutter detection in the presence of weather echoes, when it is feasible.

The following 10 discriminants can be estimated from the two pairs of signals (V_{h1}, V_{v1}) from the first scan and (V_{h2}, V_{v2}) from the second scan as follows:

$$\widehat{PR}_{H1} = 10\log\left(\frac{\left|\dfrac{1}{M}\displaystyle\sum_{m=1}^{M}V_{h1}(m)\right|^2}{\dfrac{1}{M}\displaystyle\sum_{m=1}^{M}|V_{h1}(m)|^2 - \left|\dfrac{1}{M}\displaystyle\sum_{m=1}^{M}V_{h1}(m)\right|^2}\right) \tag{5.78}$$

$$\widehat{PR}_{V1} = 10\log\left(\frac{\left|\frac{1}{M}\sum_{m=1}^{M}V_{v1}(m)\right|^2}{\frac{1}{M}\sum_{m=1}^{M}\left|V_{v1}(m)\right|^2 - \left|\frac{1}{M}\sum_{m=1}^{M}V_{v1}(m)\right|^2}\right) \tag{5.79}$$

$$\widehat{PR}_{H2} = 10\log\left(\frac{\left|\frac{1}{M}\sum_{m=1}^{M}V_{h2}(m)\right|^2}{\frac{1}{M}\sum_{m=1}^{M}\left|V_{h2}(m)\right|^2 - \left|\frac{1}{M}\sum_{m=1}^{M}V_{h2}(m)\right|^2}\right) \tag{5.80}$$

$$\widehat{PR}_{V2} = 10\log\left(\frac{\left|\frac{1}{M}\sum_{m=1}^{M}V_{v2}(m)\right|^2}{\frac{1}{M}\sum_{m=1}^{M}\left|V_{v2}(m)\right|^2 - \left|\frac{1}{M}\sum_{m=1}^{M}V_{v2}(m)\right|^2}\right) \tag{5.81}$$

$$\hat{Z}_{DR1} = 10\log\left(\frac{\left|\frac{1}{M-l}\sum_{m=1}^{M-l}V_{h1}^*(m+l)V_{h1}(m)\right|}{\frac{1}{M-l}\sum_{m=1}^{M-l}V_{v1}^*(m+l)V_{v1}(m)}\right), \quad l=0,1 \tag{5.82}$$

$$\hat{Z}_{DR2} = 10\log\left(\frac{\left|\frac{1}{M-l}\sum_{m=1}^{M-l}V_{h2}^*(m+l)V_{h2}(m)\right|}{\frac{1}{M-l}\sum_{m=1}^{M-l}V_{v2}^*(m+l)V_{v2}(m)}\right), \quad l=0,1 \tag{5.83}$$

$$\hat{\rho}_{hv1} = \frac{\left|\frac{1}{M-l}\sum_{m=1}^{M-l}V_{h1}^*(m+l)V_{v1}(m)\right| + \left|\frac{1}{M-l}\sum_{m=1}^{M-l}V_{h1}(m)V_{v1}^*(m+l)\right|}{2\sqrt{\frac{1}{M-l}\sum_{m=1}^{M-l}V_{h1}^*(m+l)V_{h1}(m)\left\|\frac{1}{M-l}\sum_{m=1}^{M-l}V_{v1}^*(m+l)V_{v1}(m)\right|}}, \quad l=0,1 \tag{5.84}$$

$$\hat{\rho}_{hv2} = \frac{\left|\frac{1}{M-l}\sum_{m=1}^{M-l}V_{h2}^*(m+l)V_{v2}(m)\right| + \left|\frac{1}{M-l}\sum_{m=1}^{M-l}V_{h2}(m)V_{v2}^*(m+l)\right|}{2\sqrt{\frac{1}{M-l}\sum_{m=1}^{M-l}V_{h2}^*(m+l)V_{h2}(m)\left\|\frac{1}{M-l}\sum_{m=1}^{M-l}V_{v2}^*(m+l)V_{v2}(m)\right|}}, \quad l=0,1 \tag{5.85}$$

$$\hat{\rho}_{12h} = \frac{\left|\frac{1}{M-l}\sum_{m=1}^{M-l}V_{h1}^*(m+l)V_{h2}(m)\right| + \left|\frac{1}{M-l}\sum_{m=1}^{M-l}V_{h1}(m)V_{h2}^*(m+l)\right|}{2\sqrt{\left|\frac{1}{M-l}\sum_{m=1}^{M-l}V_{h1}^*(m+l)V_{h1}(m)\right|\left|\frac{1}{M-l}\sum_{m=1}^{M-l}V_{h2}^*(m+l)V_{h2}(m)\right|}}, \quad l=0,1 \quad (5.86)$$

$$\hat{\rho}_{12v} = \frac{\left|\frac{1}{M-l}\sum_{m=1}^{M-l}V_{v1}^*(m+l)V_{v2}(m)\right| + \left|\frac{1}{M-l}\sum_{m=1}^{M-l}V_{v1}(m)V_{v2}^*(m+l)\right|}{2\sqrt{\left|\frac{1}{M-l}\sum_{m=1}^{M-l}V_{v1}^*(m+l)V_{v1}(m)\right|\left|\frac{1}{M-l}\sum_{m=1}^{M-l}V_{v2}^*(m+l)V_{v2}(m)\right|}}, \quad l=0,1. \quad (5.87)$$

Equations 5.78 through 5.81 represent the coherent and incoherent power ratios calculated from voltages from horizontally and vertically polarized waves collected on the first and second scans. The power ratios are expected to be large for ground clutter and small for most weather signals except for narrow-band zero-velocity weather signals. Equations 5.82 and 5.83 are the differential reflectivities of the first and second scans, respectively. Equations 5.84 and 5.85 are the co-polar correlation coefficients of the first and second scans, respectively. ρ_{hv1} and ρ_{hv2} are expected to have high (close to unity) and narrowly distributed values for weather, but have large spreads for ground clutter. Equations 5.86 and 5.87 are the correlation coefficient estimates between the two scans of the H- and V-channels, respectively. It is expected that ρ_{12h} and ρ_{12v} will be larger for ground clutter than for weather signals because the clutter tends to be stationary. Hence, a joint PDF of ρ_{hv}s and ρ_{12}s should provide a good separation between weather and clutter echoes. Some combination of these 10 discriminants can be used in a classification algorithm. An SBC is introduced and implemented for dual-pol clutter detection as follows.

5.4.3 Implementation and Evaluation of SBC Dual-Pol Clutter Detection

SBCs assume that the effect of a discriminant value on a given class is independent of the values of the other discriminants (Han et al. 2011). As described earlier, three types of radar echoes are considered: $\mathbf{C} = (c, w0, w)$. Combine the four dual-pol discriminants defined in Equation 5.78 through 5.81 to form an observational discriminant vector, $\mathbf{y} = \left(\widehat{PR}_H, \widehat{PR}_V, \hat{Z}_{DR}, \hat{\rho}_{hv}\right)$. The SBC judges whether the $\mathbf{y}$ belongs to c based on the following criteria: $\mathbf{y}$ belongs to c only if $p(c|\mathbf{y}) > p(w0|\mathbf{y})$ and $p(c|\mathbf{y}) > p(w|\mathbf{y})$, where the function p is the PDF. According to Bayes' theorem (Papoulis 1991):

$$p(C_i|\mathbf{y}) = \frac{p(\mathbf{y}|C_i)p(C_i)}{p(\mathbf{y})}, \quad (5.88)$$

where $C_i = c$, $w0$, and w.

$p(\mathbf{y}) \equiv K$ is the probability of the observational discriminants $\mathbf{y}$ and is assumed to be the same for all classes at the moment. Thus, $p(C_i|\mathbf{y})$ is proportional to $P(\mathbf{y}|C_i) p(C_i)$. Assume the probability $p(C_i)$ is the same for all classes, that is, $p(c) = p(w0) = P(w) = 1/3$. Then, Equation 5.88 becomes

$$p(C_i \mid \mathbf{y}) = \frac{1}{3K} p(\mathbf{y} \mid C_i), \tag{5.89}$$

meaning that the probability of the class C_i for a given set of observational discriminants is proportional to the conditional probability of the discriminants for the class C_i, which can be obtained from training data sets. The conditional PDF is decomposed as

$$
\begin{aligned}
p(\mathbf{y} \mid C_i) &= p\left(\widehat{pr}_\mathrm{H}, \widehat{pr}_\mathrm{V}, \hat{Z}_\mathrm{DR}, \hat{\rho}_\mathrm{hv} \mid C_i\right) \\
&= p\left(\widehat{pr}_\mathrm{H}, \widehat{pr}_\mathrm{V} \mid C_i\right) p\left(\hat{Z}_\mathrm{DR} \mid C_i\right) p\left(\hat{\rho}_\mathrm{hv} \mid C_i\right)
\end{aligned}
\tag{5.90}
$$

based on the simple assumption of class-conditional independence in SBC. Note that the joint PDF $p(\widehat{PR}_\mathrm{H}, \widehat{PR}_\mathrm{V} \mid C_i)$ is used for the dual-pol power ratios. The PDFs $p(\hat{Z}_\mathrm{DR} \mid C_i)$ and $p(\hat{\rho}_\mathrm{hv}|C_i)$ are shown in Figure 5.9 for $C_i = c$, $w0$, and w, respectively.

Using Equations 5.89 and 5.90 along with the PDFs shown in Figure 5.9, a classification decision is made for the one with the largest probability, and the results are as follows. Figure 5.11 shows the clutter maps obtained by using the SBC dual-polarization (DP) algorithm as compared with the ground truth for the three testing data. The SBC-DP algorithm performs well in detecting clutter even when the clutter is mixed with weather. This is because PR is large when clutter is mixed with $w0$, allowing for the separation of clutter and weather.

To quantify the performance, the number of true positives (TP), false negatives (FN), false positives (FP), true negatives (TN), the probability of detection P_D, probability of false alarm P_{FA}, and critical success index (CSI) (Schaefer 1990) are listed in Table 5.1 for the three controlled data sets. The terms in Table 5.1 are defined as follows:

$$P_D = \frac{TP}{TP + FN} \tag{5.91}$$

$$P_{FA} = \frac{FP}{FP + TN} \tag{5.92}$$

$$CSI = \frac{TP}{TP + FN + FP}. \tag{5.93}$$

It is evident from Table 5.1 that the probability of detection is larger than 84%, and the false alarm rate is less than 2% in all three cases. It can be expected that the SBC dual-polarization dual-scan algorithm will produce an even lower P_{FA} with higher P_D if there are many narrow-band zero-velocity weather signals.

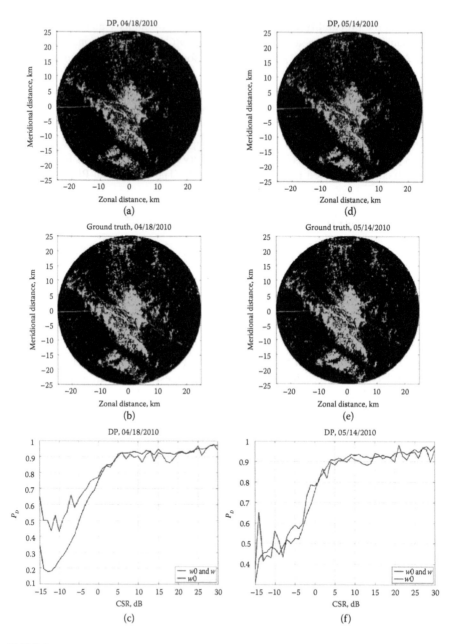

FIGURE 5.11 SBC-DP clutter maps (a, d, g) compared with the ground truth (b, e, h) for the three testing data. P_D as a function of CSR (dB) is shown in (c, f, i) for the three testing data. The blue line represents the performance of the SBC-DP in detecting ground clutter in the presence of both w and $w0$, whereas the red line represents the performance in the presence of only $w0$.   *(Continued)*

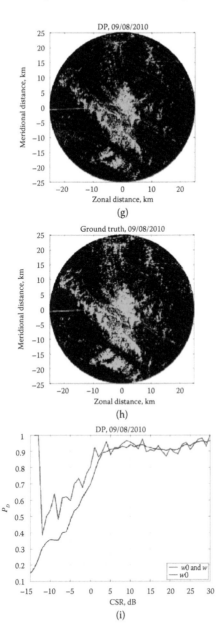

FIGURE 5.11 (Continued) SBC-DP clutter maps (a, d, g) compared with the ground truth (b, e, h) for the three testing data. P_D as a function of CSR (dB) is shown in (c, f, i) for the three testing data. The blue line represents the performance of the SBC-DP in detecting ground clutter in the presence of both w and $w0$, whereas the red line represents the performance in the presence of only $w0$.

TABLE 5.1

The Number of TP, FN, FP, TN, P_D, P_{FA}, and CSI for the Three Control Data Sets

	TP	FN	FP	TN	P_D (%)	P_{FA} (%)	CSI
04/18/2010	33,911	5399	1752	102,938	86.27	1.67	0.83
05/14/2010	33,655	5756	1162	103,427	85.39	1.11	0.83
09/08/2010	34,927	6229	666	102,178	84.86	0.65	0.84

Note: CSI, critical success index; FN, false negative; FP, false positive; P_D, probability of detection; P_{FA}, probability of false alarm; TN, true negative; TP, true positive.

5.5 CLUTTER MITIGATION

Once clutter gates are identified using the detection algorithms discussed above, the clutter effects need to be mitigated to obtain improved weather moment data. Siggia and Passarelli (2004) suggested Gaussian model adaptive processing (GMAP) to improve clutter cancellation. Similar to the GMAP, bi-Gaussian model adaptive processing (BGMAP), in which clutter filtering is performed in the spectrum domain, is discussed here. The cost functions for estimating the model parameters are provided, and the fitting results of BGMAP are presented to assess its effectiveness in mitigating clutter.

5.5.1 BI-GAUSSIAN SPECTRUM MODEL AND COST FUNCTIONS

The power spectral density of time-series voltages $V_X(m) = w_{x,m} + c_{x,m} + n_{x,m}$ ($x = h$ or v; $m = 1, 2, ..., M$) consists of a weather spectrum centered at the mean Doppler velocity of the weather signal (v_{rw}), a clutter spectrum centered at the mean Doppler velocity of ground clutter (v_{rc}), and a noise floor. The expected spectra can be written as follows:

$$S(v) = \frac{P_w}{\sigma_{vw}\sqrt{2\pi}} \exp\left[-\frac{(v - v_{rw})^2}{2\sigma_{vw}^2}\right] + \frac{P_c}{\sigma_{vc}\sqrt{2\pi}} \exp\left[-\frac{(v - v_{rc})^2}{2\sigma_{vc}^2}\right] + \frac{P_n}{2v_N}. \quad (5.94)$$

Here, P_w is the weather power, v_{rw} is the mean Doppler velocity of weather, and σ_{vw} is the spectrum width of weather; P_c is the clutter power, v_{rc} is the mean Doppler velocity of clutter, and σ_{vc} is the spectrum width of clutter. It is assumed that $\sigma_{vc} < \sigma_{vw}$. P_n is the noise power and v_N is the Nyquist velocity.

Intuitively, one would think of using the LS approach to estimate the parameters in Equation 5.94. That is, to find the parameters that minimize the cost function (the difference between the expected power spectrum [Equation 5.94] $S_m = S(v_m)$ and the estimates $\hat{S}_m$ in the log scale so that the fit is not dominated by one or a few points with large values):

$$J_{LS} = \sum_{m=1}^{M} \left(\ln S_m - \ln \hat{S}_m\right)^2. \quad (5.95)$$

However, the cost function of Equation 5.95 defined with an equal weight for all the frequency points may not necessarily be optimal. A set of weights (W_m) can be used to improve the parameter estimation. To find an optimal estimate, the maximum a posteriori (MAP) approach is used to derive a cost function as follows.

As discussed in last chapter, the total scattered wave field is Gaussian distributed based on the central limit theorem. Therefore, the complex signal (I/Q), which is proportional to the wave field, is also Gaussian distributed. Hence, the power is exponentially distributed. Using Equation 4.30 and letting the intensity $I = \hat{S}_m$ and $\langle I \rangle = S_m$, we have the estimated power at each spectral line following the exponential distribution:

$$p\left(\hat{S}_m \mid S_m\right) = \frac{1}{S_m} \exp\left(-\frac{\hat{S}_m}{S_m}\right). \tag{5.96}$$

In Equation 5.96, $\hat{S}_m$ is the estimated power corresponding to the mth spectral line and S_m is the expected value of $S(v_m)$ given in Equation 5.94. The optimal S_m estimate is that the posterior probability $p(S_m \mid \hat{S}_m)$ is maximized. Using Bayes' theorem (Papoulis 1991), we obtain

$$p\left(S_m \mid \hat{S}_m\right) = \frac{p\left(\hat{S}_m \mid S_m\right)p(S_m)}{p\left(\hat{S}_m\right)} = \frac{p\left(\hat{S}_m \mid S_m\right)p(S_m)}{\displaystyle\int_{P_n/2v_N}^{+\infty} p\left(\hat{S}_m \mid S_m\right)p(S_m)\,dS_m}. \tag{5.97}$$

Substituting Equation 5.96 into Equation 5.97 and assuming that the spectrum power distributions are independent, we comprehend that in order to maximize $p\left(S_1, S_2, ..., S_m \mid \hat{S}_1, \hat{S}_2, ..., \hat{S}_m\right)$, we only need to minimize the cost function:

$$J_{MAP} = \sum_{m=1}^{M}\left(\ln S_m + \frac{\hat{S}_m}{S_m}\right). \tag{5.98}$$

The subscript *MAP* signifies *maximum a posteriori*. Thus, if noise power P_n can be estimated, we can find the six variables P_w, v_{rw}, σ_{vw}, P_c, v_{rc}, and σ_{vc} in Equation 5.94 by minimizing the cost function J_{MAP} of Equation 5.98.

5.5.2 EXAMPLES OF BGMAP FITTING TO SEPARATE WEATHER AND CLUTTER SPECTRA

BGMAP was tested with simulations and the fitting results are shown in Figure 5.12. Time-series data were simulated using the spectrum method in Zrnić (1977). In the simulation, it was assumed that CSR = 0 dB, SNR = 30 dB, $M = 33$, $P_w = P_c = 40$ dB, $v_{rc} = 0$, $\sigma_{vc} = 0.5$ m/s, and $\sigma_{vw} = 2$ m/s.

In Figure 5.12a, the estimated weather power, radial velocity, and spectrum width are equal to 41.9 dB (1.9 dB error), 5.8 m/s (−0.2 m/s error), and 3.3 m/s

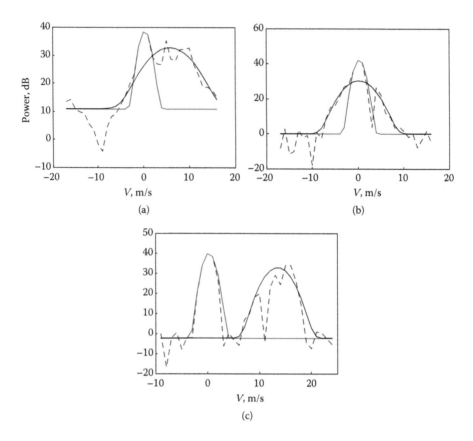

FIGURE 5.12   The blue line is the simulated observed spectra, the red line is the fitted clutter + noise Gaussian spectrum, and the black line is the fitted weather + noise Gaussian spectrum; v_{rw} is equal to (a) 6, (b) 0, and (c) 15 m/s.

(1.3 m/s error), respectively. In Figure 5.12b, the estimated weather moments are 37.7 dB (−2.3), 0.2 m/s (0.2), and 2.2 m/s (0.2); in Figure 5.12c, the estimated weather moments are 39.2 dB (−0.8), 13.5 m/s (1.5), and 1.7 m/s (−0.3). As expected, there is a relatively large (but acceptable) bias of −2.3 dB in the weather power estimate in Figure 5.12b when the weather Doppler velocity is zero. All the velocity and spectrum estimates are very good.

5.6   SPECTRUM-TIME ESTIMATION AND PROCESSING

The multilag correlation estimator and the clutter detection and filtering are combined to improve high quality dual-polarization weather radar data. One such combination, called *spectrum-time estimation and processing* (STEP), was developed and tested with OU-PRIME data (Cao et al. 2012b). The STEP algorithm contains three steps: (i) clutter detection, (ii) BGMAP clutter filtering, and (iii) multilag correlation estimation of weather moments. The procedure of implementing STEP is shown in the flow diagram of Figure 5.13.

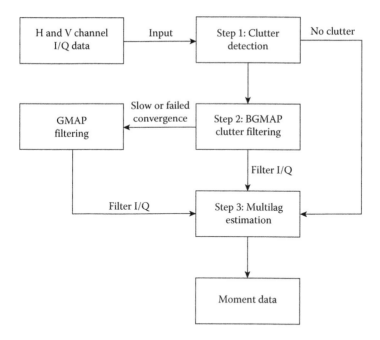

FIGURE 5.13 Flow chart of the STEP algorithm.

Figure 5.14 compares the original moments of contamination (shown in the left column) with the results processed by STEP (shown in the right column). Six rows of figures, from top to bottom, give the results of radar reflectivity Z_H, radial velocity v_h, spectrum width σ_h, correlation coefficient ρ_{hv}, differential phase ϕ_{DP}, and differential reflectivity Z_{DR}, respectively. For radar reflectivity, the clutter contamination is evident with a very high value in the original result, whereas it is gone in the STEP result. For radial velocity, the bias is clearly seen in the original result but disappears in the STEP result, which gives a smooth velocity field in the clutter region. For spectrum width, the original result shows larger values in the region of low SNR and smaller values in the clutter region. The STEP result reduces the estimation for the low-SNR region and increases the estimation for the clutter region. The improved spectrum width image looks smoother and more consistent within the storm region.

For polarimetric radar variables, especially for the co-polar correlation coefficient, the improvement is substantial. Lower values are seen for both clutter and low-SNR regions in the original result. The values for these two regions have been increased in the STEP result, indicating that precipitation is present in these regions. Now the image of correlation coefficient is smooth enough that the radar echo can be easily classified as precipitation. For differential phase and differential reflectivity, the clutter contamination is clearly shown in the original results. The STEP results give a much cleaner image for both moments. The recovered moments in the contaminated regions are generally consistent with those results in the uncontaminated region. Therefore, the STEP algorithm can extensively improve the moment estimation for polarimetric weather radar applications.

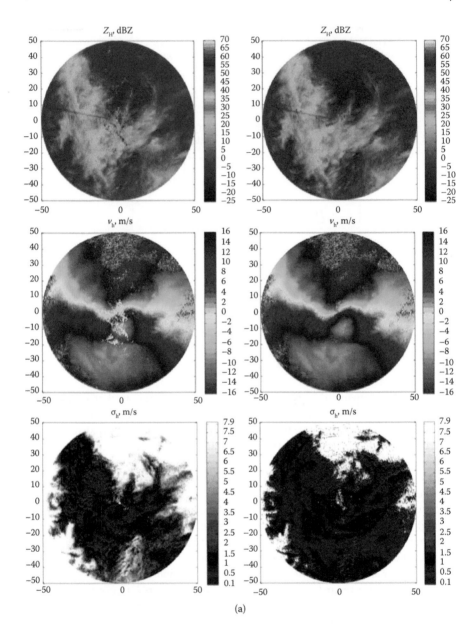

(a)

FIGURE 5.14 (a) Comparison of contaminated moments. Six rows show the radar reflectivity Z_H, radial velocity v_h, spectrum width σ_h, correlation coefficient ρ_{hv}, differential phase ϕ_{DP}, and differential reflectivity Z_{DR}, respectively. Data were collected by OU-PRIME: weather signals originated on October 21, 2009, 1803 UTC, 3.5° elevation; clutter signals come from OU-PRIME data, January 13, 2011, 2319 UTC, 0.05° elevation angle.   *(Continued)*

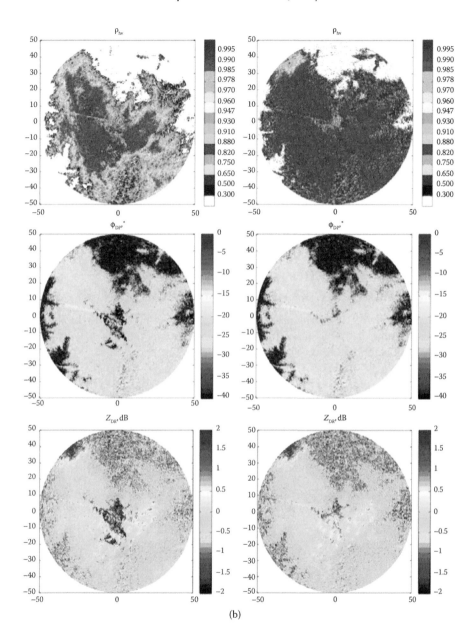

FIGURE 5.14 (Continued) (b) Comparison of STEP-processed moments. Six rows show the radar reflectivity Z_H, radial velocity v_h, spectrum width σ_h, correlation coefficient ρ_hv, differential phase ϕ_DP, and differential reflectivity Z_DR, respectively. Data were collected by OU-PRIME: Weather signals originated on October 21, 2009, 1803 UTC, 3.5° elevation; clutter signals come from OU-PRIME data, January 13, 2011, 2319 UTC, 0.05° elevation angle. (From Cao, Q., et al., 2012b. *IEEE Transactions on Geoscience and Remote Sensing*, 50, 4670–4683.)

STEP is an advanced signal-processing framework that takes advantage of the latest advances in clutter detection, filtering, and moment estimation. The major feature of the STEP algorithm is that it combines signal processing in both the time and spectral domains, making it effective in mitigating clutter and noise effects on moment estimation. It is worth noting that STEP's performance depends on the polarimetric radar system parameters and weather conditions and needs to be done adaptively for optimal results of high quality PRD. In the case of people working with moment data, quality control can be done by speckle filtering (Lee et al. 1999) and use of a moving average as well as nonweather echo identification and removal. A simple approach is to perform a running median filter first and then a running average to get smooth images.

Readers who are interested in NEXRAD moment data (Level II data) and products (Level III data) can study Appendix 5B for information on getting, converting, and reading the data.

APPENDIX 5A: DERIVATION OF COST FUNCTION FOR OPTIMAL SPECTRAL PARAMETER ESTIMATION

In Equation 5.97, $P(S_m)$ is a prior probability for S_m, which is unknown. In the absence of other information, it is assumed that $p(S_m)$ is uniform over the range $[P_n/2v_N + \infty]$. Thus Equation 5.97 can be rewritten as follows:

$$p\left(S_m \mid \hat{S}_m\right) = \frac{p\left(\hat{S}_m \mid S_m\right)}{\displaystyle\int_{P_n/2v_N}^{+\infty} p\left(\hat{S}_m \mid S_m\right) d S_m}. \tag{5A.1}$$

Substituting Equation 5.96 into Equation 5A.1, we obtain

$$p\left(S_m \mid \hat{S}_m\right) = \frac{\dfrac{1}{S_m}\exp\left(-\dfrac{\hat{S}_m}{S_m}\right)}{\displaystyle\int_{P_n/2v_N}^{+\infty} \dfrac{1}{S_m}\exp\left(-\dfrac{\hat{S}_m}{S_m}\right) d S_m}. \tag{5A.2}$$

In this book, the noise level $P_n/2v_N$ is predetermined before optimization using the method described by Hildebrand and Sekhon (1974). Thus, searching for S_m is equivalent to searching for the set of parameters P_w, V_{rw}, σ_{vw}, P_c, v_{rc}, and σ_{vc} that can maximize $p(S_m \mid \hat{S}_m)$. The denominator of Equation A.2 does not depend on the set of parameters and therefore plays no role in the optimization process (Kay 1998). Thus, maximizing $p(S_m \mid \hat{S}_m)$ is equivalent to maximizing $\dfrac{1}{S_m}\exp\left(-\dfrac{\hat{S}_m}{S_m}\right)$. Equation 5A.2 can be rewritten as

$$p\left(S_m \mid \hat{S}_m\right) = \frac{1}{g\left(\hat{S}_m, P_n/2v_N\right)S_m}\exp\left(-\frac{\hat{S}_m}{S_m}\right). \tag{5A.3}$$

In Equation 5A.3, g represents a function of $\hat{S}_m$ and $P_n / 2v_N$. In this book, it is assumed that the spectral power distribution of each spectral line is independent. In this case, the joint probability can be written as a product:

$$p\left(S_1, S_2, ..., S_m \mid \hat{S}_1, \hat{S}_2, ..., \hat{S}_m\right) = \prod_{m=1}^{M} \frac{1}{g\left(\hat{S}_m, P_n/2v_N\right) S_m} \exp\left(-\frac{\hat{S}_m}{S_m}\right). \quad (5A.4)$$

If we take the logarithm of both sides of Equation 5A.4, we obtain

$$\ln p\left(S_1, S_2, ..., S_m \mid \hat{S}_1, \hat{S}_2, ..., \hat{S}_m\right) = \sum_{m=1}^{M}\left(-\ln g\left(\hat{S}_m, P_n/2v_N\right)\right) + \sum_{m=1}^{M}\left(-\ln S_m - \frac{\hat{S}_m}{S_m}\right). \quad (5A.5)$$

Because the first term does depend on the spectrum parameters, the MAP estimation is to maximize the second term, that is, to minimize the cost function of

$$J_{\text{MAP}} = \sum_{m=1}^{M}\left(\ln S_m + \frac{\hat{S}_m}{S_m}\right). \quad (5A.6)$$

APPENDIX 5B: NEXRAD RADAR DATA ACCESS

NEXRAD data can be downloaded from NOAA's National Centers for Environmental Information (NCEI), formerly the National Climatic Data Center (NCDC) at http://www.ncdc.noaa.gov/data-access/radar-data/nexrad-products. The NCEI archive includes the base data (i.e., Level II data) and the derived products (i.e., Level III data). The Level II products (assuming the radar is post–dual-pol upgrade) include radar reflectivity (Z_H), differential reflectivity (Z_{DR}), co-polar cross-correlation coefficient (ρ_{hv}), differential phase (Φ_{DP}), radial velocity (v_r), and spectrum width (σ_v). Level II radar data from NCDC is in msg31 radar format.

The Level II NEXRAD radar can be converted to netCDF format to be read into MATLAB®. NCAR's RadX software package allows conversion between several different common radar formats, including CfRadial, DORADE, UF (universal format), netCDF, and msg31. The RadX software package can be built on Linux or MacOSX platforms.

Here is a list of important RadX command line options ('**RadxConvert –h**' to get the full list):

[-cf_classic] output classic-style netCDF file
[-const_ngates] force number of gates constant for all rays; added gates will be filled with missing values
[-dorade] convert to DORADE format (used for solo3—a popular tool to unfold aliased radial velocity data)
[-disag] disaggregate a volume into separate elevation scans (DORADE files always disaggregate)
[-f ?] set file name/paths (use * to select all files)
[-v] print verbose debug messages

Here is a list of example RadX command line options to convert NEXRAD data in a directory:

Convert from Level II/msg31 to netCDF (one file for each volume)
RadxConvert -cf_classic -v -const_ngates -f *
Convert from Level II/msg31 to netCDF (one file for each elevation in a volume):
RadxConvert -cf_classic -disag -v -const_ngates -f *
Convert from Level II/msg31 to DORADE (always one file for each elevation):
RadxConvert –dorade -v -const_ngates -f *
Convert from DORADE to netCDF (one file for whole volume)
RadxConvert -cf_classic -ag -v -const_ngates -f *
Convert from DORADE to netCDF (one file per elevation):
RadxConvert -cf_classic -v -const_ngates -f *

MATLAB natively reads netCDF (.nc) files. You must use 'ncread' to read netcdf data into MATLAB. For example, **REF=ncread(filename,'REF')** would read reflectivity data in MATLAB. A sample MATLAB plotting code is provided on the book website.

Problems

5.1 Explain why the multilag correlation estimators were developed and when they should be used in terms of SNR and spectrum width.

5.2 Show that the two-lag estimator is better than the lag-1 estimator for the co-polar cross-correlation coefficient in the presence of noise and in a narrow spectrum case.

5.3 Compare the power ratio (p_r) in Equation 5.71 and CPA in Equation 5.72 and describe their similarities and differences based on signal statistics and clutter detection application.

5.4 It takes about 30 s for WSR-88D to complete a low-level PPI scan. How large would you expect the scan-to-scan correlation coefficient estimate $\left(\hat{\rho}_{12}\right)$ for weather signals to be, compared with that for ground clutter signals? Explain why there is such a difference.

5.5 Write down the power spectrum in a bi-Gaussian form in the presence of weather, clutter, and noise and explain why the bi-Gaussian fitting is better than the notch filter method in mitigating clutter.

6 Applications in Weather Observation and Quantification

With the fundamentals and principles of weather radar polarimetry covered in Chapters 2 to 4 and our knowledge of the estimation and improvement of PRD in Chapter 5, we are now ready to discuss how to use PRD for weather studies. This chapter deals with the applications of PRD in weather observation, classification, quantification, and forecast. It describes polarimetric radar signatures for specific weather types, fuzzy logic classification of radar echoes, QPE, retrieval of rain DSD, attenuation correction, and microphysics parameterization for improving weather quantification.

6.1 OBSERVATION OF POLARIMETRIC RADAR SIGNATURES

There are a variety of types of clouds and precipitation that can be better observed with polarimetric radars than with single polarization radars. Multiparameter polarimetric radar data contain rich information about clouds and precipitation, and each polarimetric variable has specific signatures for different weather or nonweather echoes. We use PRD images to develop intuition about the polarimetric radar signatures of stratiform precipitation, mesoscale convection, squall lines followed by stratiform rain, and severe storms such as downbursts and tornados.

6.1.1 STRATIFORM PRECIPITATION

Figure 6.1 shows plane position indicator (PPI) images for a winter precipitation event collected by the polarimetric KOUN radar on January 20, 2007, at 1157 UTC. The storm was horizontally uniform with a relatively low reflectivity value, below 30 dBZ. The differential reflectivity is mostly small, less than 1 dB. The co-polar correlation coefficient is high, close to 1, except for those data points at the edges of the storm (which have low SNR), those data points in the southwest where the radar beam reached the melting layer, and those data points of clutter contamination near the radar. The differential phase is small and does not increase much along each beam, meaning that the storm had very small drops with small polarimetric signatures.

Another important radar signature for stratiform rain is the bright band due to snow melting. It is shown in Figure 6.1e through h with the PPI images at an elevation of 3.5 degrees. Similar to the bright band in radar reflectivity, there is an enhanced layer in Z_{DR}, but the height is a little lower (smaller ring) than the bright band in Z_H.

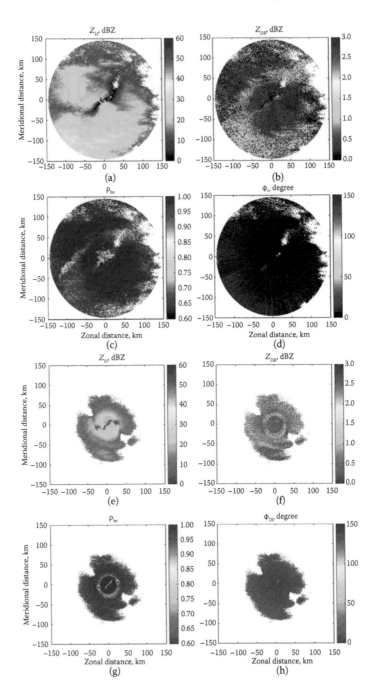

FIGURE 6.1 PRD images of winter precipitation collected by KOUN on January 20, 2007, at 1157 UTC at an elevation of 0.41 degrees and 3.5 degrees, respectively; (a, e) reflectivity factor, (b, f) differential reflectivity, (c, g) co-polar correlation coefficient, and (d, h) differential phase.

Corresponding to the enhanced layer in Z_{DR}, the co-polar correlation coefficient is reduced (a ring of ρ_{hv} values) in the melting layer because of the random differential scattering phase in the Mie scattering regime.

6.1.2 MESOSCALE CONVECTION

Figure 6.2 shows PRD images of a mesoscale convection observed by KOUN on February 10, 2009, at 2058 UTC (top four panels) and 2353 UTC (bottom four panels). At 2058 UTC, the storm system had just initiated with multiconvective cells. Z_{DR} was generally small except at the convective cores; ρ_{hv} is close to unity in the rain regions, and Φ_{DR} is also small unless the beams pass through the convective cores. Two hours later, the system had matured and became organized as a squall line where large $Z_{DR}s$ (>2 dB) are evident at the leading edge with large drops present due to size sorting or long growth time and small $Z_{DR}s$ (<1 dB) are evident in the transitional region. Again, ρ_{hv} is close to unity in the rain regions, but Φ_{DP} is large (>100 degrees) for the beam along the rain band.

6.1.3 STORM COMPLEX

Figure 6.3 shows PRD images of a storm complex observed by KOUN on August 18, 2007, at 2101 UTC. There were local convections embedded in the large system, which was a part of Tropical Storm Erin. Note that Z_{DR} and Φ_{DP} are generally small. This means that the cyclone consisted of small drops, which is a typical characteristic of tropical rain, since it lacks a sufficient ice-phase growing process.

6.1.4 SEVERE STORMS

Figure 6.4 shows NEXRAD PRD images for three downburst cases that occurred in dry, intermediate, and wet environments. Radar observables included are radar reflectivity, radial velocity, Z_{DR}, and ρ_{hv}. These events occurred in Denver, Colorado, on August 16, 2013; in Norman, Oklahoma, on June 15, 2011; and in Bladensburg, Maryland, on June 22, 2012.

Overall, the polarimetric signatures of variable Z_{DR} and low ρ_{hv} are similar for the intermediate and wet cases. By contrast, radar reflectivity is much lower in the dry case. On average, Z_{DR} is also lower in the dry case. For ρ_{hv}, values are generally higher in the dry case compared to the intermediate and wet cases. The most apparent divergent signature is present in the intermediate case; however, the downburst in the intermediate case was much closer to the radar compared to the other two. All three events had significant surface wind damage reported. The wind damage reports are from the public storm reports that were aggregated by the Storm Prediction Center (SPC) from the NWS Weather Forecast Offices.

Figure 6.5 shows PRD images of a tornado that occurred on May 10, 2010, and was observed by KOUN and C-band OU-PRIME at 2101 UTC. Whereas the KOUN reflectivity barely shows the tornadic features, the OU-PRIME PRD clearly reveals the detailed polarimetric radar signatures. Because the high resolution OU-PRIME has a half-degree beamwidth (vs. a one-degree beamwidth for KOUN)

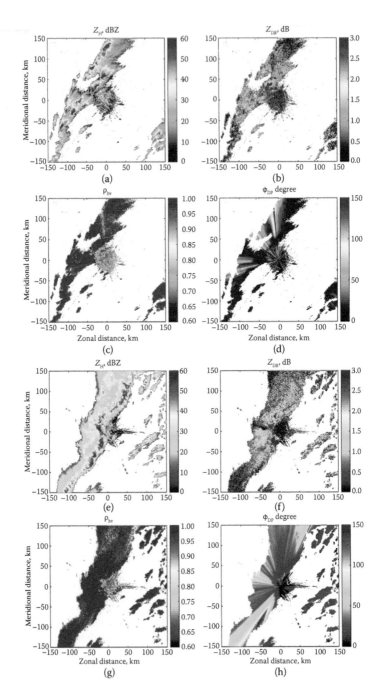

FIGURE 6.2 PRD images of a mesoscale convection observed by KOUN on February 10, 2009, at 2058 and 2353 UTC. (a, e) Reflectivity factor, (b, f) differential reflectivity, (c, g) co-polar correlation coefficient, and (d, h) differential phase.

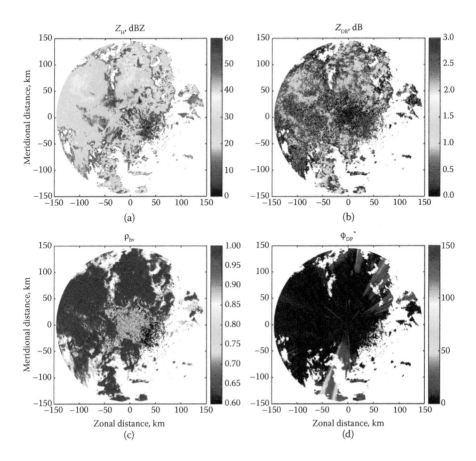

FIGURE 6.3   PRD images of a storm complex observed by KOUN on August 18, 2007, at 2101 UTC during Tropical Storm Erin. (a) Reflectivity factor, (b) differential reflectivity, (c) co-polar correlation coefficient, and (d) differential phase.

and OU-PRIME was much closer to the tornado (5 km vs. 10 km for KOUN), the hook echo and debris polarimetric signatures were seen much more clearly in the OU-PRIME radar images. The debris in the hook region cause low and fluctuating Z_{DR} and ρ_{hv} values. This is expected because the debris appear in the Mie scattering regime and have random orientations, causing a random differential scattering phase. On the leading side, the large Z_{DR} form an arc called the Z_{DR} *arc*. There are some negative values for C-band Z_{DR} on the back side of the storm because of differential attenuation. Due to different radar response and non-Rayleigh scattering effects, the Z_{DR} (ρ_{hv}) measured by the C-band OU-PRIME is larger (smaller) than that of the S-band KOUN measurements, indicating different information content in the two sets of PRD. This is due to the same scatterers appearing electrically larger in the C-band radar than in the S-band radar—hence, the OU-PRIME polarimetric signatures are more pronounced.

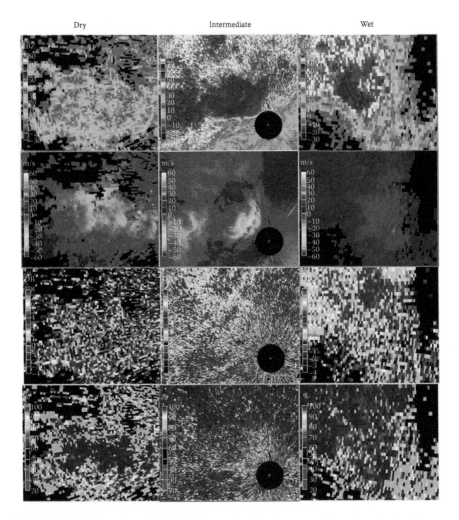

FIGURE 6.4 Examples of PRD for three downbursts representing three typical down-burst environments (dry, intermediate, and wet) that occurred on August 16, 2013, in Denver June 15, 2011, in Norman and June 22, 2012, in Bladensburg, respectively. Radar observables from top to bottom panels are radar reflectivity, radial velocity, Z_{DR}, and ρ_{hv}, respectively.

The typical values of polarimetric radar variables for different hydrometeors were first summarized by Doviak and Zrnić (1993, Table 8.1), and the ranges of the variables have now been more completely assigned in hydrometeor classification (Park et al. 2009), which is discussed next.

6.2 HYDROMETEOR CLASSIFICATION

In the previous section, we gained an understanding of polarimetric radar signa-tures for specific types of storms/echoes. Although it is important to know these polarimetric signatures and to use them for the detection of severe weather, it would

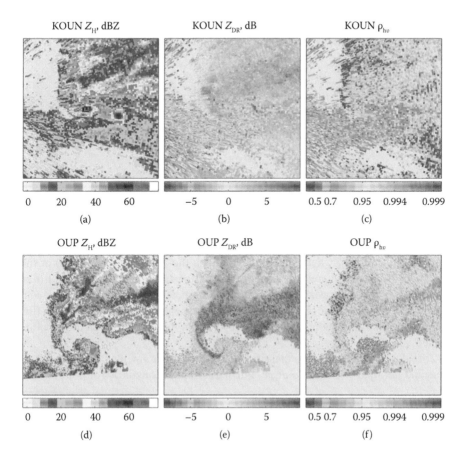

FIGURE 6.5 High-quality polarimetric radar measurements with OU-PRIME (1.0° tilt) at 2242 UTC in comparison with those of KOUN (0.7° tilt) for the tornadic storm on May 10, 2010, at 2245 UTC. First row (a, b, c): KOUN measurements for Z_H, Z_{DR}, and ρ_{hv}; second row (d, e, f): OU-PRIME measurements for Z_H, Z_{DR}, and ρ_{hv}.

be more powerful to systematically combine the multiparameter PRD properties to classify echoes. In Chapter 5, we discussed the SBC for clutter detection from polarimetric radar signals. Here, we describe the fuzzy logic algorithm for echo classification with the PRD.

6.2.1 CLASSIFICATION BACKGROUND

The first use of PRD for echo identification was applied to hail detection using Z_H and Z_{DR} (Aydin et al. 1986). Previously, the detection of hail presence was to use a reflectivity threshold with a single-polarization radar. Typically, hail would be likely to present when the reflectivity factor is larger than ($Z_H > 54$ dBZ). This single parameter threshold is obviously very rough and not accurate, because some small hailstones can present when the reflectivity factor is about 40 dBZ, which is much smaller than the threshold. Hence, Aydin et al. (1986) combined reflectivity and differential reflectivity to separate hail from rain by defining a function called the Z_{DR}-*derived hail signal*:

$$H_{DR} = Z_H - F(Z_{DR}) \tag{6.1}$$

with

$$F(Z_{DR}) = \begin{cases} 60 & Z_{DR} > 1.74 \\ 19Z_{DR} + 27 & 0 < Z_{DR} \leq 1.74. \\ 27 & Z_{DR} \leq 0 \end{cases} \tag{6.2}$$

As shown in Figure 6.6, Equation 6.2 is plotted on the simulated Z_{DR}–Z_H scatter plot from rain DSDs, indicating that rain appears to the upper left of the hail function. By contrast, when (Z_H, Z_{DR}) appears on the lower right ($H_{DR} > 0$) of the line, hail is likely to be present. Hence, the function H_{DR} is a better indicator than $Z_H > 54$ dBZ. However, when more than two parameters of PRD are available and more than two species of radar echoes need to be separated from each other, this single-line function is not convenient to use, and a more sophisticated method is needed.

Consider the PRD for echo classification: the multiparameter PRD are used as input and multiple classes are output. Furthermore, the boundaries of polarimetric

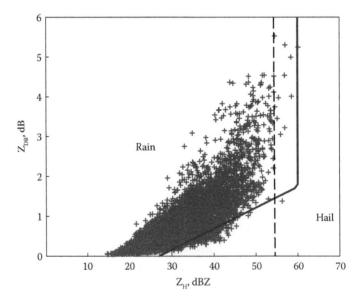

FIGURE 6.6 Hail function (solid line) plotted on simulated reflectivity and differential reflectivity data from measured rain DSDs.

variables for each class are vague and the measurements contain error. A rigorous theory/result is difficult to establish and is sometimes not necessary. A fuzzy logic approach is a natural match for the hydrometeor classification algorithm (HCA) from PRD. Compared with the SBC used for clutter detection introduced in Chapter 5, the fuzzy logic approach is easy to develop, modify, and implement (Han et al. 2011).

The fuzzy logic classification approach was originally introduced to the radar meteorology community by a joint effort between the Norman and Boulder radar communities (Straka et al. 2000; Vivekanandan et al. 1999; Zrnić et al. 2001) and was first tested on the S-Pol radar at NCAR. Later, many studies improved the fuzzy logic classification with more complicated membership functions and decision criteria. Among them are two typical algorithms. One is the algorithm developed by Colorado State University (CSU) (Lim 2005; Liu and Chandrasekar 2000), which has been transplanted into the commercial radars produced by the company Vaisala. Unlike other algorithms that apply empirically derived membership functions, a notable feature of the CSU algorithm is that its membership functions are trained in a neuro-fuzzy logic system. The other typical algorithm is the HCA developed by the NSSL (Ryzhkov et al. 2005b; Schuur et al. 2003). This algorithm is primarily designed for operation on the dual-polarization NEXRAD. The updated version of the algorithm was described by Park et al. (2009) and ultimately implemented on NEXRAD. The updated algorithm accounted for the measurement error, beam broadening effect, location of the melting layer, and precipitation type (convective vs. stratiform).

The fuzzy logic HCA has now become a mature technique for the classification of radar echoes and has been widely used in weather radar research and applications. The latest development of the HCA includes echo classification from the PRD with attenuation presented at higher frequency bands such as the C- and X-bands (Snyder et al. 2010). The fuzzy logic HCA has also been extended to detect three-body scattering signatures and improve radar data quality (Mahale et al. 2014).

6.2.2 Fuzzy Logic Approach

The fuzzy logic classification contains three steps, as shown in Figure 6.7: (i) the fuzzification process, (ii) aggregation, and (iii) defuzzification. In the first step (fuzzification), membership functions are established for each class, based on experience and the statistics of known observations. Within each class, each radar variable has a specific range of measurements. As shown in Figure 6.7, the membership functions for rain and hail classes are represented by a trapezoidal function. The ranges of measurements, such as Z_H, Z_{DR}, ρ_{hv}, and so on, are different for classes of rain and hail. There is an overlapping region between the two classes for each radar variable, which indicates the difficulty of separating the two classes using a single radar measurement. However, each radar measurement adds information that will later be used for classification.

Each membership function is defined as a polarimetric variable that has a value restricted within the range X_1–X_4 for a specific class, as shown in Figure 6.8. If the observation is beyond this range, the contribution of this variable is zero for

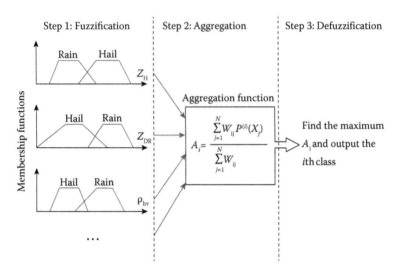

FIGURE 6.7   Illustration of the fuzzy logic method using an example of two classes.

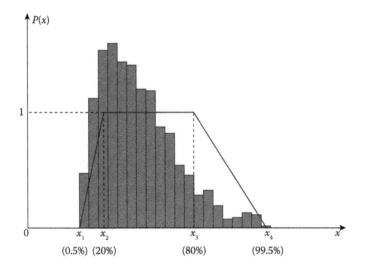

FIGURE 6.8   The trapezoidal membership function, where X is an arbitrary polarimetric variable.

the specific class. Otherwise, the weighting varies from 0 to 1, depending on the range in which the observation falls. Their control parameters (e.g., X_1–X_4) are generally found empirically or trained from observations of known classes. From the statistics of the observed data, X_1 is chosen at the 0.5th percentile; X_2 at the 20th percentile; X_3 at the 80th percentile; and X_4 at the 99.5th percentile.

In the second step (the aggregation process), the membership values are found for a set of PRD measurements and are summed up to obtain the aggregation value.

This value represents the total contribution of all the measurements, indicating how likely the input parameters will result in an output class. For radar echo classification, the aggregation value radar echo class is found as follows:

$$A_i = \frac{\sum_{j=1}^{N} W_{ij} P^{(i)}(X_j)}{\sum_{j=1}^{N} W_{ij}}, \tag{6.3}$$

where $P^{(i)}(X_j)$ is a membership function that characterizes the distribution of the jth radar variables X_j for the ith class, and W denotes a weighting that represents the significance of the radar variables for the classification.

In the third step (defuzzification), the aggregation values for each class are compared. The class that has the largest aggregation value is assigned to the corresponding set of PRD.

6.2.3 CLASSIFICATION RESULTS FROM PRD

Park et al. (2009) use six radar measurements as input: Z_H, Z_{DR}, ρ_{hv}, K_{DP}, SD(Z_H), and SD(Φ_{DP}); they use 10 classes as outputs: (1) ground clutter or abnormal propagation (GC/AP); (2) biological scatterers (BS); (3) dry aggregated snow (DS); (4) wet snow (WS); (5) crystals (CR); (6) graupel (GR); (7) big drops (BD); (8) light and moderate rain (RA); (9) heavy rain (HR); and (10) a mixture of rain and hail (RH). SD(Z_H) is the standard deviation estimated from four data points in the range and is a texture parameter of the Z_H field. SD(Φ_{DP}) is that of the Φ_{DP} field and is estimated over eight gates. The parameters for the membership function are given in Table 6.1.

The functions for the Z_{DP} parameters are as follows (Park et al. 2009):

$$f_1(Z_H) = -0.50 + 2.50 \times 10^{-3} Z_H + 7.50 \times 10^{-4} Z_H^2 \tag{6.4}$$

$$f_2(Z_H) = 0.68 - 4.81 \times 10^{-2} Z_H + 2.92 \times 10^{-3} Z_H^2 \tag{6.5}$$

$$f_1(Z_H) = 1.42 + 6.67 \times 10^{-2} Z_H + 4.85 \times 10^{-4} Z_H^2 \tag{6.6}$$

and

$$g_1(Z_H) = -44.0 + 0.8 Z_H \tag{6.7}$$

$$g_2(Z_H) = -22.0 + 0.5 Z_H. \tag{6.8}$$

The weighting W in the cost function is given in the following Table 6.2. It is evident that the contribution of each input parameter is different for the various classes. Please refer to Park et al. (2009) for more detailed descriptions.

TABLE 6.1
Parameters of the Membership Functions for 10 Classes

	GC/AP	BS	DS	WS	CR	GR	BD	RA	HR	RH
					$P[Z_H \text{ (dBZ)}]$					
x_1	15	5	5	25	0	25	20	5	40	45
x_2	20	10	10	30	5	35	25	10	45	50
x_3	70	20	35	40	20	50	45	45	55	75
x_4	80	30	40	50	25	55	50	50	60	80
					$P[Z_{DR} \text{ (dB)}]$					
x_1	−4	0	−0.3	0.5	0.1	−0.3	$f_2 - 0.3$	$f_1 - 0.3$	$f_1 - 0.3$	−0.3
x_2	−2	2	0.0	1.0	0.4	0.0	f_2	f_1	f_1	0.0
x_3	1	10	0.3	2.0	3.0	f_1	f_3	f_2	f_2	f_1
x_4	2	12	0.6	3.0	3.3	$f_1 + 0.3$	$f_3 + 1.0$	$f_2 + 0.5$	$f_2 + 0.5$	$f_1 + 0.5$
					$P[\rho_{hv}]$					
x_1	0.5	0.3	0.95	0.88	0.95	0.90	0.92	0.95	0.92	0.85
x_2	0.6	0.5	0.98	0.92	0.98	0.97	0.95	0.97	0.95	0.9
x_3	0.9	0.8	1.00	0.95	1.00	1.00	1.00	1.00	1.00	1.00
x_4	0.95	0.83	1.01	0.985	1.01	1.01	1.01	1.01	1.01	1.01
					$P[LK_{DP}]$					
x_1	−30	−30	−30	−30	−5	−30	$g_1 - 1$	$g_1 - 1$	$g_1 - 1$	−10
x_2	−25	−25	−25	−25	0	−25	g_1	g_1	g_1	−4
x_3	10	10	10	10	10	10	g_2	g_2	g_2	g_1
x_4	20	10	20	20	15	20	$g_2 + 1$	$g_2 + 1$	$g_2 + 1$	$g_1 + 1$
					$P[SD(Z_{DR}) \text{ (dB)}]$					
x_1	2	1	0	0	0	0	0	0	0	0
x_2	4	2	0.5	0.5	0.5	0.5	0.5	0.5	0.5	0.5
x_3	10	4	3	3	3	3	3	3	3	3
x_4	15	7	6	6	6	6	6	6	6	6
					$P[SD(\Phi_{DP}) \text{ (°)}]$					
x_1	30	8	0	0	0	0	0	0	0	0
x_2	40	10	1	1	1	1	1	1	1	1
x_3	50	40	15	15	15	15	15	15	15	15
x_4	60	60	30	30	30	30	30	30	30	30

Source: Park, H. S., et al., 2009. *Weather and Forecasting*, 24, 730–748.

Note: BD, big drops; BS, biological scatterers; CR, crystals; DS, dry aggregated snow; GC/AP, ground clutter or abnormal propagation; GR, graupel; HR, heavy rain, RA, light and moderate rain; RH, a mixture of rain and hail; WS, wet snow.

It is worth noting that there are some common-sense restrictions for the classification results, which depend in part on identification of the melting layer. For example, the classification results should not include snow, crystal, or graupel below the melting layer, whereas there should not be pure rain, clutter, or biological scatterers above the melting layer.

TABLE 6.2

Significance Weights of Input Parameters for 10 Classes

	Variable					
Class	Z_H	Z_{DR}	ρ_{hv}	LK_{DP}	$SD(Z_H)$	$SD(\Phi_{DP})$
GC/AP	0.2	0.4	1.0	0.0	0.6	0.8
BS	0.4	0.6	1.0	0.0	0.8	0.8
DS	1.0	0.8	0.6	0.0	0.2	0.2
WS	0.6	0.8	1.0	0.0	0.2	0.2
CR	1.0	0.6	0.4	0.5	0.2	0.2
GR	0.8	1.0	0.4	0.0	0.2	0.2
BD	0.8	1.0	0.6	0.0	0.2	0.2
RA	1.0	0.8	0.6	0.0	0.2	0.2
HR	1.0	0.8	0.6	1.0	0.2	0.2
RH	1.0	0.8	0.6	1.0	0.2	0.2

Source: Park, H. S., et al., 2009. *Weather and Forecasting*, 24, 730–748.

For the sake of convenience, the HCA is simplified to include only three input radar measurements: Z_H, Z_{DR}, and ρ_{hv} (Cao et al. 2012a). The simplified classification algorithm was applied to the convective storm of May 13, 2005 (first presented as DSD data in Figure 2.4 in Chapter 2). This is a squall line followed by widespread stratiform precipitation. Figure 6.9a through c show the PRD collected by KOUN radar at 0811 UTC for the images of Z_H, Z_{DR}, and ρ_{hv}, respectively. The squall line moves from northwest to southeast. Several convective cores, which have large values of reflectivity, are clearly evident within the squall line, from north to southwest. A line of gust front signature, along and in front of the leading edge, is also distinct in the images. Behind the squall line, there is a large region of stratiform rain, where ρ_{hv} is close to 1. The melting layer heights range from 2.6 to 3.1 km. Considering that the beam width of KOUN radar is about 1° and the elevation angle is 0.5° in Figure 6.9, measurements at the far range (>100 km) might come from the particles within or above the melting layer.

Figure 6.9d gives the classification result of radar echoes. As shown in the figure, the region close to the radar site is contaminated by ground clutter. Birds or insects might present in the region in front of the squall line. The heavy rain occurred at the convective cores and their adjacent areas. The "big drop" rain exists at the leading edge of the convection. The melting snow and the rain–hail mixture are found within the melting layer, which is observed beyond 100 km in the north and west directions. For some measurements (e.g., beyond 150 km), the radar might detect dry snow, which exists above the melting layer. Although the algorithm in this module is simple, most of the classifications seem reasonable. Additional control variables could improve the performance of the fuzzy logic classification. Problem 6.2 at the end of this chapter provides the opportunity to gain experience in HCA using PRD.

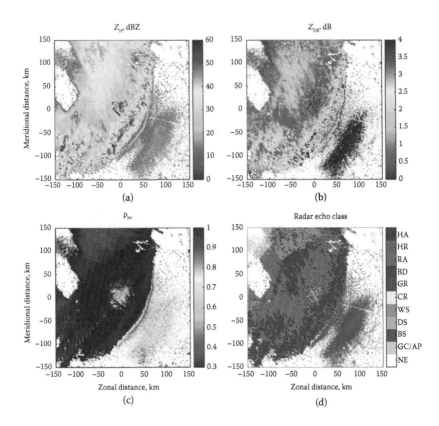

FIGURE 6.9 PRD and HCA results from S-band KOUN radar at 0811 UTC on May 13, 2005: (a) Z_H, dBZ; (b) Z_{DR}, dB; (c) ρ_{hv}; and (d) echo classification result.

HCA has now become a mature technique and has been applied to improve radar data quality (Mahale et al. 2014). Figure 6.10 shows an example of polarimetric radar observation and identification of TBSS, phenomena of the multiple-scattering effect shown in Figure 4.1d. The top panels are polarimetric radar measurements of reflectivity (Z_H) and the co-polar correlation coefficient (ρ_{hv}), which has low values for TBSS. Using hydrometeor classification, the TBSS area is identified (bottom right: HCA result) and removed, along with BS, to obtain the quality-controlled reflectivity field (bottom left).

Most HCAs use the fuzzy logic method, but other classification methods can also be used. Readers who are interested in the SBC are referred to our recent work on separation of convective and stratiform precipitation (Bukovčić et al. 2015).

6.3   QUANTITATIVE PRECIPITATION ESTIMATION

Thus far we have discussed the qualitative use of PRD by either looking at the polarimetric radar signatures or classifying echo types. We now move to the quantitative use of PRD. In this section, we study QPE from polarimetric

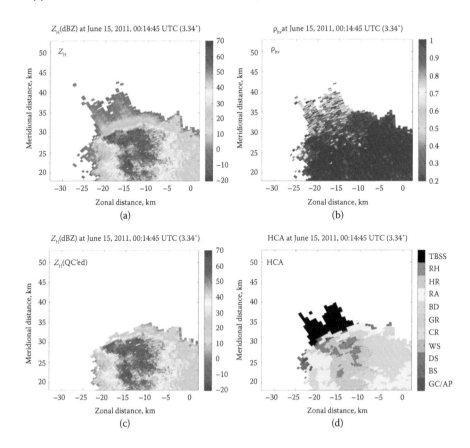

FIGURE 6.10 Polarimetric radar observation of the three-body scattering signatures and their mitigation with HCA : (a) raw reflectivity measurements, (b) raw co-polar correlation coefficient measurements, (c) quality-controlled reflectivity field based on HCA, and (d) HCA results. (From Mahale, V. N., et al., 2014. *Journal of Applied Meteorology and Climatology*, 53, 2017–2033.)

radar measurements. There are three approaches to obtain a dataset for developing rain estimators: (i) radar-gauge comparison, (ii) DSD modeling and parameter simulation, and (iii) calculation of rain rate and radar reflectivity from measured DSDs. Once the dataset is established, least squares fitting is used to find the relationship between the rain rate and the radar variable(s). Here, we focus on using 2DVD-measured DSDs and the simulated radar variables for developing rain estimators.

6.3.1 RADAR RAIN ESTIMATORS

Before we discuss QPE from PRD, let's introduce some background knowledge about single-polarization radar rain estimation. A typical single-polarization weather radar provides three measurements: reflectivity factor (Z), Doppler velocity (v_r), and spectrum width (σ_v). Only the reflectivity (Z) contains information about the rain rate (R).

TABLE 6.3
NEXRAD Single-Polarization Rain Estimators

Relation	Name	Optimal for	Also Recommended for
$Z = 300\,R^{1.4}$	Convective	Summer deep convective precipitation	Other nontropical convective
$Z = 200\,R^{1.6}$	Marshall-Palmer	General strat. prec.	
$Z = 250\,R^{1.2}$	Tropical	Tropical convective systems	
$Z = 130\,R^{2.0}$	East—cool stratiform	Winter stratiform precipitation—east of continent	Orographic rain—east
$Z = 75\,R^{2.0}$	West—cool stratiform	Winter stratiform precipitation—west of continent	Orographic rain—west

Hence, the purpose of radar rain estimation is to establish and apply power-law ($Z = \alpha R\beta$) relations to obtain the rain rate from radar reflectivity. Because Z and R are not linearly related and there is large variability in rain DSDs, the coefficient (α) and the exponent (β) in the power-law relation also vary, depending on rain types, seasons, and climatologic locations (Doviak and Zrnic 1993). Table 6.3 shows five $Z - R$ relations that were used in single-polarization NEXRAD (http://www.roc.noaa.gov/ops-/z2r_osf5.asp).

There are over 200 reported $Z - R$ relations (Rosenfeld and Ulbrich 2003). Too many relations essentially mean no relation (because it is not possible to know which one to use for radar rain estimation). There are many factors that cause this variability in $Z–R$ relations, including data selection, sampling volume discrepancy, measurement errors, and fitting procedures, as well as the difference in rain microphysics. Rosenfeld and Ulbrich (2003) gave a complete review of the various $Z–R$ relations and summarized the microphysical processes that might cause the variability in the $Z–R$ relations. The reason for this situation is that a $Z–R$ relation depends on DSDs that vary and, in general, need two or more parameters (freedoms) to characterize. Hence, a single $Z–R$ relation does not exist and cannot represent the natural variability of rain DSDs. This was one of the original motivations to develop radar polarimetry—to measure the differential reflectivity in addition to reflectivity in order to improve QPE (Seliga and Bringi 1976).

6.3.2 POLARIMETRIC RADAR RAIN ESTIMATORS

Compared with single-polarization radar, which measures the single parameter of reflectivity, polarimetric radar provides multiparameter PRD that contain DSD information and allow for a better representation of rain microphysics, hence improving rain estimation. In addition to reflectivity, differential reflectivity depends on the raindrop shape, which in turn is related to drop size. Specific differential phase is related to rain rate by a relation that is closer to linear than is the $Z-R$ relation. Hence, it is expected that QPE can be improved with added polarimetric measurements.

Polarimetric radar rain estimators are the relations between rain rate and polarimetric radar variables. These relations are also assumed in power-law form, for example $R(Z_h, Z_{dr}) = aZ_h^b Z_{dr}^c$, $R(K_{DP}) = aK_{DP}^b$, and $R(K_{DP}, Z_{dr}) = aK_{DP}^b Z_{dr}^c$. They are an extension of the $Z-R$ (equivalent form: $Z = \alpha R^\beta \Leftrightarrow R(Z_h) = aZ_h^b$) relation concept by either adding or changing to another polarimetric variable. The polarimetric estimators are established based on observed and/or simulated datasets. Because radar and gauge measurements have different sampling volume/time and error structures, it is most common to develop the polarimetric estimators from simulated datasets. To generate a rain rate and polarimetric dataset using Equation 2.4 for rain rate and Equations 4.59 through 4.62, and 4.106 for the radar variables Z_H, Z_{DR}, and K_{DP}, respectively, we need to know the DSD and the scattering amplitudes, which require the drop shape–size (axis ratio) relation.

Earlier studies used simulated DSDs by randomly generating gamma DSD parameters (N_0, μ, Λ) within certain ranges (Chandrasekar and Bringi 1988; Sachidananda and Zrnić 1987). These simulated DSDs, however, may not be representative of natural rain DSDs. Recent studies have tended to use measured DSDs because disdrometers are becoming more popular and the 2DVD can make accurate DSD measurements. DSD data used here were collected in Oklahoma by the NCAR and OU 2DVDs between May 2, 2005, and January 27, 2007. There are 36,879 oneminute DSDs, in which there are 12,756 DSDs whose rain rates are larger than 0.1 mm/hr. These 12,756 DSDs are used to calculate the rain rate and polarimetric variables for developing rain estimators.

Raindrop shape, described by the axis ratio, is the other important property for deriving QPE from polarimetric measurements. Generally, there are three kinds of raindrop axis ratio relations: (i) the equilibrium shape relation (Chuang and Beard 1990; Green 1975; Pruppacher and Beard 1970), (ii) the relation that includes the effect of oscillation on raindrop shape (which causes more spherical shapes) (Beard and Tokay 1991; Beard et al. 1983; Pruppacher and Pitter 1971), and (iii) derived raindrop axis ratio relations from previous observations or relations (Brandes et al. 2002), which are now widely accepted. It was found that the simulated Z_{DR} using the experimental shape model is 0.2 dB smaller than the corresponding value calculated using the equilibrium shape model.

Using the 2DVD-measured DSDs and the scattering amplitudes for raindrops with an empirically derived axis ratio relation (Equation 2.16), rain rate and polarimetric variables were generated and used regression fits to obtain the following estimators:

$$R(Z_h) = 0.0196Z_h^{0.688} \tag{6.9}$$

$$R(Z_h, Z_{dr}) = 0.0082Z_h^{0.918} Z_{dr}^{-3.49} \tag{6.10}$$

$$R(K_{DP}) = 44.5K_{DP}^{0.788} \tag{6.11}$$

$$R(K_{DP}, Z_{dr}) = 126.1K_{DP}^{0.956} Z_{dr}^{-2.04}. \tag{6.12}$$

To improve the fitting performance, regression fits were used in the logarithm domain with proper weighting. For a single radar variable, we minimize the errors in both the x- and y-axes to obtain $R(Z_h)$ and $R(K_{DP})$. For the two variable cases, each data point is weighted by the rain rate in using "lscov" in MATLAB to obtain the $R(Z_h, Z_{dr})$ and $R(K_{DP}, Z_{dr})$. Using the rain rate weights, the standard deviation of the estimation errors is reduced by about 30%. Otherwise, the fitted relations don't represent heavy rain well because the light rain data points dominate the fitting. The performances of these estimators are evaluated by plotting the estimates versus the DSD-calculated rain rates in Figure 6.11. The standard deviations and correlation coefficients are listed in Table 6.4.

As expected, the two-parameter estimators $R(Z_h, Z_{dr})$ and $R(K_{DP}, Z_{dr})$ have much better performance with a lower standard deviation of errors and higher correlation coefficient than these single-parameter estimators. The estimators with K_{DP} appear to perform better than their counterparts with Z_h, because the power of K_{DP} is closer to 1 than is the power of Z_h. Note that these performance results are for simulated data from the same measured DSDs. The performance of

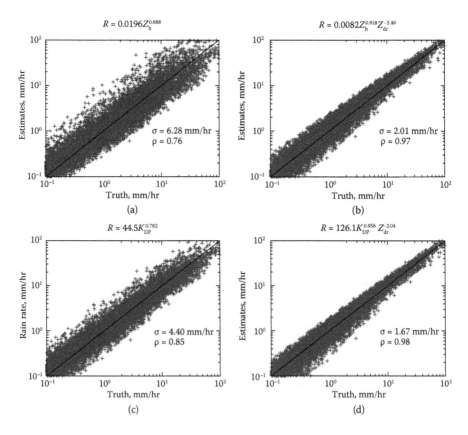

FIGURE 6.11   Radar rain estimates versus calculated truth for different estimators: (a) $R(Z_h)$, (b) $R(Z_h, Z_{dr})$, (c) $R(K_{DP})$, and (d) $R(K_{DP}, Z_{dr})$.

TABLE 6.4
Performance of Polarimetric Radar Estimators

Estimators	RMSE std$(\hat{R} - R)$, mm/hr	Correlation Coefficient
$R(Z_h)$	6.28	0.76
$R(Z_h, Z_{dr})$	2.01	0.97
$R(K_{DP})$	4.40	0.85
$R(K_{DP}, Z_{dr})$	1.67	0.98

Note: RMSE, root mean square error.

TABLE 6.5
Summary Results of Polarimetric Radar Rain Estimators for Subtropical Rain in Florida in 1998

	$R(Z_h)$	$R(K_{DP})$	$R(K_{DP}, Z_{dr})$	$R(Z_h, Z_{dr})$
Equilibrium Axis Ratios				
Mean bias factor	0.95	1.17	1.10	0.80
Bias factor range	2.55	2.67	2.43	1.82
Correlation coefficient	0.87	0.86	0.89	0.92
RMSE	7.8	8.2	7.4	8.8
RMSE (bias removed)	7.7	8.1	7.3	6.3
Empirical Axis Ratios				
Mean bias factor	0.94	0.97	0.92	0.97
Bias factor range	2.53	2.57	2.38	1.79
Correlation coefficient	0.87	0.87	0.89	0.92
RMSE	7.9	8.0	7.4	6.4
RMSE (bias removed)	7.7	7.8	7.1	6.3

Source: Brandes, E. A., et al., 2002. *Journal of Applied Meteorology*, 41, 674–685.
Note: RMSE, root mean square error.

polarimetric radar rain estimators can be very different with real data from radar-gauge comparisons due to the differences in sampling volume and measurement error characteristics.

As shown in Table 6.5, Brandes et al. (2002) applied and evaluated two sets of polarimetric radar rain estimators for subtropical rain in Florida during the summer of 1998. Rainfall estimates were made for 25 events from NCAR S-Pol radar measurements using polarimetric radar rain estimators and were compared with observations from two rain gauge networks. Rain accumulations for each event were used for comparison to reduce error. Biases, defined as the gauge/radar ratio, and root mean square errors (RMSEs) were calculated and listed in Table 6.5, in addition to correlation coefficients. As expected, the RMSEs are larger and the correlation coefficients are generally smaller than those in Table 6.4 for the simulated data, which is understandable due to the differences in sampling volume and error statistics. Table 6.5 shows that the estimators developed from an empirical axis ratio relation (Equation 2.16) yield smaller biases and errors than those developed from an equilibrium drop shape relation. The estimator $R(Z_h, Z_{dr})$ has the best performance, having the smallest errors and highest correlation coefficient.

6.3.3   RAIN ESTIMATION RESULTS

Using the four derived radar estimators of Equations 6.9 through 6.12 described in Subsection 6.3.2, rain estimates are made from polarimetric radar measurements of the squall-line case, shown in Figure 6.9. Figure 6.12 displays the rain estimate

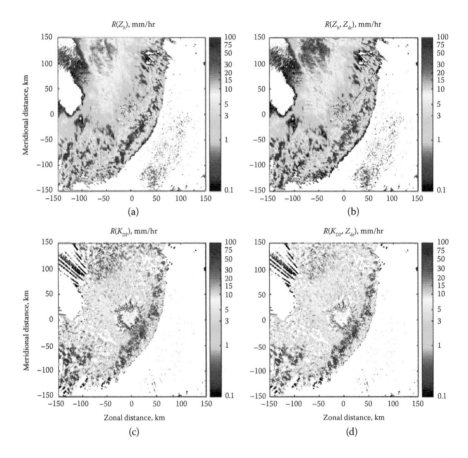

FIGURE 6.12   QPE results using different radar rain estimators: (a) $R(Z_h) = 0.0196 Z_h^{0.688}$, (b) $R(Z, Z_{dr}) = 0.0082 Z^{0.918} Z_{dr}^{-3.49}$, (c) $R(K_{DP}) = 44.5 K_{DP}^{0.788}$, and (d) $R(K_{DP}, Z_{dr}) = 126.1 K_{DP}^{0.956} Z_{dr}^{-2.04}$.

results of the four estimators in Equations 6.9 through 6.12. Rain estimation was performed for the region with $\rho_{hv} > 0.8$ to remove clutter and biological scatterers. The estimator $R(Z_h) = 0.0196Z_h^{0.688}$ (i.e., $Z_h = 303R^{1.45}$) is close to the default estimator ($Z = 300R^{1.4}$) applied by NEXRAD for midlatitude rain (Fulton et al. 1998). The other three are polarimetric estimators. The polarimetric estimators have a clear improvement in the region of strong convection. The estimator $R(Z_h)$ obviously overestimates the rainfall rate for convective cores because the Z_h value was significantly enlarged by a few large drops and/or the contamination of melting hail. Because the increase in drop size also causes the increase in Z_{dr}, the $R(Z_h, Z_{dr})$ estimator effectively reduces the overestimation of $R(Z_h)$. The $R(K_{DP})$ estimator, less sensitive to hail contamination, also has a smaller but more reasonable estimation than $R(Z_h)$ in this region. However, the relative uncertainty of the K_{DP} measurement is much larger than Z_h or Z_{dr}. It is evident that estimators with K_{DP} have poorer rain estimation in the light rain region (Figure 6.12c and d) than the estimators without K_{DP} (Figure 6.12a and b).

6.3.4 ESTIMATION ERRORS AND PRACTICAL ISSUES

There are different error structures for the different radar rain estimators. For convenient error analysis, we use the general form $R = aZ_h^b Z_{dr}^c K_{DP}^d$. Taking a differential, we have

$$\delta R = \frac{\partial R}{\partial Z_h}\delta Z_h + \frac{\partial R}{\partial Z_{dr}}\delta Z_{dr} + \frac{\partial R}{\partial K_{DP}}\delta K_{DP}$$

$$\frac{\delta R}{R} = b\frac{\delta Z_h}{Z_h} + c\frac{\delta Z_{dr}}{Z_{dr}} + d\frac{\delta K_{DP}}{K_{DP}}. \tag{6.13}$$

Because the measurement errors can be considered to be independent of each other, Equation 6.13 shows that the relative error of rain estimates is proportional to the errors of radar variable estimates and to their power coefficients. This explains why rain estimators with K_{DP} perform poorly in light rain regions, because the error (~0.2 degree km^{-1}) of K_{DP} estimates translates to about 100% for the relative error, and hence large errors for light rain estimates. Equation 6.13 also means that error may be increased when a new measurement is added to the rain estimators. Therefore, more measurements do not necessarily yield more accurate rain estimation, unless the error effects are handled properly (Brandes et al. 2004a).

Because polarimetric estimators have different error structures and perform differently in estimating different types of rain, it is better to apply these relations appropriately in different situations. For example, when rain is intense and/or mixed with hail, K_{DP} generally has a better representation of rain physics. In that case, $R(K_{DP})$ normally has a small error for rain estimation. However, when rain is light, the measurement error of K_{DP} may be large. Therefore, it is not appropriate to apply $R(K_{DP})$. Ryzhkov et al. (2005b) suggested a "synthetic" approach, applying $R(Z_h, Z_{dr})$, $R(K_{DP})$, and $R(K_{DP}, Z_{dr})$, with a minor difference from the power-law forms, in three different ranges of rainfall rate estimated from the $Z - R$ relation. For $R(Z_h) < 6$ mm hr^{-1}, $R(Z_h, Z_{dr})$ is applied; for $R(Z_h) > 50$ mm hr^{-1}, $R(K_{DP})$ is applied; and $R(K_{DP}, Z_{dr})$ is applied for other cases.

Because different estimators are applicable for different types of precipitation, it is natural and reasonable to apply the estimators based on the hydrometeor classification results discussed in Section 6.2. Giangrande and Ryzhkov (2008) introduced a classification-based rain estimator, which consists of three relations:

$$R(Z_h) = 1.70 \times 10^{-2} Z_h^{0.714}$$

$$R(Z_h, Z_{dr}) = 1.42 \times 10^{-2} Z_h^{0.77} Z_{dr}^{-1.67} \qquad (6.14)$$

$$R(K_{DP}) = 44.0 |K_{DP}|^{0.822} \operatorname{sign}(K_{DP}).$$

The usage of the set of relations is as follows: $R = 0$ if the nonmeteorological echo is classified; $R = R(Z_h, Z_{dr})$ if rain, including light/intermediate/heavy rain and big drops, is classified; $R = R(K_{DP})$ if rain/hail is classified; $R = 0.6R(Z)$ if wet snow is classified; and $R = 2.8R(Z)$ if dry snow or crystals are classified. As expected, the classification-based rain estimator has robust performance.

Polarimetric radar estimators have improved QPE accuracy from about the 35% RMSEs for single polarization reflectivity measurement to about 10%~15% RMSEs for polarization radar data (Balakrishnan et al. 1989; Brandes et al. 2002; Matrosov et al. 2002; Ryzhkov and Zrnić 1995; Ryzhkov et al. 2005a). This is because polarimetric measurements contain information about the DSD variability and allow for better data quality, less contamination, and separation of different precipitation types so that proper estimators can be used accordingly. Hence, the reduction of rain estimation error is substantial at near ranges (<100 km). At farther ranges (>150 km), however, the error reduction is marginal due to the inherent problems of radar remote sensing such as non-uniform beam filling and overshooting.

Although polarimetric estimators improve QPE, there are limitations to this empirical approach. A set of relations need to be derived to estimate rain rate, but other rain variables (e.g., number concentration) of interest are not available. Moreover, those relations are normally dependent on radar frequency. Because the DSD information is the main reason for the improved QPE, it is logical to directly retrieve rain DSDs for QPE and precipitation microphysics studies. This approach is more flexible for the application of measurements on different radar platforms. We discuss DSD retrievals next.

6.4   DSD RETRIEVAL

As noted, QPE depends on rain microphysics (DSDs), and polarimetric measurements contain DSD information. It is natural and more useful and accurate to directly retrieve rain DSDs from PRD rather than determine the QPE from the derived estimators. The retrieved DSDs can then be used to calculate rain physics and physical process parameters, including rain rate, number concentration, mean drop diameter, water content, evaporation rate, accretion rate, and so on. Hence, rain DSD retrieval provides far more detailed information than just QPE.

6.4.1 SELECTION OF MEASUREMENTS AND DSD MODEL

Dual-polarization radar provides polarimetric measurements of radar reflectivity (Z_H), differential reflectivity (Z_{DR}), specific differential phase shift (K_{DP}), and the co-polar correlation coefficient (ρ_{hv}). They are DSD-weighted integral parameters and contain DSD information, but the number of independent pieces of information is limited (no more than four). For example, ρ_{hv} is close to 1 for all rain, especially at S-band frequencies. Although ρ_{hv} is a very useful parameter for qualitative use in hydrometeor classification, it is of little use in providing quantitative information about rain DSDs, especially at S-band frequencies at which ρ_{hv} is always close to unity for rain. K_{DP} measurements/estimates have a large relative error, especially for light rain. Furthermore, K_{DP} is estimated from the range derivative of the differential phase (Φ_{DP}) over more than 10 gates and does not have the same resolution volume as the Z_H and Z_{DR} measurements. This makes the usage of K_{DP} in DSD retrieval difficult, and sometimes the use of K_{DP} has a negative impact (introduces error) rather than a positive impact (unless the error covariance is accurately accounted for in retrievals, such as with the variational approaches to be discussed in Chapter 7).

Excluding ρ_{hv} and K_{DP} leaves the polarimetric radar measurements of Z_H and Z_{DR} for rain DSD retrieval. This means that a two-parameter DSD model is needed to facilitate a retrieval of the two DSD parameters from the two polarimetric measurements of Z_H and Z_{DR}, although a real measured DSD is normally represented by a 41-bin spectrum. The most popular two-parameter DSD model is the exponential distribution, which has been widely used. Recent disdrometer observations, however, show that natural rain DSDs are not exponentially distributed. In other words, the DSDs do not appear in a straight line in a semilog plot. As shown in Figure 2.4, both convex and concave shapes present for the measured DSDs. This result has led to a search for other, more representative two-parameter DSD models.

In analyzing DSD data collected during PRECIP-98 in Florida, Zhang et al. (2001) found that there was a strong correlation between the shape (μ) and slope (Λ) parameters and introduced a constrained-gamma (C-G) DSD model. The C-G model was used for DSD parameter retrieval from polarimetric radar measurements, yielding reliable results. The C-G model was further refined with more DSD observations from Oklahoma, giving

$$\mu = -0.0201\Lambda^2 + 0.902\Lambda - 1.718. \tag{6.15}$$

The $\mu - \Lambda$ relation in Equation 6.15 was derived from the truncating moment fitting (TMF) of the 2DVD-measured DSDs processed with a sorting and averaging method based on two parameters (SATP) (Cao et al. 2008). The SATP method was introduced to reduce sampling error and ameliorate the problem of overrepresentation of light rain. It results in a more representative constraining shape–slope relation for the C-G DSD model.

Figure 6.13a shows the scatterplot for the $\mu - \Lambda$ relation, and Figure 6.13b for the $\sigma_m - D_m$ relation. It is shown that the Oklahoma DSD data with SATP processing yields a $\mu - \Lambda$ relation with a lower slope than that for the Florida data. Note that moment errors can cause a correlation among fitted DSD parameters (Zhang et al. 2003). To verify the refined $\mu - \Lambda$ relation, we examine the mean mass-weighted diameter (D_m) and standard deviation of the mass-weighted diameter distribution (σ_m), because both can be directly derived from observations and are independent of sorting and fitting procedures. If the relation in Equation 6.15 represents rain physics, the $\sigma_m - D_m$ relation derived from observations and from Equation 6.15 should be consistent. Figure 6.12b shows the results of these calculations. Crosses denote calculations of D_m and σ_m from observed 1-min DSDs. The solid line, which represents the result derived from Equation 6.15, is in the middle of the data points and agrees well with direct calculations of D_m and σ_m from observations. This shows the equivalence between the $\mu - \Lambda$ relation and the $\sigma_m - D_m$ relation; both serve as a constraint for DSD shape to reduce the three-parameter gamma model to a two-parameter C-G model (Zhang 2015). The equivalence is showed in Appendix 6B.

The two-parameter C-G DSD models can be expressed by

$$N(D) = N_0 D^\mu \exp(-\Lambda D) \tag{6.16}$$

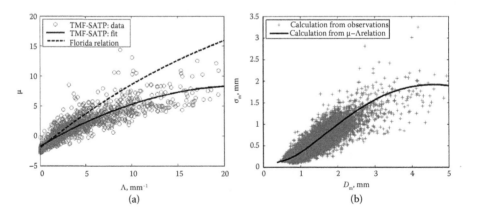

FIGURE 6.13 Scatterplots of developing the constrained-gamma model. (a) $\mu - \Lambda$ relation: Circles denote the SATP (sorting and averaging method based on two parameters) DSD data fitted by the truncating moment fitting method. The solid line is the mean curve fitted to circle points by the second order polynomial fitting. The dashed line corresponds to the Florida relation (Zhang et al. 2001). (b) $\sigma_m - D_m$ relation: Circles denote that D_m and σ_m are calculated from 14,200 min of observed DSDs. The solid line denotes that D_m and σ_m are calculated from gamma DSDs with $\mu - \Lambda$ constrained by Equation 6.15. (From Cao, Q., et al., 2008. *Journal of Applied Meteorology and Climatology*, 47, 2238–2255.)

and

$$\mu = a\Lambda^2 + b\Lambda + c. \tag{6.17}$$

It reduces to the exponential (EXP) model when $\mu = 0$. Once a two-parameter DSD model is selected, the direct DSD retrieval is to find the two DSD parameters (N_0, Λ) from the polarimetric radar measurements of Z_H and Z_{DR}.

6.4.2 Retrieval Procedure

If we ignore model and measurement errors, finding the two DSD parameters is straightforward. Ignoring raindrop canting and using the DSD model of Equations 6.16 and 6.17 in the definition of polarimetric radar variables of Z_{DR} and Z_H, respectively, we obtain

$$Z_{dr} = \frac{Z_{hh}}{Z_{vv}} = \frac{\int \left| s_a(\pi, D) \right|^2 D^{\mu(\Lambda)} \exp(-\Lambda D) dD}{\int \left| s_b(\pi, D) \right|^2 D^{\mu(\Lambda)} \exp(-\Lambda D) dD} = g_1(\Lambda) \tag{6.18}$$

and

$$\frac{N_0}{Z_{hh}} = \frac{\pi^4 |K_w|^2}{4\lambda^4} \frac{1}{\int \left| s_a(\pi, D) \right|^2 D^{\mu(\Lambda)} \exp(-\Lambda D) dD} = g_2(\Lambda). \tag{6.19}$$

Using the T-matrix-calculated backscattering amplitudes for the empirically derived raindrop shape model of Equation 2.16, the ratios of Equations 6.18 and 6.19 are calculated and plotted in dB scale in Figure 6.14. Both the EXP and C-G (Equation 6.15) DSD models are used for the calculations.

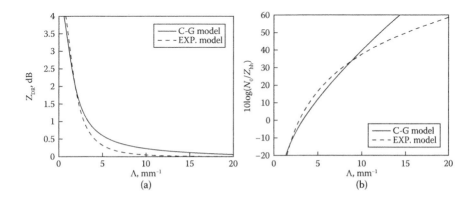

FIGURE 6.14 Polarimetric radar variables versus the DSD slope parameter for both the constrained-gamma and exponential models: (a) Differential reflectivity and (b) ratio between N_0 and reflectivity factor.

Then, the procedure to obtain the C-G DSD parameters is as follows:[*]

1. From the measured Z_{DR}, find Λ as shown in Figure 6.13a and then from the $\mu - \Lambda$ relation, obtain μ.
2. Use the retrieved μ and Λ as well as the N_0/Z_{hh} ratio in Figure 6.13b to obtain N_0.

To conveniently use the C-G (Equation 6.15) DSD retrieval, the DSD parameters of Λ and N_0 are expressed in terms of Z_{DR} and Z_{hh}, respectively, as follows:

$$\Lambda = 0.0125Z_{DR}^{-3} - 0.3068Z_{DR}^{-2} + 3.3830Z_{DR}^{-1} + 0.1790 \tag{6.20}$$

and

$$N_0 = Z_{hh} \times 10^{0.00285\Lambda^3 - 0.0926\Lambda^2 + 1.409\Lambda - 3.764}. \tag{6.21}$$

The same procedure applies to the EXP DSD model, with $\mu = 0$.

To show the performance of the C-G DSD retrieval, we first apply it to measured DSD data. From the DSD data, the polarimetric radar variables of Z_H and Z_{DR} are calculated, and then the calculated Z_H and Z_{DR} are used to determine the DSD parameters using the procedure described above. The results are shown in Figure 6.15.

This kind of DSD retrieval essentially fits measured DSDs with the C-G and EXP DSD models using two radar variables (Z_H and Z_{DR}), which is similar to the moment fit of DSDs to a distribution model, as described in Chapter 2. This is because Z_h is close to the 6th moment of DSD, and Z_{dr} is approximately proportional to the ratio of the 6th and 5.4th moments (Zhang et al. 2001). Because the radar variables are all proportional to higher moments that are dominated by large drops and contain very little information about small drops, an accurate DSD model is needed to facilitate an accurate DSD retrieval. It is clear that the C-G model results fit with the DSD data much better than the EXP model. Further, the $\mu - \Lambda$ relation can be adjusted as needed.

The retrieval procedure is then applied to the real PRD, and the retrieved results are shown in Figure 6.16.

As shown in the figure, the rain rates with both the C-G and EXP models are similar to those of the $R(Z_h, Z_{dr})$ rain estimator in Figure 6.12b. This is reasonable because the same radar measurements of Z_H and Z_{DR} are used in either the DSD retrieval or the estimator. However, DSD retrieval also provides other microphysics parameters, including mass-weighted diameter (D_m) and drop number concentration (N_t), that are shown in Figure 6.16 and that cannot obtained from the empirical rain estimators.

To verify the retrieval algorithms, rain microphysical parameters calculated from radar-retrieved DSD parameters using different DSD models are compared. Direct calculations with disdrometer measurements are also presented in Figure 6.17 for reference. Radar variable (Z_H, Z_{DR}) calculations from the DSD data and their KOUN radar measurements are shown on the left, microphysics parameters on the right. Both the C-G and EXP DSD model retrievals agree well with the 2DVD measurements for rain

---

[*] In the case of $Z_{DR} < 0.3$ dB, the derived relations in Equations 6.27 and 6.28 are suggested to use for finding DSD parameters.

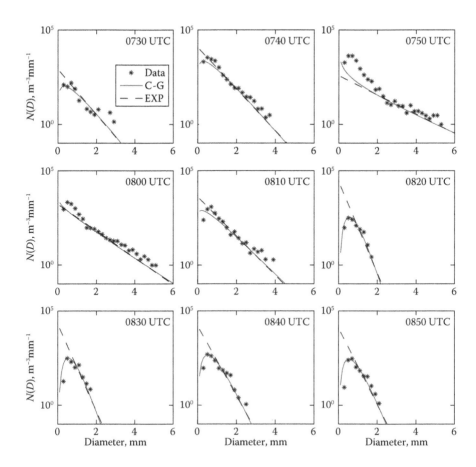

FIGURE 6.15   Examples of DSD retrieval from calculated polarimetric radar variables for 2DVD-measured DSD data collected in Oklahoma on May 13, 2005.

rate, mass-weighted diameter, and number concentration, and the C-G model retrieved results agree with observation better than the EXP model. The radar retrievals from reflectivity with the M–P model have the worst performance; its problems include that it (i) overestimates N_t except for heavy convective rains and has a small dynamic range, (ii) overestimates W for stratiform rain (after 0830 UTC), and (iii) underestimates D_m for stratiform rain. Clearly, the polarization radar-based C-G and EXP models characterize rain microphysics more accurately than the reflectivity-based M–P model.

6.4.3   APPLICATION OF DSD RETRIEVAL IN PARAMETERIZATION OF RAIN MICROPHYSICS

Using the C-G DSD model with a constraining $\mu - \Lambda$ relation like Equation 6.15, the rain state parameters of rain rate and number concentration can be retrieved from the polarization radar measurements of the radar reflectivity factor (Z_H) and differential reflectivity (Z_{DR}). In addition, the rain physical processes such as evaporation rate (R_e),

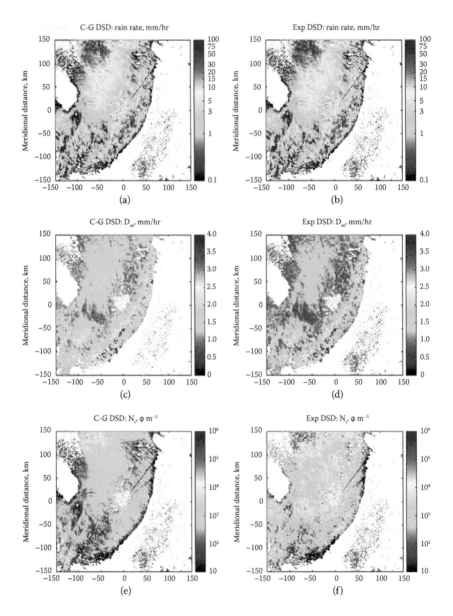

FIGURE 6.16 Spatial distribution of DSD retrieval results for (a) and (b) rain rate (R), (c) and (d) mass-weighted diameter (D_m), and (e) and (f) number concentration (N_t) with the C-G model (left column) and the EXP model (right column).

accretion rate (R_c), and mass-weighted terminal velocity (V_t) can also be determined (Zhang et al. 2006). This shows the importance and potential of using PRD in microphysics parameterization in numerical weather prediction (NWP). Because the C-G model is a two-parameter model, it is consistent with a two-moment microphysics parameterization (Milbrandt and Yau 2005b, 2006), which is more accurate than

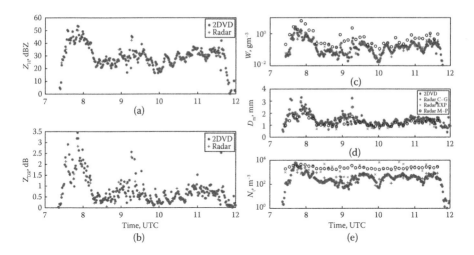

FIGURE 6.17 Comparisons of rain microphysics and physical process parameters between 2DVD measurements and PRD retrievals. (a) Reflectivity (Z_H, dBZ), (b) differential reflectivity (Z_{DR}, dB), (c) rain rate (R, mm/hr), (d) mass-weighted diameter (D_m, mm), (e) number concentration (N_t, # m^{-3}).

the single-moment parameterization (Lin et al. 1983). Following Kessler's parameterization procedure (1969), the microphysical process parameters are derived (Appendix of Zhang et al. [2006]) by integration of the gamma DSD. Assuming an evaporation coefficient $E_e = 1$ and accretion coefficient $E_c = 1$, we obtain R_e (g m^{-3} s^{-1}) for a unit vapor saturation deficit ($m_e = 1$ g m^{-3}) from Equation 6A.3, R_a(in g m^{-3} s^{-1}) or a unit cloud water content ($m_c = 1$ g m^{-3}) from Equation 6A.6, and V_{tm} (in m/s) from Equation 6A.7, along with rain water content (W in g m^{-3}) as follows:

$$W = \frac{\rho_w \times 10^{-3}\pi}{6} N_0 \int_{D_{min}}^{D_{max}} D^{\mu+3} \exp(-\Lambda D)\,dD$$

$$= \frac{\rho_w \times 10^{-3}\pi}{6} N_0 \Lambda^{-(\mu+4)} \left[\gamma(\Lambda D_{max}, \mu+4) - \gamma(\Lambda D_{min}, \mu+4)\right], \tag{6.22}$$

$$R_e = 6.78 \times 10^{-4} W\Lambda^{7/5} \frac{\left[\gamma(\Lambda D_{max}, \mu+13/5) - \gamma(\Lambda D_{min}, \mu+13/5)\right]}{\left[\gamma(\Lambda D_{max}, \mu+4) - \gamma(\Lambda D_{min}, \mu+4)\right]} \tag{6.23}$$

$$R_c = \frac{3 \times 10^{-3} W}{2} \sum_{i=0}^{4} c_l \Lambda^{-l+1} \frac{\left[\lambda(\Lambda D_{max}, \mu+l+3) - \gamma(\Lambda D_{min}, \mu+l+3)\right]}{\left[\gamma(\Lambda D_{max}, \mu+4) - \gamma(\Lambda D_{min}, \mu+4)\right]} \tag{6.24}$$

$$V_{tm} = \sum_{l=0}^{4} c_l \Lambda^{-l} \frac{\left[\gamma(\Lambda D_{max}, \mu+l+4) - \gamma(\Lambda D_{min}, \mu+l+4)\right]}{\left[\gamma(\Lambda D_{max}, \mu+4) - \gamma(\Lambda D_{min}, \mu+4)\right]}, \tag{6.25}$$

where γ is the incomplete gamma function, and D_{min} and D_{max} are the raindrops' minimal and maximal diameters. D_{min} was set to 0.1 mm. D_{max}, the size of the largest drop, can be estimated from radar reflectivity or differential reflectivity (Brandes et al. 2003).

These equations allow for detailed study of precipitation microphysics for convective and stratiform precipitation and their evolution (Brandes et al. 2004b). Figure 6.18 compares retrieved microphysical process parameters using the C-G and EXP DSD models with those from the M–P DSD model. Calculations with the disdrometer observations are shown for reference. If the disdrometer results can be considered as "truth," the M–P model overestimates evaporation and accretion for stratiform rain by up to a factor of 10 and underestimates them for strong convection. This might be the reason that the parameterization coefficients in a single moment scheme are usually reduced by one-half (or more) in an attempt to improve weather model forecasts (Miller and Pearce 1974; Sun and Crook 1997). The M–P model also underestimates the mass-weighted terminal velocity for stratiform rain because the droplet size is underestimated. The M–P model yields a smaller dynamic range for all microphysical process parameters, which could be a reason (in addition to grid resolution) that cloud models have difficulty resolving fine-scale storm features. It is apparent that the C-G DSD model gives a more accurate estimation of rain microphysical processes than does the M–P model. It is because of the use of reflectivity and differential reflectivity (i.e., two parameters) in the retrieval process that the C-G rain model more closely represents the disdrometer-measured (natural) raindrop spectra than the single-parameter M–P model. Therefore, the C-G parameterization scheme (Equations 6.23 through 6.25) can be used to improve two-moment models that forecast two microphysical parameters (e.g., W and D_m).

For computational convenience, expressions for rainfall rate (R in mm hr^{-1}), rain water content (W in g m^{-3}), and D_m (in mm) can be expressed in terms of polarimetric measurements of Z_H and Z_{DR} as follows:

$$R = 0.0113 Z_h \times 10^{\left(-0.0553 Z_{DR}^3 + 0.382 Z_{DR}^2 - 1.175 Z_{DR}\right)} \qquad (6.26)$$

$$W = 1.023 \times 10^{-3} Z_h \times 10^{\left(0.0742 Z_{DR}^3 + 0.511 Z_{DR}^2 - 1.511 Z_{DR}\right)} \qquad (6.27)$$

$$D_m = 0.0657 Z_{DR}^3 - 0.332 Z_{DR}^2 + 1.090 Z_{DR} + 0.689, \qquad (6.28)$$

where Z_h is in linear units (mm^6 m^{-3}) and Z_{DR} is in dB. The similar relations in Equations 6.26 through 6.28 have been verified for tropical rain in Florida (Brandes et al. 2003, 2004a, 2004b) and for southern Great Plains precipitation in Oklahoma (Cao et al. 2008).

Because W and D_m can be determined from NWP model-predicted state variables such as the water mixing ratio and number concentration, it is convenient to express the microphysical processes in terms of the two parameters of W and D_m as

$$R_e = 4.142 \times 10^{-3} W \times 10^{\left(-0.0421 D_m^3 + 0.315 D_m^2 - 0.958 D_m\right)} \qquad (6.29)$$

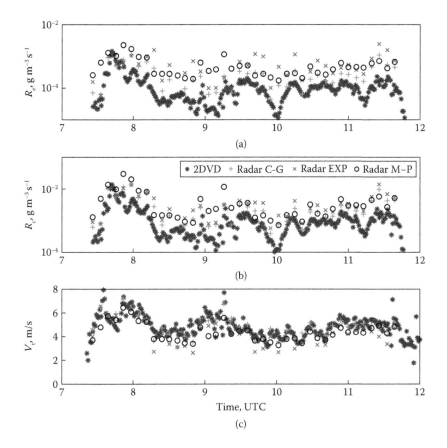

FIGURE 6.18 Comparisons of rain physical process parameters from radar retrievals with different DSD models: (a) evaporation rate (R_e, g m^{-3} s^{-1}), (b) accretion rate (R_c, g m^{-3} s^{-1}), and (c) mass-weighted mean terminal velocity (V_t, m/s).

$$R_a = 7.036 \times 10^{-3} W \times 10^{\left(0.000445 D_m^3 + 0.00132 D_m^2 - 0.0744 D_m\right)} \tag{6.30}$$

$$V_{tm} = 0.139 D_m^3 - 1.343 D_m^2 + 5.245 D_m - 0.176. \tag{6.31}$$

Equations 6.29 through 6.31 are alternative forms of the integrals like Equations 6.23 through 6.25, obtained by fitting log(R_e/W), log(R_c/W), and V_{tm} to polynomial functions of D_m, as shown in Figure 6.19. The discrete points are model calculations based on the CG model with the $\mu - \Lambda$ relation (Equation 6.15). There is very little deviation from the fitted curves, suggesting that Equations 6.26 through 6.31 accurately represent the two-parameter C-G model and its microphysics parameterization. As expected, the concentration parameter ratios (R/Z_h, W/Z_h, R_e/W, and R_c/W) decrease as Z_{DR} or D_m increases. This is because the denominators are proportional to the higher DSD moments, which are dominated by large drops. The total surface

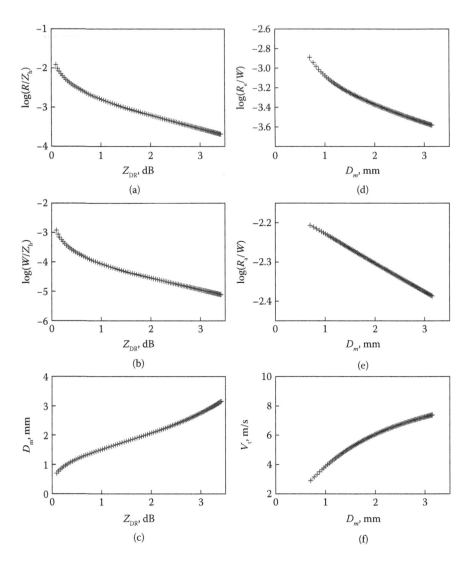

FIGURE 6.19 Parameterization of rain microphysics states and physical processes for the two-parameter C-G DSD model: (a) R/Z_h ratio, (b) R_e/W ratio, (c) W/Z_h ratio, (d) R_a/W ratio, (e) mass-weighted mean diameter (D_m in mm), and (f) mass-weighted mean terminal velocity (V_t in m/s).

area and cross section associated with evaporation and accretion are smaller for DSDs dominated by large raindrops than for small drops at the same W. It is obvious that microphysical process parameters computed from the C-G DSD depend on both rainwater content and droplet size, which can and should be captured by polarimetric measurements.

6.5 ATTENUATION CORRECTION

Attenuation is a fundamental issue encountered in communication and remote sensing (Crane 1996). Both clear air and precipitation causes wave attenuation at microwave frequencies, but wet precipitation cause much more attenuation than clear air or dry precipitation. The impact of attenuation on weather radar measurements can be substantial, especially for radars that operate at higher frequencies such as C- and X-bands. Figure 6.20 shows examples of S-band (a, c) and X-band (b, d) radar reflectivity measurements. It is clear that the attenuation causes reduced X-band reflectivity measurements at the farther side of the storms (see lower-left corner of the upper-right panel) or even the complete disappearance (below noise level)

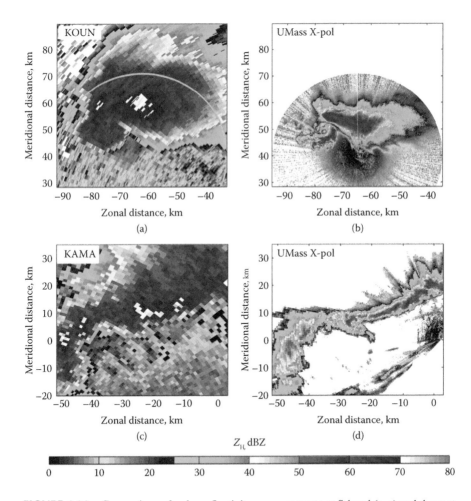

FIGURE 6.20 Comparison of radar reflectivity measurements at S-band (a, c) and those at C-band (b, d) to show the impact of attenuation. (From Snyder, J. C., et al., 2010. *Journal of Atmospheric and Oceanic Technology*, 27, 1979–2001.)

of echo (see panel d). In addition to the reduced reflectivity measurements, attenuation also has severe negative impacts on polarimetric radar measurements, including Z_{DR} bias due to differential attenuation and increased errors in PRD due to the reduced SNR. To use the PRD with attenuation, either the attenuation effects need to be included in the forward operator or the attenuation needs to be corrected before meteorological application. We focus on direct and simple attenuation corrections in this section.

6.5.1 Correction Methods

As described in Section 4.4.2, attenuation-included radar variables are related to the intrinsic (unattenuated) radar variables by subtracting the two-way PIA, as shown in Equations 4.137 and 4.140. Attenuation correction consists of finding the PIA_H and PIA_{DP} and adding them back into the measured Z_H and Z_{DR}, which are propagation-included radar variables, Z'_H and Z'_{DR}, to obtain the intrinsic (attenuation-corrected) Z_H and Z_{DR} estimates. Equations 4.137 and 4.140 are rewritten as

$$Z_H = Z'_H + 2\int_0^r A_H(\ell)d\ell \equiv Z'_H + PIA_H \qquad (6.32)$$

$$Z_{DR} = Z'_{DR} + 2\int_0^r A_{DP}(\ell)d\ell \equiv Z'_{DR} + PIA_{DP} \qquad (6.33)$$

There are different ways to find and distribute PIA_H and PIA_{DP} for attenuation correction.

6.5.1.1 DP Method

Simple DP attenuation correction is straightforward and is called the *DP method*. The DP method is based on the fact that both the attenuation and the propagation phase are integral effects of forward scattering. Hence, the attenuation (A_H) and differential attenuation (A_{DP}) are correlated with the specific differential phase (K_{DP}). Figure 6.21 shows the simulation results of A_H and A_{DP} versus K_{DP}, based on the calculation from 2DVD-measured DSDs. The top panels are for S-band at 2.8 GHz; the middle panels are for C-band at 5.5 GHz; and the bottom panels are for X-band at 3 GHz. Whereas the attenuation and differential attenuation at S-band are very small and can be ignored except for a beam along a squall-line of strong convection, those at C-band and X-band are substantial. Scattered data points are fitted to linear lines to give

$$A_H = cK_{DP} \qquad (6.34)$$

$$A_{DP} = dK_{DP}. \qquad (6.35)$$

We discuss attenuation correction in this section and leave simultaneous microphysics retrieval and attenuation correction for Chapter 7. The fitting relations and coefficients are provided in Figure 6.21.

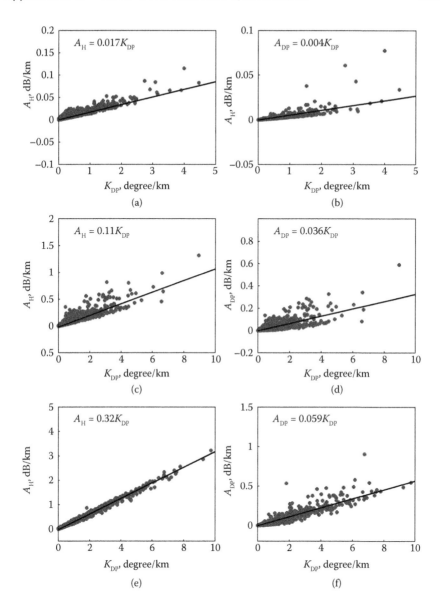

FIGURE 6.21   Attenuation (A_H: left column) and differential attenuation (A_{DP}: right column) versus specific differential phase for (a, b) S-band, (c, d) C-band, and (e, f) X-band.

Substituting Equations 6.34 and 6.35 into Equations 6.32 and 6.33 and using the differential phase relation of $\phi_{DP} = 2\int_0^r K_{DP}(\ell)d\ell$, we have the simple DP correction as

$$Z_H(r) = Z_H'(r) + c\phi_{DP}(r) \qquad (6.36)$$

$$Z_{DR}(r) = Z'_{DR}(r) + d\phi_{DP}(r). \qquad (6.37)$$

Equations 6.36 and 6.37 are applied to the measured $Z'_H(r)$ and $Z'_{DR}(r)$ at every range gate (r) by adding the estimated PIA of $PIA_H(r) = c\phi_{DP}(r)$ and $PIA_{DP}(r) = d\phi_{DP}(r)$ to obtain attenuation-corrected $Z_H(r)$ and $Z_{DR}(r)$. Note that the initial differential phase $\phi_{DP}(0)$, which can be azimuth-dependent, needs to be estimated and subtracted to obtain the propagation differential phase $\phi_{DP}(r)$ over the range of $[0, r]$. The key of this DP attenuation is to determine the attenuation coefficients c and d, because the coefficient can vary/scatter depending on the type of rain and rain microphysics. In practice, it is possible for A_{DP} to vary by a factor of 0.5 to 2.0 of the numbers given in Figure 6.21.

6.5.1.2 Z-PHI Method

The Z-PHI method is a variant of the DP method described above. Whereas the PIA_H and PIA_{DP} are estimated and added to the polarimetric radar measurements at every range in the DP method, in the Z-PHI method, the PIAs are estimated for the last gate (r_N) (in practice, the last few gates of data are used and averaged), that is, $PIA_H(r_N)$ and $PIA_{DP}(r_N)$. Then, the total attenuation is distributed to every gate based on reflectivity measurements. Because the total attenuation is estimated from $\phi_{DP}(r_N)$ and distributed based on the reflectivity Z_H, it is called the *Z-PHI method* (Testud et al. 2000). The Z-PHI method was originally developed for attenuation correction with a single measurement of reflectivity by Hitschfeld and Bordan (1954) and was called the *H-B method*. The single reflectivity measurement can be made by a TRMM precipitation radar (PR) or a single-polarization weather radar. The H-B method used the relation between the attenuation and reflectivity given below:

$$A = aZ_h^b. \qquad (6.38)$$

Rewrite Equation 4.137 in linear form,

$$Z'_h(r) = Z_h(r)\exp\left[-0.46\int_0^r A(\ell)d\ell\right] \equiv Z_h(r)g^{1/b}(r), \qquad (6.39)$$

where

$$g(r) = \exp\left[-0.46b\int_0^r A(\ell)d\ell\right]. \qquad (6.40)$$

Differentiating Equation 6.40 and using Equations 6.38 and 6.39 yields

$$\frac{dg(r)}{dr} = -0.46bg(r)A(r)$$

$$= -0.46bg(r)aZ_h^b(r)$$

$$= -0.46baZ_h'^b(r).$$

That is,

$$dg(r) = -0.46ba Z_h'^b(r)dr.$$

Performing integration gives

$$g(r) - g(0) = -0.46ba \int_0^r Z_h'^b(\ell)d\ell$$

$$g(r) = 1 - 0.46ba \int_0^r Z_h'^b(\ell)d\ell. \tag{6.41}$$

Substituting Equation 6.41 into Equation 6.39 and solving for the intrinsic (attenuation-corrected) reflectivity $Z_h(r)$ yields

$$Z_h(r) = \frac{Z_h'(r)}{g^{1/b}} = \frac{Z_h'(r)}{\left[1 - 0.46ba \int_0^r Z_h'^b(\ell)d\ell\right]^{1/b}}. \tag{6.42}$$

The attenuation correction given by Equation 6.42 is unfortunately unstable, because the $A - Z$ relation contains a modeling error, and the reflectivity measurement contains an estimation error. When these errors accumulate, they lead to divergence. To obtain a convergent, stable result, instead of using the predetermined coefficient a, the coefficient is calculated from the total attenuation $PIA_H(r_N)$ using Equation 6.41, that is,

$$g(r_N) = \exp[0.23b PIA_H(r_N)] = 1 - 0.46ba \int_0^{r_N} Z_h'^b(\ell)d\ell.$$

Solving for a gives

$$a = \frac{1 - \exp[0.23b PIA_H(r_N)]}{0.46b \int_0^{r_N} Z_h'^b(\ell)d\ell}. \tag{6.43}$$

Hence, Equations 6.42 and 6.43 constitute the formulation for the Z-PHI method for attenuation correction. The same approach can be applied for differential attenuation correction for differential reflectivity by replacing (a, b) with (a_d, b_d), Z_h with Z_{dr}, and PIA_H with PIA_{DP}. The Z-PHI method was introduced to correct C-band polarimetric radar measurements (Testud et al. 2000).

6.5.1.3 Self-Consistent with Constraint Method

As discussed in previous sections, the attenuation and differential attenuation coefficients (c, d) in Equations 6.34 and 6.35 are needed in the DP and Z-PHI methods for attenuation correction. However, these coefficients vary greatly depending on precipitation microphysics, causing inaccuracy in attenuation. To address this issue, the self-consistent with constraint (SCWC) method was introduced by

Bringi et al. (2001). Instead of using the prefixed coefficients c and d, the coefficients are found through optimization, in which the difference between the reconstructed $\phi_{DP}^{(e)}(r,c)$ and measured $\phi_{DP}^{(m)}(r)$ is minimized. This is done by estimating PIA_H in Equation 6.43 from $\Delta\phi_{DP}^{(m)}$, that is,

$$a(c) = \frac{1 - \exp\left[0.23b \times c\Delta\phi_{DP}^{(m)}\right]}{0.46b \int_0^{r_N} Z_h'^b(\ell)d\ell} \tag{6.44}$$

and reconstructing the differential phase of $\phi_{DP}^{(e)}(r,c)$ as

$$\phi_{DP}^{(e)}(r,c) = \int_0^r \frac{aZ_h^b(\ell,c)}{c}d\ell. \tag{6.45}$$

Then, the total absolute difference

$$\chi = \sum_{n=1}^N \left| \phi_{DP}^{(e)}(r_n,c) - \phi_{DP}^{(m)}(r_n) \right| \tag{6.46}$$

is minimized to find the optimal coefficient c_{opt}, which can vary from ray to ray. The same minimization procedure can be applied to find d_{opt} to correct the differential reflectivity.

Yet another way to obtain d_{opt} is to use the attenuation-corrected Z_H and a linear relation between the intrinsic values of Z_H and Z_{DR} to determine $Z_{DR}(r_N)$ (Park et al. 2005). Then, the optimal d_{opt} is estimated from

$$d_{opt} = \frac{Z_{DR}(r_N) - Z_{DR}'(r_N)}{\phi_{DP}^{(m)}(r_N) - \phi_{DP}^{(m)}(0)}. \tag{6.47}$$

Once the optimal coefficients c_{opt} and d_{opt} are found, the total attenuation PIA_H and differential attenuation PIA_{DP} can be obtained as well as the attenuation-corrected reflectivity and differential reflectivity.

6.5.1.4 Dual-Frequency Method

As noted earlier, rain attenuation is negligible for S-band radar measurement. We also noted earlier that the national WSR-88D network, which covers the United States, operates at S-band. *PIA* at C-band or X-band can be estimated from the dual-frequency ratio (DFR) or dual-wavelength ratio (DWR), defined as the difference between the S- and C-/X-band horizontal reflectivity (Tuttle and Rinehart 1983). The DF method can also be applied jointly with the Z-PHI method by using PLA_H at the end of a ray as a constraint. The specific form used is that from Zhang et al. (2004a), termed the *adjusted Hitschfeld–Bordan method*. The DF method is limited in its application because it requires (1) radar data from systems having two different frequencies, (2) data free of hail contamination, and (3) the assumption of the same intrinsic reflectivity at the two frequencies. Otherwise, a more complicated approach is needed to correct the attenuation (see Chapter 7 for discussion).

6.5.2 Example Results

Attenuation correction for reflectivity measurements at X-band using the methods mentioned above was performed in Snyder et al. (2010). The results are shown in Figures 6.22 and 6.23. The weather event being measured was a supercell that occurred at 0055:37 UTC on May 30, 2004, and was observed by UMass X-pol and polarimetric KOUN S-band radars. Figure 6.22a and b illustrate the X-pol-measured differential phase and reflectivity. The rest of the panels are attenuation-corrected reflectivity fields using the DP, Z-PHI, SCWC, and DF methods, respectively. As expected, the attenuation-corrected reflectivity fields appear more reasonable, and there is no more decay as the range increases.

To quantify the performance, the S-band reflectivity measurement was used as the reference, and the mean absolute error (MAE), bias, and bias-corrected MAE (BCMAE) were calculated for different upper-reflectivity thresholds and shown in Table 4 of Snyder et al. (2010). Whereas the SCWC provides attenuation-corrected reflectivity with a slightly smaller bias, the bias-corrected errors are about the same for the three (DP, Z-PHI, and SCWC) methods.

Figure 6.23 shows the attenuation correction result for differential reflectivity. The raw differential reflectivity measurement Z'_{DR} is shown in Figure 6.23a. Figure 6.23b through d displays the attenuation-corrected differential reflectivity Z_{DR} with the DP, Z-PHI, and SCWC methods, respectively. After the corrections, there is no apparent trend in Z_{DR} with range except for that with the Z-PHI method, which underestimates differential attenuation for this case.

In this section, basic and simple attenuation correction methods based on empirical relations are described. They work well for pure rain regions and for rough estimation. When melting hail/snow is present, however, the non-Rayleigh scattering (resonance) and backscattering phase difference become important, yielding a nonmonotonic increase in ϕ_{DP} and with large fluctuation errors due to the reduced co-polar correlation coefficient. In that case, further research is required. The hail/melting snow contaminated regions should be excluded using a ρ_{hv} threshold or a classification result, and then a multiparameter optimization scheme should be used to jointly minimize the difference between the analyzed and measured Z_H, Z_{DR}, and ϕ_{DP} to achieve optimal attenuation correction results (see Chapter 7 for discussion).

APPENDIX 6A: RAIN ESTIMATION BASED ON BAYES' THEOREM

Let rain state variable x and radar variable y follow a joint Gaussian distribution with a PDF of

$$p(x,y) = \frac{1}{2\pi \cdot \sigma_x \sigma_y \sqrt{1-\rho_{xy}^2}}$$

$$\times \exp\left(-\frac{1}{2(1-\rho^2)}\left[\frac{(x-\mu_x)^2}{\sigma_x^2} - \frac{2\rho(x-\mu_x)(y-\mu_y)}{\sigma_x\sigma_y} + \frac{(y-\mu_y)^2}{\sigma_y^2}\right]\right), \quad (6A.1)$$

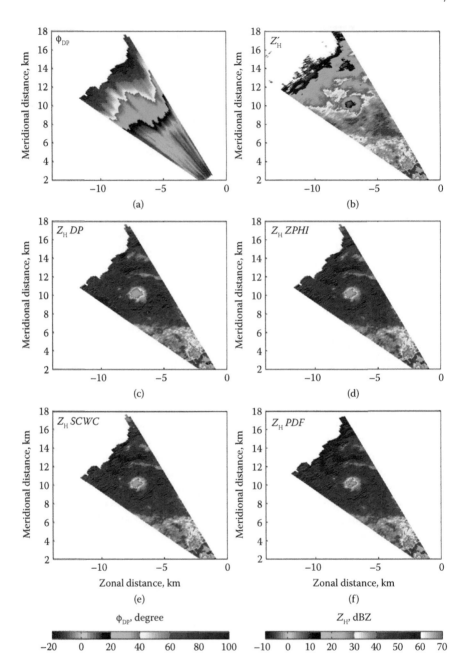

FIGURE 6.22 UMass X-Pol (a) ϕ_{DP}, (b) Z'_H, and attenuation-corrected Z_H using the (c) dual-polarization, (d) Z-PHI, (e) self-consistent with constraint, and (f) dual frequency techniques from 0055:37 UTC 5/30/2004. (From Snyder, J. C., et al., 2010. *Journal of Atmospheric and Oceanic Technology*, 27, 1979–2001.)

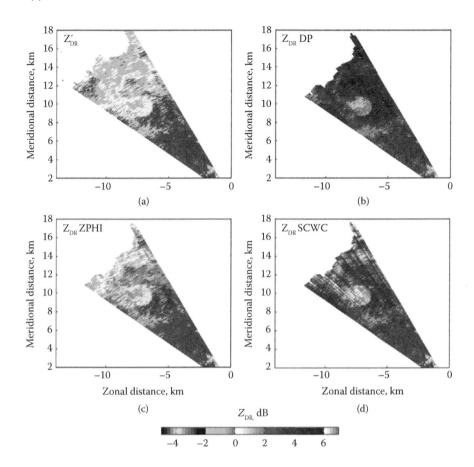

FIGURE 6.23 Same dataset as in Figure 6.22, but for differential reflectivity correction: (a) Z'_{DR} and attenuation-corrected Z_{DR} using the (b) dual-polarization, (c) Z-PHI, and (d) self-consistent with constraint methods. (From Snyder, J. C., et al., 2010. *Journal of Atmospheric and Oceanic Technology*, 27, 1979–2001.)

where (μ_x, μ_y), (σ_x, σ_y), and ρ_{xy} are their means, standard deviations, and correlation coefficient, respectively.

From Bayes' theorem (Papoulis 1991), we obtain the PDF of x under the condition of measurement y as

$$p(x|y) = \frac{p(x,y)}{p(y)} = \frac{1}{\sqrt{2\pi}\sigma_{x|y}} \exp\left(-\frac{(x-\mu_{x|y})^2}{2\sigma_{x|y}^2}\right), \quad (6A.2)$$

where

$$\mu_{x|y} = \mu_x + \rho_{xy}\frac{\sigma_x}{\sigma_y}(y - \mu_y) \quad (6A.3)$$

and

$$\sigma_{x|y} = \sigma_x \left(1 - \rho_{xy}^2\right)^{1/2}. \tag{6A.4}$$

Equation 6A.3 can be used to derive the R–Z relation. Let $x = \ln(R)$ and $y = \ln(Z)$; Equations 6A.3 and 6A.4 become

$$\mu_{\ln(R)|\ln(Z)} = \mu_{\ln(R)} + \rho_{\ln(R),\ln(Z)} \frac{\sigma_{\ln(R)}}{\sigma_{\ln(Z)}}\left(\ln(Z) - \mu_{\ln(Z)}\right) \tag{6A.5}$$

$$\sigma_{\ln(R)|\ln(Z)} = \sigma_{\ln(R)}\left(1 - \rho_{\ln(R),\ln(Z)}^2\right)^{1/2} \tag{6A.6}$$

We consider the local mean as the logarithm of the rain estimate $\mu_{\ln(R)|\ln(Z)} = \ln\left(\hat{R}\right)$ and obtain

$$
\hat{R} = \exp\left(\mu_{\ln(R)} + \rho_{\ln(R),\ln(Z)} \frac{\sigma_{\ln(R)}}{\sigma_{\ln(Z)}}\left(\ln(Z) - \mu_{\ln(Z)}\right) + \frac{\sigma_{\ln(R)}^2}{2}\left(1 - \rho_{\ln(R),\ln(Z)}^2\right)\right)
$$

$$
= \exp\left(\mu_{\ln(R)} - \rho_{\ln(R),\ln(Z)} \frac{\sigma_{\ln(R)}}{\sigma_{\ln(Z)}}\mu_{\ln(Z)} + \frac{\sigma_{\ln(R)}^2}{2}\left(1 - \rho_{\ln(R),\ln(Z)}^2\right)\right) Z^{\rho_{\ln(R),\ln(Z)} \frac{\sigma_{\ln(R)}}{\sigma_{\ln(Z)}}} \tag{6A.7}
$$

$$
= aZ^b,
$$

where

$$a = \exp\left(\mu_{\ln(R)} - \rho_{\ln(R),\ln(Z)} \frac{\sigma_{\ln(R)}}{\sigma_{\ln(Z)}}\mu_{\ln(Z)} + \frac{\sigma_{\ln(R)}^2}{2}\left(1 - \rho_{\ln(R),\ln(Z)}^2\right)\right) \tag{6A.8}$$

$$b = \rho_{\ln(R),\ln(Z)} \frac{\sigma_{\ln(R)}}{\sigma_{\ln(Z)}}. \tag{6A.9}$$

Using a similar procedure, we can derive the rain estimators for the reflectivity and differential reflectivity measurements. We let $x = \ln(R)$, $y = \ln(Z)$, and $z = \ln(Z_{dr})$, and we derive the rain estimator as follows:

$$\hat{R} = \exp\left(\mu_{\ln(R)} + b\left(\ln(Z) - \mu_{\ln(Z)}\right) + c\left(\ln(Z_{dr}) - \mu_{\ln(Z_{dr})}\right) + \frac{\sigma_{\ln(R)|\ln(Z),\ln(Z_{dr})}^2}{2}\right) \tag{6A.10}$$

$$= aZ^b Z_{dr}^c,$$

where

$$a = \exp\left(\mu_{\ln(R)} - b_Z\mu_{\ln(Z)} - b_D\mu_{\ln(Z_{dr})} + \frac{\sigma_{\ln(R)|\ln(Z),\ln(Z_{DR})}^2}{2}\right) \tag{6A.11}$$

$$b = \frac{\sigma_{\ln(R)}}{\sigma_{\ln(Z)}} \frac{\left(\rho_{\ln(R),\ln(Z)} - \rho_{\ln(R),\ln(Z_{dr})}\rho_{\ln(Z),\ln(Z_{dr})}\right)}{\left(1 - \rho^2_{\ln(Z),\ln(Z_{dr})}\right)} \tag{6A.12}$$

$$c = \frac{\sigma_{\ln(R)}}{\sigma_{\ln(Z_{dr})}} \frac{\left(\rho_{\ln(R),\ln(Z_{dr})} - \rho_{\ln(R),\ln(Z)}\rho_{\ln(Z),\ln(Z_{dr})}\right)}{\left(1 - \rho^2_{\ln(Z),\ln(Z_{dr})}\right)} \tag{6A.13}$$

and the standard deviation of the estimate is

$\sigma_{\ln(R)|\ln(Z),\ln(Z_{DR})}$

$$= \sigma_{\ln(R)} \left(\frac{1 + 2\rho_{\ln(R),\ln(Z)}\rho_{\ln(R),\ln(Z_{dr})}\rho_{\ln(Z),\ln(Z_{dr})} - \rho^2_{\ln(R),\ln(Z)} - \rho^2_{\ln(R),\ln(Z_{dr})} - \rho^2_{\ln(Z),\ln(Z_{dr})}}{1 - \rho^2_{\ln(Z),\ln(Z_{dr})}} \right)^{1/2}. \tag{6A.14}$$

APPENDIX 6B: EQUIVALENCE AMONG CONSTRAINED GAMMA DSD MODELS

As discussed in Section 6.3, a two-parameter rain DSD model is needed to facilitate an observation-based retrieval because both the network of dual-polarization WSR-88D radars and the Global Precipitation Measurement dual-frequency radar each provide two independent measurements. Since the $\mu - \Lambda$ relation was introduced to reduce the three parameters of the gamma distribution model to the two-parameters of the C–G model for rain DSDs (Zhang et al. 2001), there have been many papers, including Williams et al. (2014), publishing new constraints and debating the validity of the C-G model in remote sensing applications. However, the C-G DSD models with different constraints are essentially equivalent, shown as follows.

For a gamma DSD of $N(D) = N_0 D^\mu \exp(-\Lambda D)$, the nth moment of the DSD (ignoring truncation effects for simplicity) is $M_n = \int_0^\infty D^n N(D) dD = N_0 \Lambda^{-(\mu+n+1)} \Gamma(\mu+n+1)$. A general mean characteristic size can be defined as a ratio between the $(n + 1)$th and the nth moments as follows:

$$D_n \equiv M_{n+1}/M_n = (\mu + n + 1)/\Lambda. \tag{6B.1}$$

The standard deviation (or spectrum width) of the distribution $p_n(D) = D^n N(D)/M_n =$ is

$$\sigma_n \equiv \left[\int_0^\infty (D - D_n)^2 D^n N(D) dD / M_n \right]^{1/2} = (\mu+n+1)^{1/2}/\Lambda. \tag{6B.2}$$

If the σ_n and D_n are related by

$$\sigma_n = aD_n^b, \tag{6B.3}$$

then a relation between μ and Λ can be derived by substituting Equations 6B.1 and 6B.2 into Equation 6B.3. After simplification, we obtain

$$\mu = a^{\frac{-2}{2b-1}} \Lambda^{\frac{2(b-1)}{2b-1}} - (n+1).$$
(6B.4)

This means that the width–size ($\sigma_n - D_n$) relation (6B.3) is equivalent to the shape–slope ($\mu - \Lambda$) relation (Equation 6B.4) for a gamma distribution. Both serve as a constraint of a gamma DSD and reduce the three parameters of the gamma distribution to two parameters.

In the case of mass-weighted mean diameter D_m for a rain DSD, we have $n = 3$ in Equations 6B.1 through 6B.4, yielding $D_m \equiv D_3 = (\mu + 4)/\Lambda$ and $\sigma_m = (\mu + 4)^{1/2}/\Lambda$. Therefore, the $\sigma_m - D_m$ relation of $\sigma_m = aD_m^b$ (Williams et al. 2014) becomes

$$\mu = a^{\frac{-2}{2b-1}} \Lambda^{\frac{2(b-1)}{2b-1}} - 4,$$
(6B.5)

which is essentially a μ–Λ relation, just in a different functional form, compared with the previously introduced μ–Λ relations in quadratic form (Cao et al. 2008; Zhang et al. 2001, 2003).

Problems

6.1 Describe the polarimetric radar signatures of a melting layer in terms of Z_H, Z_{DR}, and ρ_{hv} and compare these with those of dry snow and rain. Explain why there are these differences based on the cloud/precipitation physics discussed in Chapter 2 and the wave scattering theory covered in Chapter 3.

6.2 Download polarimetric radar data to perform echo classification. Use the three measurements of Z_H, Z_{DR}, and ρ_{hv} as the input and the parameters of the membership functions in Table 6.1 to categorize the radar echo into the ten classes: GC/AP, BS, DS, WS, CR, GR, BD, RA, HR, and RH. There are some restrictions that can be used in classification according to the location of the melting layer. For the region below the melting layer, only the classes GC/AP, BS, BD, RA, HR, and HA are allowed. For the region above the melting layer, only DS, CR, GR, and HA are allowed. For the region within the melting layer, only GC/AP, BS, DS, WS, GR, BD, and HA are allowed. Plot the radar measurements and the classification results. Check and interpret the classification using your common sense knowledge of storm physics.

6.3 Use the rain rate (R) and reflectivity (Z_h) values you calculated in Problem 4.4b from DSD data to derive the radar rain estimators by fitting the data to the power-law relations of $R = aZ^b$ and $Z = cR^d$, respectively, with a least square fit in the logarithmic domain. Analyze and discuss the difference between the two approaches. In addition, derive an R(Z) relation using Equations 6A.7, 6A.8, and 6A.9 provided in

Appendix 6A based on the Bayes theorem and compare it with your least square fit results. Explain why there are differences among different approaches and discuss how to accurately derive a rain estimator from real data.

6.4 Perform rain DSD retrieval and use the KOUN radar data for Z_H and Z_{DR} from Problem 4.5c to retrieve rain DSDs

    a.   Assuming the DSD follows the M–P DSD model, solve for the parameter Λ from the observed Z_H.

    b.   Assuming the DSD follows the C-G DSD model, solve for the parameters N_0 and Λ from the observed Z_H and Z_{DR}.

    c.   Calculate the total number concentration (N_t), rainfall rate (R), and mass-weighted mean diameter (D_m) from both the DSD data provided in Problem 2.4 and the retrieved DSD parameters.

    d.   Plot the retrieved N_t, R, and D_m for both the M–P and C-G DSD models along with those calculated from the DSD data. Compare the results and discuss your findings.

    e.   Perform the DSD retrieval for the data provided in Problem 6.2. Plot and analyze the results of N_0, Λ, R, and D_m. Compare the rain rate with that obtained using the empirical relations in Equations 6.9 and 6.10.

6.5 Use the retrieved DSD parameters that you obtained in Problem 6.4e for S-band and calculate Z_H, Z_{DR}, A_H, and A_{DP} at X-band. Simulate measured reflectivity and differential reflectivity Z_H' and Z_{DR}' by including the PIA and differential attenuation, PIA_H and PIA_{DP}. Plot the Z_H' and Z_{DR}' images and compare them with the S-band measurements.

7 Advanced Methods and Optimal Retrievals

In Chapter 6, we described the simple and common applications of PRD in weather observation and quantification, in which measurement errors and error structures, spatial and temporal information, and NWP model constraints were not optimally utilized. This chapter introduces advanced methods for optimally retrieving cloud and precipitation microphysics for weather quantification and forecasting. It describes simultaneous attenuation correction and DSD retrieval, the statistical retrieval of rain DSDs, variational analysis, and the challenges and potential of assimilating PRD into NWP models to improve weather forecasts.

7.1 SIMULTANEOUS ATTENUATION CORRECTION AND DSD RETRIEVAL

As discussed in Sections 6.3 and 6.4, rain estimation and DSD retrieval are successful for S-band nonattenuated PRD. PRD at higher frequencies, however, suffer substantial attenuation, and attenuation correction is needed before meteorologists can correctly interpret/use the data, as described in Section 6.5. Because rain attenuation depends on the rain microphysics, the parameters/coefficients (b,c,d) used in the attenuation correction procedures also depend on rain DSDs. Therefore, it is more natural and better to simultaneously correct attenuation while retrieving DSDs, a method that corrects the attenuation problems in both dual-polarization and dual-frequency radar measurements. Recent advancements also necessitate this simultaneous correction and retrieval: the WSR-88D network has been upgraded with dual-polarization capability, and NASA has just launched the Global Precipitation Measurement (GPM) satellite with dual-frequency precipitation radar.[*] Attenuation correction and DSD retrieval with dual-pol measurements and with dual frequency share the same methodology, described in this section.

7.1.1 ANALOGY BETWEEN DUAL-POLARIZATION AND DUAL-FREQUENCY RADAR TECHNIQUES

Figure 7.1 shows examples of ray plots of dual-polarization and dual-frequency radar measurements. The left panels show dual-polarization measurements of reflectivity at horizontal and vertical polarizations (Figure 7.1a) and differential reflectivity (Figure 7.1c); the right panels show dual-frequency reflectivities (Figure 7.1b) and the DWR (Figure 7.1d). The measurements should be understood as the estimates of the radar variables, with the prime symbol (′) indicating attenuation/propagation-included measurements.

---

* http://www.nasa.gov/mission_pages/GPM/main/index.html

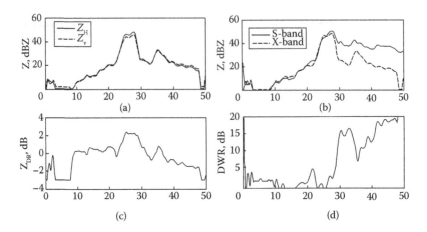

FIGURE 7.1 Examples of ray plots of (a, c) dual-polarization and (b, d) dual-frequency radar measurements.

In either dual-polarization or dual-frequency radar retrieval from rain DSDs, two radar measurements of either (Z_H, Z_{DR}) or (Z, DWR) are used to retrieve two-parameter (N_0, Λ) DSDs. In the case of no attenuation, dual-pol DSD retrieval is very straightforward, solving the size parameter Λ from Z_{DR} and the concentration parameter N_0 from Z_H, as described in Sections 6.4 and 6.5. In cases where attenuation is present, however, attenuation and differential attenuation need to be corrected before the DSD retrieval can be applied. Because the attenuations depend on the cloud/precipitation microphysics, attenuation estimation and correction rely on DSD information. Hence, attenuation correction and DSD retrieval need to be performed simultaneously. This simultaneous attenuation and DSD retrieval applies to both dual-pol and dual-frequency methods (Meneghini and Liao 2007). Here, we focus on the retrieval from PRD.

7.1.2 FORMULATION: INTEGRAL EQUATION METHOD

Assume that rain DSDs are represented by the constrained gamma model in Equations 6.16 and 6.17, in which there are only two free DSD parameters: N_0 and Λ. Using the C-G model in the expression of the attenuation-included reflectivity, we have from Equation 6.39

$$Z'_{h,v}(r)$$

$$= Z_{h,v}(r)e^{-0.46\int_{r1}^{r} A_{H,V}(\ell)d\ell}$$

$$= \frac{4\lambda^4}{\pi^4 |K_w|^2} \int |s_{hh}(\pi,D)|^2 N_0 D^\mu \exp(-\Lambda D)dDe^{-0.46\int_{r1}^{r}\left\{8.686\lambda \int Im[s_{hh,vv}(0,D)]N_0 D^\mu \exp(-\Lambda D)dD\right\}d\ell}$$

$$= \frac{4\lambda^4}{\pi^4 |K_w|^2} N_0(r) \int |s_{hh}(\pi,D)|^2 D^\mu \exp(-\Lambda D)dDe^{-0.46\int_{r1}^{r} N_0(\ell)\left\{8.686\lambda \int Im[s_{hh,vv}(0,D)]D^\mu \exp(-\Lambda D)dD\right\}d\ell}$$

$$= C_Z N_0(r) I_{Z_h,Z_v}(\Lambda)e^{-0.46\int_{r1}^{r} N_0(\ell)I_{AH,AV}(\ell)d\ell},$$

$$(7.1)$$

where

$$C_Z = \frac{4\lambda^4}{\pi^4 |K_w|^2} \tag{7.2}$$

$$I_{Z_h,Z_v}(\Lambda) = \int |s_{hh,vv}(\pi,D)|^2 D^\mu \exp(-\Lambda D)dD \tag{7.3}$$

$$I_{A_H,A_V}(\Lambda) = 8.686\lambda \int \text{Im}[s_{hh,vv}(0,D)]D^\mu \exp(-\Lambda D)dD. \tag{7.4}$$

Writing the attenuation-included reflectivity (Equation 7.1) in decibel form for horizontal polarization, we have

$$Z_H'(r) = 10\log\{C_Z N_0(r)I_{Z_h}[\Lambda(r)]\} - 2\int_{r_1}^r N_0(\ell)I_{A_H}[\Lambda(\ell)]d\ell. \tag{7.5}$$

Taking the difference between the horizontal reflectivity and vertical reflectivity yields the attenuation-included differential reflectivity:

$$Z_{DR}'(r) = 10\log\{I_{Z_h}[\Lambda(r)]/I_{Z_v}[\Lambda(r)]\} \tag{7.6}$$
$$- 2\int_{r_1}^r N_0(\ell)\{I_{A_H}[\Lambda(\ell)] - I_{A_H}[\Lambda(\ell)]\}d\ell.$$

Thus, Equations 7.5 and 7.6 constitute the integral equation formulation for simultaneous attenuation correction and DSD retrieval. For a ray of N range gates, there are N measurements of (Z_H', Z_{DR}') and N pairs of DSD parameters (N_0, Λ). Hence the $2N$ DSD parameters can be solved from the $2N$ measurements/equations with the C-G DSD model. Equations 7.5 and 7.6 can be solved with recursion methods as follows:

1. *Forward recursion:* The solution starts from the initial gate of $r = r_1$ and works toward the last gate, $r = r_N$. It includes the following steps:
   a. At the initial gate of $r = r_1$, the PIA can be ignored. Hence, the second terms in Equations 7.5 and 7.6 disappear. $\Lambda(r_1)$ is solved from $Z_{DR}'(r_1)$ with Equation 7.6, and $N_0(r_1)$ is obtained from $Z_H'(r_1)$ using Equation 7.5 in the same procedure described in Section 6.4.
   b. Once (N_0, Λ) are known for the first gate, the attenuation terms for the measurements at the second gate are calculated and included in Equations 7.5 and 7.6 to solve for (N_0, Λ) from (Z_H', Z_{DR}') at the second gate.
   c. Repeat (b) until the last gate, $r = r_N$.

2. *Backward recursion:* The solution starts from the last gate of $r = r_N$ and works toward the initial gate, $r = r_1$. The PIA_H and PIA_{DP} are estimated and included in Equations 7.5 and 7.6, yielding

$$Z_H'(r) = 10\log\{C_Z N_0(r)I_{Z_h}[\Lambda(r)]\} - PIA_H + 2\int_r^{r_N} N_0(\ell)I_{A_H}[\Lambda(\ell)]d\ell \tag{7.7}$$

$$Z'_{DR}(r) = 10\log\left\{I_{Z_h}[\Lambda(r)]/I_{Z_v}[\Lambda(r)]\right\}$$

$$- PIA_{DP} + 2\int_r^{r_N} N_0(\ell)\left\{I_{A_H}[\Lambda(\ell)] - I_{A_H}[\Lambda(\ell)]\right\}d\ell. \tag{7.8}$$

a. Starting from the last gate, $r = r_N$, the attenuation within the gate can be ignored. $\Lambda(r_N)$ is solved from $Z'_{DR}(r_N)$ with Equation 7.8, and $N_0(r_N)$ is obtained from $Z'_H(r_N)$ using Equation 7.7.

b. Use $N_0(r_N)$ and $\Lambda(r_N)$ to calculate the attenuation terms for the last gate (the last terms in Equations 7.7 and 7.8) and then solve for $N_0(r_{N-1})$ and $\Lambda(r_{N-1})$ from $Z'_H(r_N)$ and $Z'_{DR}(r_N)$ at the gate, r_{N-1}.

c. Repeat (b) until the initial gate, $r = r_1$.

In principle, both the forward and backward recursions allow for DSD retrieval. In practice, however, the forward recursion tends to be *unstable* due to errors and error accumulation. The backward recursion is *stable* because the PIA_H and PIA_{DP} constraints are used. Hence, the backward recursion is normally used in practical applications.

7.1.3 EXAMPLE OF A SIMULTANEOUS RETRIEVAL

To test the performance of the integral equation formulation discussed in Section 7.1.2, X-band dual-polarization measurements of rain are simulated, and the simulated data are used to simultaneously estimate attenuation and retrieval rain DSDs and microphysical parameters, as shown in Figure 7.2.

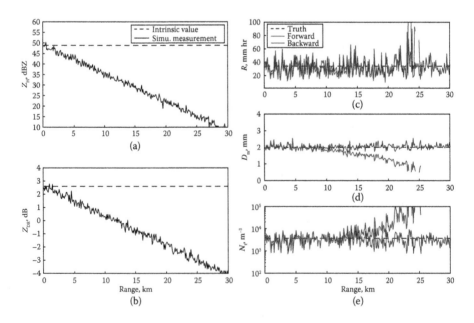

FIGURE 7.2 Simulated X-band polarimetric radar measurements (a, b) and retrievals (c, d, e). Input parameters are $N_0 = 8000$ m⁻³mm⁻¹ and $\Lambda = 2.0$ mm⁻¹.

A ray with a constant rain DSD of $N_0 = 8000$ m^{-3}mm^{-1} and $\Lambda = 2.0$ mm^{-1} is assumed. Then, true rain microphysical parameters and intrinsic radar variables are calculated to obtain $R = 34.4$ mm hr^{-1}, $D_m = 2.0$ mm, $Z_H = 48.8$ dBZ, and $Z_{DR} = 2.60$ dB. PIA_H and PIA_{DP} are then calculated and added to the radar variables along with 1 dB error for Z_H and 0.2 dB error for Z_{DR}, to obtain the simulated radar measurements shown in the left column of Figure 7.2. The decreasing trend is due to the attenuation for Z_H and differential attenuation for Z_{DR}, whereas the random fluctuations are due to the added errors. The retrieval results using both the forward and backward recursion procedures are shown in the right column of Figure 7.2, along with the truth. It is clear that the backward recursion gives stable results, whereas the results of the forward recursion diverge and do not even yield valid retrievals after range $r > 25$ km when the PIA is significantly large ($PIA_H > 30$ dB).

In this section, the method of simultaneous attenuation and DSD retrieval is discussed. This method should be investigated further for wide applications in rain DSD retrieval from PRD. By taking DSD variation into account during the retrieval/correction procedure, this method is more accurate than separate attenuation correction when DSD model and measurement errors are present. Dual-pol and dual-freq techniques share this same methodology and similar equations.

7.2 STATISTICAL RETRIEVAL OF RAIN DSDs

The DSD retrieval discussed in Section 6.4 is a deterministic retrieval and shows the potential to improve QPE and microphysics parameterization. A deterministic retrieval, however, is limited in handling measurement/model errors, cannot use the prior statistical information of DSD data except for DSD model constraints such as a $\mu - \Lambda$ relation, and does not quantify the performance of the retrieval. For example, if the measurement of Z_{DR} is close to zero or is negative due to a measurement error, there is no solution for Λ from Equation 6.18, and it is therefore impossible to retrieve a DSD. This is shown graphically in Figure 6.14a, in which the $Z_{DR} - \Lambda$ line does not cross the x-axis. Another disadvantage of the deterministic retrieval is that it does not provide information about the error covariance of the retrieval, which is important to know when we want to optimally utilize the retrieval results. To overcome these deficiencies, a statistical retrieval based on Bayesian theory was introduced in Cao et al. (2010).

7.2.1 STATISTICAL RETRIEVAL METHOD: BAYESIAN APPROACH

A statistical retrieval treats both the rain state (DSD) parameters and polarimetric radar measurements as random variables that are characterized by PDFs. Let x be a state vector, representing DSD parameters, and let y be a measurement vector, representing PRD. The retrieval is to find the PDF of x in the posterior condition of the measurement vector of y. Based on Bayes' theorem, the posterior conditional PDF $p_{post}(\mathbf{x}|\mathbf{y})$ is given by

$$p_{post}\left(\mathbf{x}|\mathbf{y}\right) = \frac{p_f\left(\mathbf{y}|\mathbf{x}\right) \cdot p_{pr}\left(\mathbf{x}\right)}{\int p_f\left(\mathbf{y}|\mathbf{x}\right) \cdot p_{pr}\left(\mathbf{x}\right) \cdot d\mathbf{x}}, \tag{7.9}$$

where $p_{pr}(\mathbf{x})$ is the prior PDF of state x and $p_f(\mathbf{x}|\mathbf{y})$ is the forward conditional PDF of the observation y, given state x. Once the posterior conditional PDF of the state vector is found for a given observation y, the expected value $\langle \mathbf{x} \rangle$ and the standard deviation σ_x are then calculated by integrating over the entire range of state x:

$$\langle \mathbf{x} \rangle = \frac{\int \mathbf{x} \cdot p_f(\mathbf{y}|\mathbf{x}) \cdot p_{pr}(\mathbf{x}) \cdot d\mathbf{x}}{\int p_f(\mathbf{y}|\mathbf{x}) \cdot p_{pr}(\mathbf{x}) \cdot d\mathbf{x}}, \tag{7.10}$$

$$\sigma_x = \sqrt{\frac{\int (\mathbf{x} - <\mathbf{x}>)^2 \cdot P_f(\mathbf{y}|\mathbf{x}) \cdot P_{pr}(\mathbf{x}) \cdot d\mathbf{x}}{\int P_f(\mathbf{y}|\mathbf{x}) \cdot P_{pr}(\mathbf{x}) \cdot d\mathbf{x}}}, \tag{7.11}$$

As in the deterministic retrieval, state variables are DSD parameters and measurements are reflectivity and differential reflectivity. Hence, we have $\mathbf{x} = [N_0', \Lambda']^t$ and $\mathbf{y} = [Z_H, Z_{DR}]^t$. The DSD parameters are transformed to $N_0' = \log_{10} N_0$ and $\Lambda' = \Lambda^{0.25}$ so that they are more Gaussian-distributed.

7.2.2 PRIOR DISTRIBUTION OF DSD PARAMETERS

As shown in Equation 7.9, the prior PDF of the state vector (DSD parameters) and the forward conditional PDF of the observation vector (PRD) for a given state vector are required to obtain/retrieve the posterior conditional PDF of the state parameters. To obtain the prior distribution of the DSD parameters, 2DVD-measured DSDs are fitted to the gamma distribution model using the truncated moment fit (Vivekanandan et al. 2004), which utilizes the second, fourth, and sixth DSD moments.

The distributions of the estimated DSD parameters N_0 and Λ are found to be greatly skewed and to have large dynamic ranges and nonlinear relations with physics parameters. In order to reduce their dynamic ranges and mitigate the nonlinear effects, the DSD parameters are transformed to N_0' and Λ' for use in the Bayesian retrieval. The occurrence frequency of N_0' and Λ' is shown in Figure 7.3a and b. The dynamic ranges of N_0' and Λ' for light rains are reduced significantly so that they are close to being Gaussian-distributed, whereas N_0 and Λ are not; and moderate/heavy rains (e.g., $0 < \Lambda < 3$) are now better represented than they were before. It is evident that most DSDs have an N_0' value between 3 and 5 (i.e., N_0 is about 10^3–10^5 m^{-3} mm^{-1}) and a Λ' value between 1.1 and 1.6 (i.e., Λ is about 1.5–6). Figure 7.3c shows the contour plot of the joint occurrence of the DSD-estimated N_0' and Λ' after discretization of N_0' over the interval 0.1 and of Λ' over the interval 0.05. The joint PDF of N_0' and Λ' is equal to the normalization of this distribution, which has been saved as a lookup table for each discrete (N_0', Λ') pair for efficient computation in Equations 7.9 through 7.11.

7.2.3 THE FORWARD CONDITIONAL DISTRIBUTION

To facilitate the Bayesian retrieval, the forward conditional PDF $p_f(\mathbf{x}|\mathbf{y})$ is also needed to link the state variables with the measurements, as in Equations 7.9 through 7.11. Considering that the sampling errors of radar estimates are generally

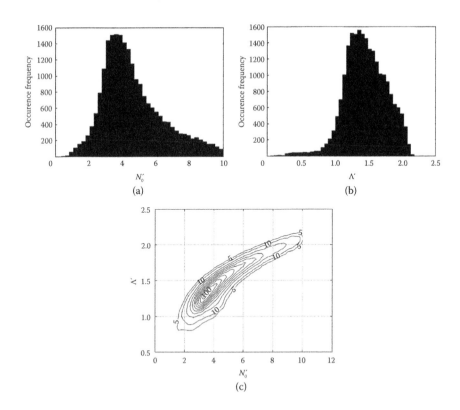

FIGURE 7.3 Occurrences of estimated DSD parameters based on 2DVD data: (a) Occurrence of N'_0, (b) occurrence of Λ', and (c) contour plot of the joint occurrence frequency of the estimated DSD parameters N'_0 and Λ'.

Gaussian-distributed, the forward conditional PDF is assumed to follow a bivariate Gaussian distribution as

$$P_f\left(Z_H, Z_{DR} | \Lambda', N'_0\right) = \frac{1}{2\pi \cdot \sigma_{Z_H} \sigma_{Z_{DR}} \sqrt{1-\rho^2}}$$

$$\times \exp\left(-\frac{1}{2(1-\rho^2)}\left[\frac{\left(Z_H - \langle Z_H \rangle\right)^2}{\sigma_{Z_H}^2} - \frac{2\rho\left(Z_H - \langle Z_H \rangle\right)\left(Z_{DR} - \langle Z_{DR} \rangle\right)}{\sigma_{Z_H}\sigma_{Z_{DR}}} + \frac{\left(Z_{DR} - \langle Z_{DR} \rangle\right)^2}{\sigma_{Z_{DR}}^2}\right]\right),$$

$$(7.12)$$

where ρ denotes the correlation coefficient between Z_H and Z_{DR}. It is typically very low and assumed to be 0, because the measurement errors of Z_H and Z_{DR} are uncorrelated and, although their model errors may be correlated, this correlation is difficult to quantify.

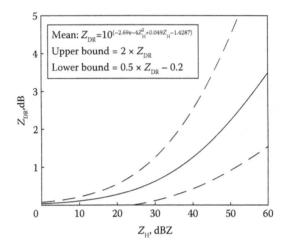

FIGURE 7.4 Sketch of Z_{DR} (dB) vs. Z_H (dBZ) from 2DVD measurements for specifying observation errors. Upper and lower bounds (dashed lines) are given according to the mean curve. (From Cao, Q., et al., 2010. *Journal of Applied Meteorology and Climatology*, 49, 973–990.)

In Equation 7.12, the expected values $(\langle Z_H \rangle, \langle Z_{DR} \rangle)$ of Z_H and Z_{DR} can be calculated from the state (N_0', Λ') using the forward model/operators in Equations 4.59 through 4.62. The standard deviation, however, needs to be specified. Considering that Z_H is generally more reliable than Z_{DR} and its measurement error is generally accepted as 1–2 dB, σ_{Z_H} is assumed to be constant at 2 dB and $\sigma_{Z_{DR}}$ is assumed to be a function of Z_H and Z_{DR}, as shown in Figure 7.4. Within the upper and lower bounds, the Z_{DR} error $\sigma_{Z_{DR}}$ is assumed to be constant at 0.3 dB, mainly from estimation error. If observed PRD fall outside of this region, $\sigma_{Z_{DR}}$ is believed to be contaminated by atypical rain and therefore the error is larger than 0.3 dB, expressed as

$$\sigma_{Z_{DR}} = \begin{cases} 0.3[1+(Z_{DR}-Z_{DR}^{up})] & \text{above} \\ 0.3 & \text{within} \\ 0.3[1+(Z_{DR}^{low}-Z_{DR})] & \text{below.} \end{cases} \tag{7.13}$$

7.2.4 RESULTS AND EVALUATION

The Bayesian approach was tested by simulated radar measurements from DSD data. According to the measured DSDs, Z_H and Z_{DR} were simulated based on Equations 4.59 and 4.62. The simulated Z_H and Z_{DR} were used to retrieve the DSD parameters by applying Equations 7.9 and 7.10. With the retrieved mean values $\langle N_0' \rangle$ and $\langle \Lambda' \rangle$, the mean values of rain integral variables are calculated. Figure 7.5 shows one-to-one plots of retrieved values versus observations. It is evident that most of the retrieved R values are very close to the observations. The retrieved D_m is a little more scattered than the retrieved R, which is likely attributed to the inaccuracy of the

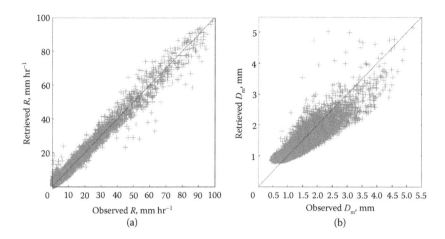

FIGURE 7.5 Scatterplots of retrieved values vs. observations: (a) R (mm hr^{-1}) and (b) D_m (mm). Crosses represent data points, and solid lines represent equal values of axes.

DSD model because some DSDs are not well represented by a gamma distribution. Even those that could be described approximately by a gamma distribution may not be accurately described by the C-G model with the $\mu - \Lambda$ relation (Equation 6.15) used in the retrieval. Bayesian retrieval with the C-G model performs well, giving accurate rain estimations with a bias of less than 8% and a standard deviation lower than 18%.

The retrieval algorithm was applied to real PRD collected by KOUN radar for the same case used in Sections 6.2 and 6.3, as shown in Figure 7.6. Figure 7.6a shows the rain rate R from the Bayesian retrieval, which is similar to the dual-pol empirical QPE results (not shown). The rain rate estimation from the empirical single-polarization estimator using Equation 6.16 is shown in Figure 7.6 for comparison. Although Figure 7.6a and b exhibits similar storm features and rainfall rates within the stratiform region, Figure 7.6b has a higher rainfall rate than Figure 7.6a for the region of strong convection. Figure 7.6c and d displays images of standard deviations of Λ' and N_0' estimates, respectively, for the Bayesian retrieval. Both SD(Λ') and SD(N_0') images have a similar trend, implying that either one can be used as an indicator of retrieval confidence. As expected, the large standard deviations appear in clutter, melting snow/hail, or biological scatterer–contaminated regions and hence cause large uncertainty in the rain retrieval.

To verify the Bayesian retrieval results, the retrieved rainfall rate and 1 hr rain accumulation are compared with in situ measurements, as shown in Figure 7.7. Empirical QPE results are also shown for comparison. Rain rates of in situ measurements include rain gauge measurements at six Mesonet sites and 2DVD measurements at the Kessler Farm Field Laboratory. Thick grey solid lines indicate in situ measurements. Thin solid lines denote radar retrievals using the Bayesian approach. As a reference, dashed lines represent radar retrievals using the empirical estimators of Ryzhkov et al. (2005a). It is evident that both Bayesian retrieval and empirical estimators give reasonable results. The Bayesian approach, however,

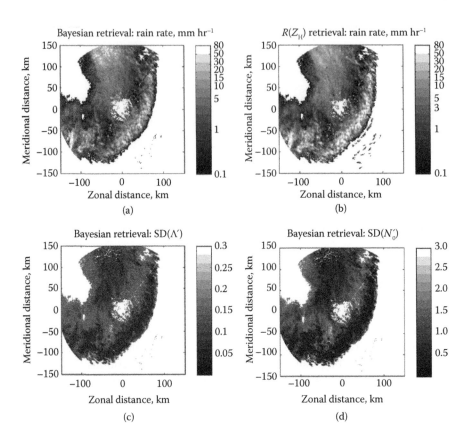

FIGURE 7.6 Retrieval results from radar observations at 0830 UTC on May 13, 2005: (a) R (mm hr^{-1}) from Bayesian retrieval, (b) R (mm/hr) from $R(Z_H)$ retrieval, (c) SD(Λ') (mm$^{-1/4}$) from Bayesian retrieval, and (d) SD(N_0')$\left[\log 10 \left(\text{mm}^{-1-\mu} \text{m}^{-3} \right) \right]$ from Bayesian retrieval.

performs better than the empirical estimators in the region of convective cores (e.g., around 0655 UTC at MINC; around 0815 UTC at SHAW), where radar echoes are often contaminated by the hail and radar-measured Z_H and Z_{DR} are therefore extremely large (e.g., $Z_H > 55$ dBZ and $Z_{DR} > 3.5$ dB).

To quantify their performance, Table 7.1 gives the bias and standard deviation of 1 hr rain accumulation retrievals versus in situ measurements. The empirical single-polarization estimator has the worst result. At the seven sites (six Mesonet and one disdrometer), the Bayesian retrieval generally has both smaller biases (up to approximately 16%) and smaller RMSE (up to approximately 22%) than the empirical estimators, except for that at CHIC.

In summary, the Bayesian approach has the following advantages: (i) both estimates and estimation errors are obtained so that the reliability of the retrieval can be assessed; (ii) measurement and model errors are considered and included in the retrieval; (iii) prior (historical) information of rain microphysics is utilized; and (iv) no fixed empirical estimator is needed.

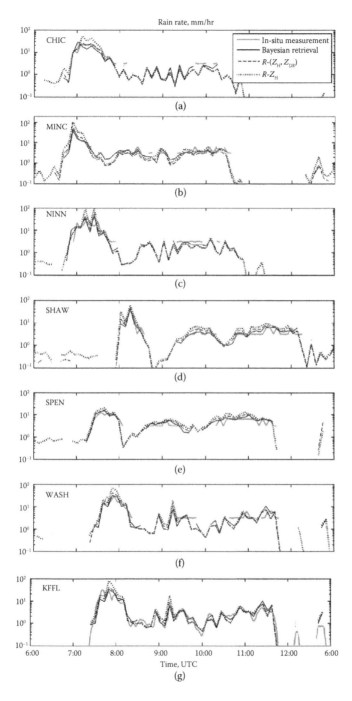

FIGURE 7.7 Comparisons of R (mm/hr) between radar retrievals and in situ measurements at seven sites: (a) CHIC, (b) MINC, (c) NINN, (d) SHAW, (e) SPEN, (f) WASH, and (g) KFFL. (From Cao, Q., et al., 2010. *Journal of Applied Meteorology and Climatology*, 49, 973–990.)

TABLE 7.1

Bias (%) and RMSE (%) of Rain Retrievals (1-hr Rain Accumulation) vs. In Situ Measurements at Sites

Retrieval Type		CHIC	MINC	NINN	SHAW	SPEN	WASH	KFFL
Bayesian	Bias	27.3	2.5	8.3	7.8	15.5	10.7	10.0
	RMSE	21.0	11.5	21.7	14.7	19.6	16.4	15.5
Empirical	Bias	1.1	1.6	12.3	22.4	28.0	19.8	20.2
$R(Z_H, Z_{DR})$	RMSE	17.3	26.4	33.5	29.9	33.9	34.4	33.3
Empirical	Bias	35.4	48.1	48.2	49.5	44.9	48.9	50.4
$R(Z_H)$	RMSE	80.4	103.5	106.8	72.3	56.0	95.3	92.5

7.3   VARIATIONAL RETRIEVAL

The simultaneous attenuation correction and DSD retrieval discussed in Section 7.1 is a deterministic approach; that is, no error effects are taken into account. Hence, multiparameter PRD may not be optimally utilized even though the DSD information is included in attenuation correction. The Bayesian method for DSD retrieval described in Section 7.2 is a statistical method that accounts for error effects, but does not account for attenuation. Neither the Bayesian method nor the simultaneous method includes spatial weather information, and neither is flexible enough to allow for the addition of more observations/constraints. Variational methods (to be described in this section) combine the Bayesian and simultaneous approaches and optimally use all available information/measurements. Those that use observations only are called *observation-based retrievals* (Cao et al. 2013; Hogan 2007), whereas those that include NWP model forecast results in the analysis are called *data assimilation* (DA) *methods* (Kalnay 2003; Xue et al. 2009). DA methods are discussed in Section 7.4. This section will present the variational formulation and then apply it to a two-dimensional variational retrieval for rain DSDs from S-, C-, and X-band PRD of Z_H, Z_{DR}, and Φ_{DP}/K_{DP} to show its applicability.

7.3.1   FORMULATION OF VARIATIONAL RETRIEVAL

7.3.1.1   General Formulation

The variational retrieval is a statistical retrieval like that described in Section 7.2. Furthermore, it takes the spatial covariance of the state variables into account. The variational approach can be formulated from the Bayesian theory (Equation 7.7) based on the assumption of the Gaussian-distributed error models (Kalnay 2003). Assuming that the state vector estimate $\mathbf{x}$ is Gaussian-distributed with a covariance matrix $\mathbf{B}$, the observation vector $\mathbf{y}$ is also Gaussian with an error covariance $\mathbf{R}$, and considering that the denominator is a constant, Equation 7.9 becomes

$$-2\ln[p(\mathbf{x}|\mathbf{y})] = [\mathbf{y} - \mathbf{H}(\mathbf{x})]^t \mathbf{R}^{-1}[\mathbf{y} - \mathbf{H}(\mathbf{x})] + [\mathbf{x} - \mathbf{x}_b]^t \mathbf{B}^{-1}[\mathbf{x} - \mathbf{x}_b] + C, \qquad (7.14)$$

where $\mathbf{x}_b$ is the mean background or the first guess of the state vector, and $\mathbf{H}(\cdots)$ is the observation operator. The superscript t denotes a transposed matrix. An optimal retrieval is to find the state vector with the maximum probability $p(\mathbf{x}|\mathbf{y})$, which is equivalent to finding the state vector such that the cost function

$$J = [\mathbf{x} - \mathbf{x}_b]\mathbf{B}^{-1}[\mathbf{x} - \mathbf{x}_b] + [\mathbf{y} - \mathbf{H}(\mathbf{x})]^t\, \mathbf{R}^{-1}\,[\mathbf{y} - \mathbf{H}(\mathbf{x})] \tag{7.15}$$

is minimized, yielding

$$\hat{\mathbf{x}} = \mathbf{x}_b + [\mathbf{B}^{-1} + \mathbf{H}'^t\mathbf{R}^{-1}\mathbf{H}']^{-1}\mathbf{H}'^t\mathbf{R}^{-1}[\mathbf{y} - \mathbf{H}(\mathbf{x}_b)], \tag{7.16}$$

where $\mathbf{H}'$ represents the Jacobian operator, a matrix containing the partial derivative of forward observation operator $\mathbf{H}$ with respect to each element of the state vector. The variational formulation (Equation 7.16) can be used in either an observation-based retrieval or a model-based analysis (i.e., data assimilation—see discussion in Section 7.4). $\mathbf{x}_b$ represents the model background in data assimilation, but represents an earlier iteration of analysis in observation-based retrieval, and $\hat{\mathbf{x}}$ is solved iteratively. Hence, once the error covariances and the forward operator are defined, Equation 7.16 can be used to find the optimal estimate of the state vector.

7.3.1.2 Formulation for DSD Retrieval from PRD

In the DSD retrieval from the polarimetric radar measurements, the C-G model presented in Section 6.4 is used and $N_0' = \log_{10} N_0$ and Λ are chosen as the two state variables, that is, $\mathbf{x} = [N_0', \Lambda]^t$. (Ideally, $[N_0', \Lambda']$ should be used, as in Section 7.2.) The observation vector is the PRD: $\mathbf{y} = [Z_H', Z_{DR}', K_{DP}]^{t*}$. Hence, we have the cost function $J(\mathbf{x})$ defined as

$$J(\mathbf{x}) = J_b(\mathbf{x}) + J_{Z_H'}(\mathbf{x}) + J_{Z_{DR}'}(\mathbf{x}) + J_{K_{DP}}(\mathbf{x}), \tag{7.17}$$

where

$$J_b(\mathbf{x}) = \frac{1}{2}(\mathbf{x} - \mathbf{x}_b)^t\mathbf{B}^{-1}(\mathbf{x} - \mathbf{x}_b) \tag{7.18}$$

$$J_{Z_H'}(\mathbf{x}) = \frac{1}{2}\left[H_{Z_H'}(\mathbf{x}) - \mathbf{y}_{Z_H'}\right]^t \mathbf{R}_{Z_H'}^{-1}\left[H_{Z_H'}(\mathbf{x}) - \mathbf{y}_{Z_H'}\right] \tag{7.19}$$

$$J_{Z_{DR}'}(\mathbf{x}) = \frac{1}{2}\left[H_{Z_{DR}'}(\mathbf{x}) - \mathbf{y}_{Z_{DR}'}\right]^t \mathbf{R}_{Z_{DR}'}^{-1}\left[H_{Z_{DR}'}(\mathbf{x}) - \mathbf{y}_{Z_{DR}'}\right] \tag{7.20}$$

$$J_{K_{DP}}(\mathbf{x}) = \frac{1}{2}\left[H_{K_{DP}}(\mathbf{x}) - \mathbf{y}_{K_{DP}}\right]^t \mathbf{R}_{K_{DP}}^{-1}\left[H_{K_{DP}}(\mathbf{x}) - \mathbf{y}_{K_{DP}}\right], \tag{7.21}$$

where J_b represents the contribution from the background. $J_{Z_H'}, J_{Z_{DR}'},$ and J_{KDP}, correspond to the measurements of Z_H', Z_{DR}', and K_{DP}, respectively. The subscripts Z_H', Z_{DR}', and K_{DP} are used to denote the terms for the corresponding radar observations.

---

* K_{DP} was used instead of Φ_{DP} for the simplicity of the covariance; however, it would be worthwhile to study the use of Φ_{DP}.

The background error covariance matrix $\mathbf{B}$ is an m-by-m matrix, where m is the size of state vector $\mathbf{x}$ and equal to the number of grids (in the 2D analysis region) times the number of state parameters. To avoid the inversion of such a large matrix as $\mathbf{B}$, a new state variable $\mathbf{v}$ is introduced, written as

$$\mathbf{v} = \mathbf{D}^{-1}\Delta\mathbf{x}, \tag{7.22}$$

with $\Delta\mathbf{x} = \mathbf{x} - \mathbf{x}_b$ and $\mathbf{DD}^{\mathrm{T}} = \mathbf{B}$ (Gao et al. 2004; Parrish and Derber 1992). Δ is the notation of the increment. $\mathbf{D}$ is the square root of matrix $\mathbf{B}$. The cost function is then rewritten as

$$J(\mathbf{v}) = \frac{1}{2}\mathbf{v}^t\mathbf{v} + \frac{1}{2}\left[H_{Z_H}(\mathbf{x}_b + \mathbf{Dv}) - \mathbf{y}_{Z_H}\right]^t \mathbf{R}_{Z_H}^{-1}\left[H_{Z_H}(\mathbf{x}_b + \mathbf{Dv}) - \mathbf{y}_{Z_H}\right]$$

$$+ \frac{1}{2}\left[H_{Z_{DR}}(\mathbf{x}_b + \mathbf{Dv}) - \mathbf{y}_{Z_{DR}}\right]^t \mathbf{R}_{Z_{DR}}^{-1}\left[H_{Z_{DR}}(\mathbf{x}_b + \mathbf{Dv}) - \mathbf{y}_{Z_{DR}}\right] \tag{7.23}$$

$$+ \frac{1}{2}\left[H_{K_{DP}}(\mathbf{x}_b + \mathbf{Dv}) - \mathbf{y}_{K_{DP}}\right]^t \mathbf{R}_{K_{DP}}^{-1}\left[H_{K_{DP}}(\mathbf{x}_b + \mathbf{Dv}) - \mathbf{y}_{K_{DP}}\right].$$

This way, the minimization of cost function J is achieved by making use of the cost function gradient $\nabla_v J$, which is given by

$$\nabla_v J = \mathbf{v} + \mathbf{D}^t\mathbf{H}_{Z_H}^t\mathbf{R}_{Z_H}^{-1}\left(\mathbf{H}_{Z_H}\mathbf{Dv} - \mathbf{d}_{Z_H}\right) + \mathbf{D}^t\mathbf{H}_{Z_{DR}}^t\mathbf{R}_{Z_{DR}}^{-1}\left(\mathbf{H}_{Z_{DR}}\mathbf{Dv} - \mathbf{d}_{Z_{DR}}\right)$$

$$+ \mathbf{D}^t\mathbf{H}_{K_{DP}}^t\mathbf{R}_{K_{DP}}^{-1}\left(\mathbf{H}_{K_{DP}}\mathbf{Dv} - \mathbf{d}_{K_{DP}}\right). \tag{7.24}$$

Here, $\mathbf{d}$ is the innovation vector of the observations (Gao et al. 2004), that is, $\mathbf{d} = \mathbf{y} - \mathbf{H}(\mathbf{x}_b)$.

The spatial influence of the observation is determined by the background error covariance matrix $\mathbf{B}$. Huang (2000) showed that the element b_{ij} of matrix $\mathbf{B}$ could be modeled using a Gaussian correlation model:

$$b_{ij} = \sigma_b^2 \exp\left[-\frac{1}{2}\left(\frac{r_{ij}}{r_L}\right)^2\right], \tag{7.25}$$

where the subscripts i and j denote two grid points in the analysis space and σ_b^2 is the background error covariance; r_{ij} is the distance between the ith and jth grid points, and r_L is the spatial decorrelation length of the background error, typically assumed to be constant (2–4 km) in storm-scale radar data analysis (Gao et al. 2004).

7.3.2 Forward Observation Operator and Iteration Procedure

Given the two DSD parameters $\mathbf{x} = [N_0', \Lambda]^t$ at each grid point, the rain DSD is determined and intrinsic radar variables of Z_H, Z_{DR}, and K_{DP} can be calculated from the precalculated values of scattering amplitudes using the T-matrix method, as well as specific attenuations at horizontal polarization (A_H) and specific differential attenuation (A_{DP}), using the formulas given in Section 4.3.

The forward operators of Z'_H and Z'_{DR} at each range gate are given by

$$Z'_H(n) = Z_H(n) - 2\sum_{i=1}^{n-1} A_H(i)\Delta r \qquad (7.26)$$

and

$$Z'_{DR}(n) = Z_{DR}(n) - 2\sum_{i=1}^{n-1} A_{DP}(i)\Delta r, \qquad (7.27)$$

where the numbers i and n denote the ith and nth range gates from the radar location, respectively, and Δr is the range resolution.

The partial derivatives of each of the polarimetric measurement variables (Z'_H, Z'_{DR}, and K_{DP}) with respect to each of the two state variables (Λ or N'_0) are then calculated and stored with 10 lookup tables of derivatives

$$\left(\frac{\partial Z_H}{\partial \Lambda}, \frac{\partial Z_{DR}}{\partial \Lambda}, \frac{\partial K_{DP}}{\partial \Lambda}, \frac{\partial A_H}{\partial \Lambda}, \frac{\partial A_{DP}}{\partial \Lambda}, \frac{\partial Z_H}{\partial N'_0}, \frac{\partial Z_{DR}}{\partial N'_0}, \frac{\partial K_{DP}}{\partial N'_0}, \frac{\partial A_H}{\partial N'_0}, \frac{\partial A_{DP}}{\partial N'_0}\right).$$

In each lookup table, the derivative values are precalculated for parameter Λ varying from 0 to 50 and parameter N'_0 varying from 0 to 15. To ensure sufficient accuracy, the range of each parameter is discretized at an interval of 0.02. As a result, each of the lookup tables has 2501×751 elements. In this way, the partial derivative value for the operator **H** is found from these tables for any given values of Λ and N'_0. Interpolation can be performed for values between the lookup table values of Λ or N'_0 to further improve the accuracy.

The iteration procedure for minimizing the cost function J is as follows:

1. Input the radar data of Z'_H, Z'_{DR}, and K_{DP}, lookup tables, and the background state parameters. For the purpose of data quality control, only the radar measurements with SNR > 1 dB are used in the analysis region.
2. The initial state vector of variational retrieval is equal to the background state vector (i.e., $\mathbf{x} = \mathbf{x}_b$), and the iteration starts with $\mathbf{v} = 0$. The radar variables Z_H, Z_{DR}, K_{DP}, A_H, and A_{DP} are calculated at each grid point using the forward operator as well as the lookup tables of scattering amplitudes. These radar variables are interpolated to observation points and the attenuations are added to yield Z'_H, Z'_{DR}, and K_{DP} and the cost functions (Equations 7.18 through 7.21).
3. The state vector is modified according to the innovation vector of the observations, and then the calculation of radar variables is repeated for the next iteration. The update of the state vector continues during the search for the minimum gradient of the cost function and until the convergence of the iteration.
4. After the minimization process converges, the analysis field of DSD parameters is obtained. Note that the analysis result from the first convergence might not be satisfactory. To improve the retrieval, the analysis result from the first convergence is used as a new background to repeat the analysis (i.e., iteration) process. This kind of repetition, which applies the previous analysis result as the background for a new analysis, is regarded as an "outer loop" of iteration. In general, several outer loops alone would give a satisfactory analysis result with a relatively small cost function.

7.3.3 APPLICATION TO PRD

The variational retrieval algorithm described above is now ready for application in rain microphysics retrieval from PRD. It was first tested and evaluated with simulated radar data and then used with real radar data.

In the case of simulated data, X-band CASA (Center for Collaborative Adaptive Sensing of the Atmosphere) polarimetric radar measurements were simulated from real data from an S-band polarimetric KOUN radar located at Norman, Oklahoma. The S-band radar measurements of Z_H and Z_{DR} are assumed to be free of precipitation attenuation and are used to retrieve rain DSDs with the approach discussed in Section 7.2. The retrieved DSDs are then used to calculate X-band radar variables that are considered the "truth." The attenuated X-band Z_H' and Z_{DR}' are then obtained from Equations 7.26 and 7.27. Finally, bias and random error are added to the attenuated Z_H' and Z_{DR}' to simulate the X-band radar observations for testing the variational retrieval algorithm. Different biases and RMS errors are used to conduct the retrieval experiments to assess the performance of the algorithm. The results are summarized in Table 7.2. The configuration of noise and bias can vary for different experiments.

The X-band radar observations were simulated using the S-band KOUN radar measurements on May 8, 2007 (1230 UTC, elevation angle 0.5°), when a convective system with widespread stratiform precipitation passed through Oklahoma in a west-to-east direction. In total, 12 experiments were designed for the tests. All these tests contain DSD model error, that is, the simulated truth assumes exponential DSDs whereas the retrieval assumes C-G DSDs. A constant background

TABLE 7.2
The Biases and RMS Errors of Variational Retrieval Used for Different Experiments

Test	Retrieval Bias/Simulated Bias			Retrieval RMSE/Simulated Error		
	Z_H (dBZ)	Z_{DR} (dB)	K_{DP} (degree/km)	Z_H (dbZ)	Z_{DR} (dB)	K_{DP} (degree/km)
1	0.091/0	0.027/0	0.004/0	0.393/0.5	0.107/0.1	0.084/0.1
2	0.083/0	0.009/0	0.006/0	0.409/1.0	0.108/0.2	0.083/0.2
3	0.178/0	0.023/0	0.012/0	0.476/1.5	0.110/0.3	0.088/0.3
4	0.267/0	0.036/0	0.019/0	0.537/2.0	0.120/0.4	0.093/0.4
5	0.440/0.125	0.115/0.025	0.020/0.025	0.597/0.5	0.159/0.1	0.084/0.1
6	0.841/0.25	0.219/0.05	0.037/0.05	0.952/0.5	0.253/0.1	0.092/0.1
7	1.575/0.5	0.411/0.1	0.067/0.1	1.687/0.5	0.445/0.1	0.114/0.1
8	2.879/1.0	0.755/0.2	0.118/0.2	3.037/0.5	0.807/0.1	0.160/0.1
9	0.448/0.125	0.113/0.025	0.022/0.025	0.606/0.75	0.157/0.15	0.085/0.15
10	0.862/0.25	0.216/0.05	0.040/00.05	0.979/1.0	0.250/0.2	0.095/0.2
11	1.604/0.5	0.408/0.1	0.071/0.1	1.724/1.25	0.443/0.25	0.117/0.25
12	2.940/1.0	0.747/0.2	0.122/0.2	3.117/1.5	0.801/0.3	0.165/0.3

$\left(N_0' = 3, \Lambda = 5\right)$ is applied in these tests. Tests 1 through 4 assume no bias but differ-ent random errors for the simulated observations. Tests 5 through 8 assume the same random errors but different biases. Tests 9 through 12 assume the same biases as Tests 5 through 8 except for different random errors. The detailed configurations of simulated data and error statistics of retrievals are shown in Table 7.1. In each cell of the table, the values to the right of the slash notation are simulated biases or RMSEs for simulated observations. Values to the left are retrieval biases or RMSEs com-puted against the simulated truth. For an optimal analysis system, the observation error covariance matrix **R** should properly characterize the expected observation errors, including their magnitudes and spatial correlations. The variational method based on the optimal estimation theory also assumes that all errors are unbiased (Kalnay 2003). However, the RMSEs and biases of real observations are difficult to accurately estimate. Therefore, mismatched errors are introduced in our tests to examine the sensitivity of the analysis to such error mismatches. The RMSEs of Z_H', Z_{DR}', and K_{DP} in the variational scheme are assumed to be 0.5 dB, 0.1 dB, and 0.1°/km, respectively. That is to say, "measurement" errors only match the truth in Test 1. In the other tests, true errors are generally larger than assumed errors in the variational scheme.

In Tests 1 through 4, retrieval RMSE values are generally less than true RMSE values. This means that the algorithm can smooth out observation errors to result in less error in the final analysis, consistent with the optimal estimation theory that the error of the final analysis should be smaller than the error of all sources of information used (Kalnay 2003). Bias increases with increasing RMS error, but in general the biases are very small, consistent with the fact that there is no systematic bias in the measurements. Tests 5 through 8 have the same random errors as Test 1 except that they contain different biases, created by adding con-stant values to all measurements. Compared to Test 1, Tests 5 through 8 show notable biases and RMS errors in the retrieval results. Except for some values of K_{DP}, all retrieval biases or RMS errors are larger than simulated measurement biases or errors in Tests 5 through 8. Test 8 shows that 1 dB bias in Z_H' measure-ments leads to about 3 dB bias and about 3 dB RMSE in the Z_H retrieval. This fact implies that the variational algorithm is more sensitive to the measurement bias than to random error. The measurement bias not only introduces a larger bias in the retrieval but also enlarges the retrieval RMS error. Hence, every effort should be made to remove the measurement biases before the variational analysis (e.g., Harris and Kelly 2001).

Tests 9 through 12 have a set of measurement biases that match those of Tests 5 through 8, but the random errors are larger. However, retrieval biases and RMSEs of Test 12 are almost the same as those of Test 8. This again shows that the algo-rithm is more sensitive to the measurement bias than to the measurement error. It is worth noting that a retrieval based on simulated data generally gives good results even though the constant background contains useless information on the measured precipitation. These results make sense because the simulated data are generally of good quality and provide complete data coverage in the analysis regions.

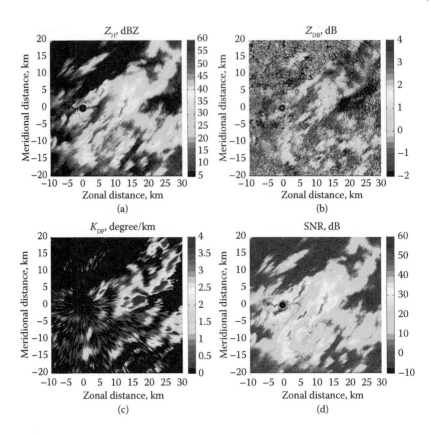

FIGURE 7.8   X-band CASA IP1 KSAO radar observations: (a) radar reflectivity; (b) differential reflectivity; (c) specific differential phase; and (d) SNR (2° elevation angle, 1950 UTC on April 24, 2011).

Figures 7.8 and 7.9 give an example of real X-band data retrieval. Figure 7.8 shows X-band KSAO radar observations, and Figure 7.9 shows a comparison of the retrieval and the KOUN observations, which have been converted to X-band. The retrieval analyzes a 40 km by 40 km region (401 by 401 grids with a space of 100 m) and assumes the observation errors are the same: 2 dB, 0.2 dB, and 0.2°/km for Z'_H, Z'_{DR}, and K_{DP}, respectively. The correlation length r_L is assumed to be 2 km. It is evident that the X-band retrieval also agrees well with the KOUN observations while capturing more details of the storm structures in Z_H, Z_{DR}, and K_{DP}. This example illustrates the validity of the retrieval algorithm for real X-band radar data applications.

A variational approach was developed and successfully applied to optimally retrieve rain DSD from attenuated PRD. It primarily uses the Z'_H, Z'_{DR}, and K_{DP} data of a single radar but can be easily extended to include observations from multiple radars to further improve the retrieval. The verification of the retrieval focuses on the accuracy of polarimetric variables calculated from the DSD. A sensitivity study shows that the algorithm is more sensitive to bias than to random error in the observations. The uncertainty of retrieval in the region of low data quality can be

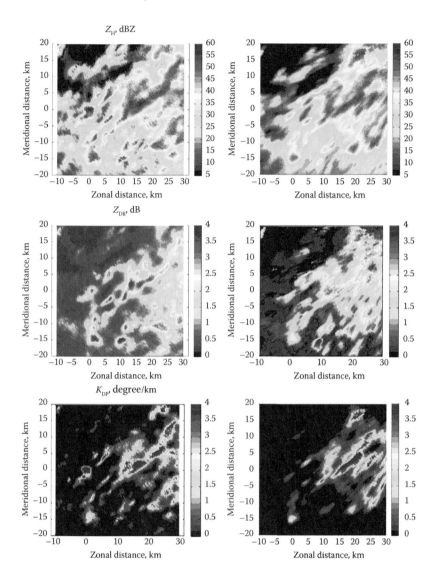

FIGURE 7.9 Comparison of retrieval results (left column) and simulation results from S-band KOUN data (right column). The retrieval is based on the X-band KSAO data shown in Figure 7.8. The three rows from top to bottom show radar reflectivity, differential reflectivity, and specific differential phase, respectively.

mitigated by using additional data from other radars that provide useful precipitation information in the same region.

The variational retrieval described above focuses on rain only (i.e., DSD parameters) and so do the experimental data sets that have excluded other species. To extend the variational method in different atmospheric conditions, multiple hydrometeor species, such as rain, snow, hail, and/or melting phase should be considered.

The increase in the number of species leads to an increase in state variables to retrieve, hence requiring more independent information. Whereas simultaneous use of multifrequency PRD is worth pursuing, model physics constraints present another avenue to explore, which is discussed next.

7.4 OPTIMAL RETRIEVAL THROUGH DA

Whereas the variational method takes into account the measurement errors, spatial correlation of weather, and attenuation, it works only when the number of independent information is larger than the number of the independent state variables. This is not true for retrieving cloud/precipitation physics when ice phase, multiple species of hydrometeors, and/or multimoment microphysics are considered. For example, the WRF double moment with six species (WDM6) scheme has 12 microphysics state variables, while the WSR-88D PRD has only four measurements that provide no more than three independent variables most of the time. It is impossible to determine all 12 state variables from only the PRD. Hence, other constraints, such as NWP model physics constraints, need to be used jointly with PRD. Specifically, the background states are obtained from the NWP model output. Those methods that include NWP model forecast results in the retrieval/analysis are called *DA methods* (Kalnay 2003; Xue et al. 2009). DA was attempted in order to optimally use PRD for improving weather forecasting by Jung et al. (2008a, 2008b), and some of the issues were identified. This section presents the challenges and potential of using PRD in weather analysis and forecast.

7.4.1 GENERAL CHALLENGES

Compared with the single-polarization reflectivity factor, PRD contain rich information that allow for successful severe weather observation, hydrometeor classification, and improved quantitative precipitation estimation. Because PRD allow for a better understanding of cloud and precipitation microphysics and improved microphysical parameterization, it is highly anticipated that PRD can improve NWP model initialization and forecasts through DA. Intuitively, this is feasible because PRD are now available nationwide, and the DA methods of variational, ensemble Kalman filter and hybrid methods are becoming mature enough for operational use. In practice, however, it has not been easy to implement PRD DA, and its progress has been limited. The reasons for this limited progress are (i) the gap between the models/parameters used in weather science and in radar polarimetry, (ii) a lack of simple forward operators, (iii) large relative errors in PRD, (iv) large variations/errors in model microphysics parameterization, and (v) highly nonlinear relations between model state variables and polarimetric variables. To succeed in PRD DA, we need to have a compatible microphysics parameterization scheme for the NWP model, accurate and fast forward operators, and accurate error characterization.

The gap between theory/model/parameters used in numerical weather prediction and weather radar engineering communities is outlined in Figure 7.10.

NWP models and radar polarimetry were developed from two fields—weather science and radar engineering, respectively. NWP models consist of a set of dynamic, thermodynamic, and microphysical equations. These equations are numerically solved

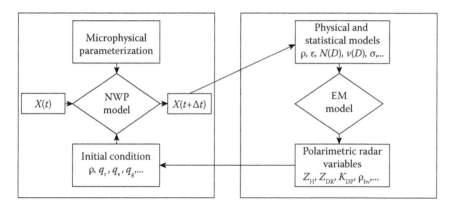

FIGURE 7.10 The relationship between the models used in meteorology and radar engineering.

using initial atmospheric states (pressure, temperature, density, and cloud physics) to arrive at predictions of future conditions. Hence, NWP meteorologists are used to dealing with state variables that are not directly measured by radar; they might not be very familiar with polarimetric radar variables, their information content, and their measurement error structure despite a few meteorological publications on these issues (Li and Mecikalski 2012; Posselt 2015). The different approaches used by NWP meteorologists and radar engineers present a problem that requires consideration.

As described in Chapters 4 and 5, polarimetric radar variables were defined based on EM wave scattering/propagation and wave statistics and are estimated from radar pulsed signals. Polarimetric radar variables cannot be written in simple functions of state variables that modelers can easily use. Weather radar engineers are used to developing technology/algorithms for making weather measurements and observation-based retrievals. Meteorologists are interested in observing, describing, and explaining the polarimetric signatures for microphysics study. Scientists having the background and experience in both NWP and radar polarimetry are in demand, and thorough studies to establish accurate relations between state and radar variables are in need.

7.4.2 Observation Operators and Errors

As in Chapter 4, polarimetric radar variables are expressed as the second moments of the scattering amplitudes, which are calculated by solving Maxwell equations for hydrometeor scattering. Microphysics information (DSD/PSD, shape, orientation, and composition) is needed to calculate the scattering amplitudes and radar variables, but not all these microphysics properties are predicted in NWP model simulation. Hence, many assumptions (DSD model, shape size relation, density, and orientation, and so on) have to be made to facilitate the calculations of radar variables from NWP model outputs. Furthermore, the most dominant species in terms of water mixing ratio (e.g., q_r, q_s, q_g,...) may not necessarily be the main contributor to polarimetric signatures. For example, hail may account for only <5% of water in a radar resolution volume but dominates polarimetric radar measurements. Moreover, the water–ice

TABLE 7.3

Typical Values of Polarimetric Radar Measurements and Errors

Variable	Range	Error	Rel. Error (%)
Z_H (dBZ)	0–70	1.0	<10
V_H (m/s)	−25–25	1.0	<10
Z_{DR} (dB)	0–4	0.2	~100
K_{DP} (degree/km)	0–3	0.2	~100
ρ_{hv}	0.9–1	0.01	~50

mixtures such as melting snow and melting hail/graupel are the main contributors to radar returns, but they are not even predicted in NWP models. Therefore, new melting species (melting snow and melting hail) have to be constructed to properly simulate polarimetric radar variables.

Jung et al. (2008a, 2010a) developed and published forward operation operators of PRD, which takes NWP model outputs to simulate PRD. The PRD simulator code has just been released on OU's ARPS website (http://arps.ou.edu/downloadpyDualPol. html). This simulator will allow for more thorough NWP simulations and comparisons for optimal use of PRD.

The development of the simulators does not solve all the problems. EM modeling errors and PRD measurement errors have not been fully characterized and documented to allow a DA expert to optimally use PRD for NWP model–based analysis. Table 7.3 shows the dynamic ranges of polarimetric radar variables of weather and the typical errors of their estimates.

The measurement errors of reflectivity and of Doppler velocity are large at ~1 dB and 1 m/s, respectively. Their relative errors are small, at ~10%. Moreover, the radial velocity is linearly related to the wind components of model states in NWP models. Hence, it is not surprising to see a big impact from assimilating v_r in NWP models. It is difficult to show the impact of assimilating Z because of its nonlinear relation with the predicted water mixing ratios. Nevertheless, the relative error is still small. In the case of polarimetric measurements, although the absolute errors of Z_{DR}, K_{DP}, and ρ_{hv} appear small (<0.2), they vary and their relative errors are huge, ~100%. This is because the intrinsic values of these variables (Z_{DR} and K_{DP}) are very small for certain species of cloud and precipitation like light rain and dry snow. The dynamic range of ρ_{hv} is very small (0.8–1.0) for weather echoes, and a 0.1 change in ρ_{hv} measurement could mean a change in hydrometeor species. Due to the non-Gaussian error distribution and nonlinear relation, these errors eventually cause bias in analysis through error propagation. Careful handling of these errors is also needed in PRD DA use.

7.4.3 Model Microphysics Uncertainty

Another issue in DA use of PRD is that there is large variability and uncertainty in microphysics parameterization and the NWP forecasts. These model variabilities limit the extent to which the PRD can contribute to the analysis of the state parameters and the improvement of model forecasts. For PRD to positively impact DA analysis, it is important that the model microphysics be compatible with the PRD information. For example, most operational NWP models use single-moment (SM) microphysics with a single-parameter DSD model, and only water mixing ratios are predicted. Hence, all radar variables are uniquely related to the water mixing ratios. That is, only the radar reflectivity factor is sufficient to determine the rainwater mixing ratio. This model allows no room for differential reflectivity and other polarimetric radar variables to improve rainwater content estimation.

Fortunately, double-moment (DM) and multimoment (MM) microphysics parameterizations have received much attention and been developed recently. Developments include the following: a hybrid scheme of SM and DM parameterization by Thompson et al. (2008), MM (one, two, and three moments) schemes by Milbrandt and Yau (2005a, 2005b), and the WDM6 scheme by Lim and Hong (2010). These DM or MM microphysics parameterizations have the flexibility to incorporate the dual-polarization information because the number concentration and particle size are independent, allowing for simulation of realistic PRD. Considering that the independent information of PRD is limited, DM is appropriate for PRD DA use at this stage.

Figures 7.11 and 7.12 show comparisons of the polarimetric radar signatures of a supercell storm between NWP model simulations with the schemes by Milbrandt and Yau and radar observations. Figure 7.11 shows the low-level polarimetric radar data of Z_H (left column) and Z_{DR} (right column). The top row shows the NWP model simulations with SM (left column) microphysics, the middle row shows the NWP results with DM microphysics, and the bottom row shows the radar measurements from a tornadic storm of May 10, 2010, observed by polarimetric KOUN radar. The storm started to split into two cells, and both of them have enhanced Z_{DR}, called Z_{DR} *arcs*, at their leading sides. Whereas the SM microphysics produces no enhanced Z_{DR} arc at the leading edge of the storm, the DM model result has a strong Z_{DR} arc as in the radar observation. The strong Z_{DR} arc is expected at the leading edge of the storm and is mainly due to size sorting, in which the small drops are blown to the trailing side, whereas the big drops have more momentum and fall faster, yielding large Z_{DR} at the low level (Kumjian and Ryzhkov 2008).

Figure 7.12 shows the midlevel of the ρ_{hv} ring signature (right column) by model simulations and radar observation. The midlevel of the ρ_{hv} ring is due to melting hailstones, which cause non-Rayleigh scattering with random scattering phase difference and hence reduced ρ_{hv}. As we can see, the ρ_{hv} and Z_H features are highly correlated with the SM microphysics (shown in a,b), which produce no ring structure. This is because all the state/observation variables are connected through a single PSD/DSD parameter. However, the DM microphysics (c,d), having two degrees of freedom, clearly shows the ring structure seen in the radar observations (bottom row). This shows the importance of a compatible microphysics parameterization scheme used in NWP model simulation in producing realistic polarimetric radar signatures so that PRD can have a positive impact on DA analysis and prediction.

The above comparisons of polarimetric signatures between model simulations and radar observations suggest that a two-moment (or multimoment) microphysics scheme should be used in the PRD DA. However, even after making this choice, there

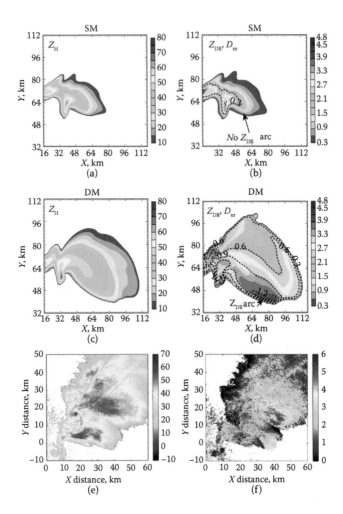

FIGURE 7.11 Low-level polarimetric radar signature of the Z_{DR} arc revealed by NWP model simulations with SM (a,b), DM (c,d), and radar measurements (right extreme column). The left column is the reflectivity factor (Z_H), and the right column is the differential reflectivity (Z_{DR}). (Simulation results are from Jung, Y., et al., 2010a. *Journal of Applied Meteorology and Climatology*, 49, 146–163.)

are still large variations among the two-moment microphysics schemes in producing polarimetric signatures. Figure 7.13 shows observed (Figure 7.13a through c) and 4 hr forecasted/simulated PRD of reflectivity (Z_H: top row), differential reflectivity (Z_{DR}: middle row), and specific differential phase (K_{DP}: bottom row) fields for a mesoscale convective system on the 0.5 elevation of various radars, at 0400 UTC on May 20, 2013. The simulations of the PRD are from the WRF model output using the forward operator (Jung et al. 2010a) with different microphysics schemes: Thompson, Morrison, and WDM6, Milbrandt and Yau, and WSM6 (Hong and Lim 2006). As shown in the figure, NWP model simulated PRD values varies among

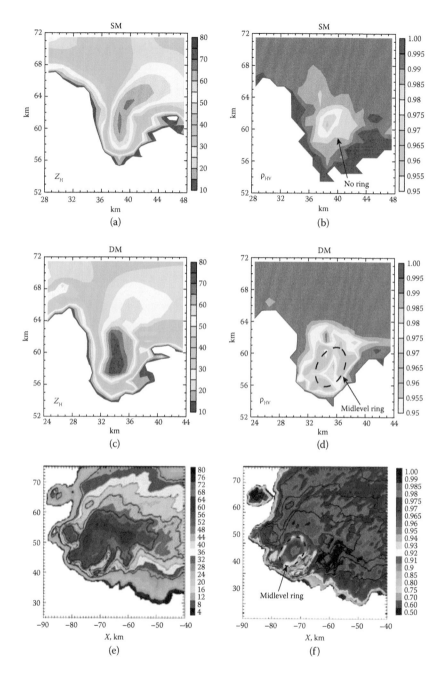

FIGURE 7.12 Midlevel polarimetric radar signature of the ρ_{hv} ring revealed by NWP model simulations with SM (a,b), DM (c,d), and radar measurements (right extreme column). The left column is the reflectivity factor (Z_H), and the right column is the co-polar correlation coefficient (ρ_{hv}). (Simulation results from Jung, Y., et al., 2010a. *Journal of Applied Meteorology and Climatology*, 49, 146–163.; measurements provided by Dr. Matthew Kumjian.)

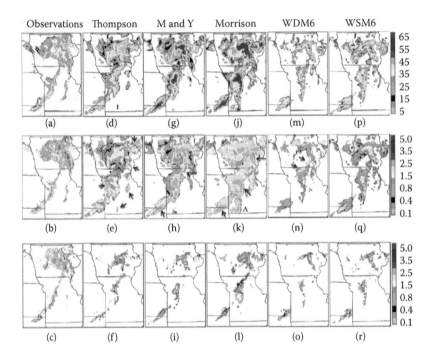

FIGURE 7.13 Mosaics of observed (a) reflectivity (dBZ), (b) differential reflectivity (dB), and (c) specific differential phase (degree/km) at a 0.5° tilt at 0400 UTC on May 20, 2013, and simulated values of 4-hour forecast at the same tilt locations from the (d–f) Thompson, (g–i) Milbrandt–Yau, (j–l) Morrison, (m–o) WDM6, and (p–r) WSM6 forecasts. (From Putnam, B., et al. 2013. *36th Conference on Radar Meteorology*, Breckenridge, CO.)

different microphysics schemes. The substantial difference in the simulated (Z_H, Z_{DR}, K_{DP}) fields with different microphysics schemes indicates significant uncertainties in the schemes' treatments of microphysical processes. There is still much work to be done to obtain converging results that would demonstrate the practical impact of PRD DA on NWP.

7.4.4 EXPECTATION FOR FUTURE PRD DA

Realizing the large relative errors in polarimetric radar measurements and the issue of uncertainty in NWP model parameterization of microphysical states and processes, a more systematic approach needs to be taken to optimally utilize PRD for improving QPE and QPF. Figure 7.14 summarizes the various components involved in optimally utilizing PRD and their interdependencies and connections. Each component is worthy of thorough study so that physics models and constraints are accurately constructed and the associated errors are correctly characterized, in order for PRD to be optimally utilized.

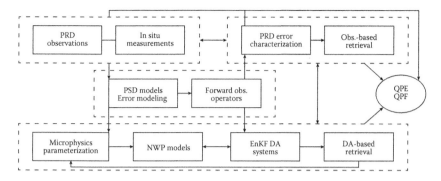

FIGURE 7.14   Schematic diagram of optimal utilization of PRD for QPE and QPF.

Problems

7.1 Discuss the advantages and disadvantages of the simultaneous attenuation correction and DSD retrieval in Section 7.1 versus that of attenuation correction and DSD retrieval done separately in Chapter 6.

7.2 Explain the similarities and differences between the deterministic DSD retrieval in Section 6.4 and the Bayesian DSD retrieval in Section 7.2. Under what condition are the two retrieval methods reconciled? Discuss when each retrieval should be used.

7.3 Describe how the variational retrieval is related to the Bayesian theory.

7.4 What are the challenges in optimally utilizing PRD for improving weather forecasts? What are the main challenges? How are you going to tackle and solve your main challenge if you are (i) a student, (ii) a scientist, or (iii) a manager/administrator?

8 Phased Array Radar Polarimetry

Although radar polarimetry with multiparameter measurements has become a mature technology in weather applications, there is a need for fast data updates for weather observations and quantification in future. The new phased array radar (PAR) technology has recently been introduced to the weather community and has received great attention. This chapter discusses PAR polarimetry for weather measurements, as well as its challenges and opportunities. A theory of PAR polarimetry is developed to establish the relation between electric fields at the PAR antenna and the fields in a resolution volume filled with hydrometeors. The approaches to correct and avoid bias and error in PAR polarimetry are examined and explored.

8.1 BACKGROUND AND CHALLENGES

Currently, it takes about 5 min for a WSR-88D radar to complete a volumetric scan, as shown in Table 8.1 (provided by Mr. Richard Ice at NOAA's Radar Operation Center [ROC]). The latest information about the WSR-88D scan strategy can be found from the NOAA-ROC website (http://www.roc.noaa.gov/WSR88D/NewRadarTechnology/NewTechDefault.aspx). The 5 min data update time is too slow for severe weather events such as tornados and downbursts, which sometimes last only a few minutes. It is desirable to have radar data with a higher temporal resolution (<1 min) so that detailed evolutions of severe storm phenomena can be revealed and tracked. This level of detail is difficult to achieve operationally with a mechanically scanning, dish-antenna radar, although a few research radars (e.g., RaXPol and Rapid Doppler on Wheels, Rapid DOW) are capable of performing faster scans for short periods of time (Pazmany et al. 2013; Wurman et al. 2001).

The need for fast data updates points to using advanced radar technology such as the PAR, which has an agile beam that steers electronically and quickly. Figure 8.1 shows that whereas a normal radar scans by mechanically rotating the dish antenna, a PAR steers its beam electronically by adjusting the phase of each element of the physically stationary antenna array.

Another motivation and driving force for using PAR technology is the MPAR (multifunction PAR) initiative, led by a joint NOAA–FAA effort (JAG/PARP report 2006) to use a single radar network to replace the four radar networks in the United States that currently serve for weather surveillance, air-traffic control, and target detection and recognition. The four networks are (1) National Weather Surveillance Radar (WSR-88D or NEXRAD), which is used for weather surveillance; (2) Terminal Doppler Weather Radar, which detects low altitude wind shear along aircraft approach and departure corridors; (3) Airport Surveillance Radar, which detects

TABLE 8.1
WSR-88D Volumetric Coverage Patterns

Quick Reference VCP Comparison Table for RPG Operators					February 2008
Slices	Tilts	VCP	Time[a]	Usage	Limitations
	14	11	5 min	Severe and nonsevere convective events. Local 11 has R_{max} = 80 nm. Remote 11 has R_{max} = 94 nm.	Fewer low elevation angles make this VCP less effective for long-range detection of storm features when compared to VCPs 12 and 212.
		211	5 min	Widespread precipitation events with embedded, severe convective activity (e.g. MCS, hurricane). Significantly reduces range-obscured V/SW data when compared to VCP 11.	All bins clutter suppression is *not* recommended. PRFs are not editable for SZ-2 (Split Cut) tilts.
	14	12	4.5 min	Rapidly evolving, severe convective events. Extra low elevation angles increase low-level vertical resolution when compared to VCP 11.	High antenna rotation rates decrease the effectiveness of clutter filtering, increase the likelihood of bias, and slightly decrease the accuracy of the base data estimates.
		212	4.5 min	Rapidly evolving, widespread severe convective events (e.g. squall line, MCS). Increased low-level vertical resolution compared to VCP 11. Significantly reduces range-obscured V/SW data when compared to VCP 12.	"All bins" clutter suppression is NOT recommended. PRFs are not editable for SZ-2 (split cut) tilts. High antenna rotation rates decrease the effectiveness of clutter filtering, increase the likelihood of bias, and slightly decrease accuracy of the base data estimates.

(Continued)

TABLE 8.1 *(Continued)*
WSR-88D Volumetric Coverage Patterns

Quick Reference VCP Comparison Table for RPG Operators					February 2008
Slices	Tilts	VCP	Time[a]	Usage	Limitations
19.5° 14.6° 9.9° 6.0° 4.3° 3.4° 2.4° 1.5° 0.5° 0.0°	9	21	6 min	Nonsevere convective precipitation events. Local 21 has R_{max} = 80 nm. Remote 21 has R_{max} = 94 nm.	Gaps in coverage above 5°.
		121	6 min	VCP of choice for hurricanes. Widespread stratiform precipitation events. Significantly reduces range-obscured V/SW data when compared to VCP 21.	PRFs are not editable for any tilt. Gaps in coverage above 5°.
		221	6 min	Widespread precipitation events with embedded, possibly severe convective activity (e.g. MCS, hurricane). Further reduces range-obscured V/SW data when compared to VCP 121.	"All bins" clutter suppression is *not* recommended. PRFs are not editable for SZ-2 (split cut) tilts. Gaps in coverage above 5°.
4.5° 3.5° 2.5° 1.5° 0.5° 0.0°	5	31	10 min	Clear air, snow, and light stratiform precipitation. Best sensitivity. Detailed boundary layer structure often evident.	Susceptible to velocity de-aliasing failures. No coverage above 5°. Rapidly developing convective echoes aloft might be missed.
		32	10 min	Clear air, snow, and light stratiform precipitation.	No coverage above 5°. Rapidly developing convective echoes aloft might be missed.

Note: MCS: mesoscale convective system; RPG: Radar Product Generator; SZ-2: Sachidanada and Zrnic code version 2; V/SW: velocity or spectrum width.

[a] VCP update times are approximate.

and tracks aircraft during their approaches and departures around airports; and (4) Air Route Surveillance Radar, which is used for the long range surveillance of aircraft. All of the radars are mechanically scanned dish-antenna systems. Each radar network has its own designated spatial and temporal resolution and serves its own mission. It is cost-effective and management/operationally efficient to

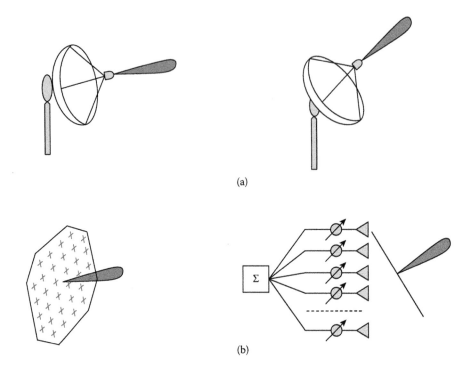

(a)

(b)

FIGURE 8.1 Conceptual sketch of (a) mechanical scan beam versus (b) the PAR electronic steering beam.

perform all of these missions and functions within a single radar network. To fulfill the requirements of all the missions, the single radar network needs to be equipped with an MPAR network that is fast, capable, and flexible in adjusting beam and scan characteristics (Weber et al. 2007).

PAR technology has been successfully utilized in the military for more than five decades (Brookner 2008). Recent advancements in microwave technology have made phased array antennas more feasible and affordable, and the weather radar community has started to invest time and resources into phased array technology. Through a joint effort by a government/university/industry team, the nation's first phased array weather radar, the National Weather Radar Testbed (NWRT), was developed in Norman, Oklahoma (Zrnić et al. 2007), as shown in Figure 8.2a. It has been demonstrated that its pulse-to-pulse beam steering capability enables accurate meteorological measurements in a shorter dwell time than with a mechanically steered beam, resulting in faster data updates (Heinselman et al. 2006, 2008). Figure 8.2b shows an example of radar observations of a strong severe thunderstorm observed by WSR-88D KTLX radar and the NWRT/PAR. It is clear that the NWRT/PAR provides detailed information on the evolution of the storm with updates every 36 s, whereas KTLX cannot do so with its 4-minute time interval. The NWRT/PAR have also been studied with regard to spaced antenna interferometry for use in transverse wind measurement and inhomogeneity detection within the beam (Zhang and Doviak 2007, 2008).

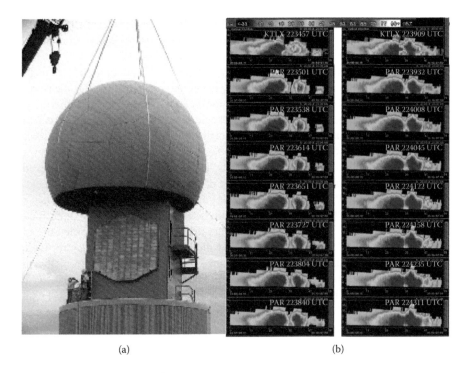

(a)   (b)

FIGURE 8.2   (a) The National Weather Radar Testbed and (b) its observations at time intervals of 36 s (second through seventh rows), compared with that by the WSR-88D KTLX radar (top row) with time separations of >4 min, observed on May 30, 2006. (Figure 8.2a courtesy of A. Zahrai. Figure 8.2b from Heinselman, P., et al., 2006. *23rd Conference on Severe Local Storms*, American Meteorological Society.)

Whereas PAR technology has the potential to advance weather observations and to serve for multiple missions, radar polarimetry also has great potential to improve weather quantification and forecasts (as shown in previous chapters). It is desirable for future weather radars to have both the polarimetry (multiparameter measurements) and the PAR fast-scan capabilities, specifically polarimetric PAR (PPAR).

PPARs have been developed for NASA and military missions, but with limited scanning angles and aperture size due to technical challenges and high cost. The SIR-C/X-SAR (Spaceborne Imaging Radar at C-/X-band Synthetic Aperture Radar) missions, conducted by a joint US/German/Italian space agency effort, are PARs with full polarization capabilities built by JPL and the Ball Communication Systems Division (Jordan et al. 1995). An Airborne Synthetic Aperture Radar mission is also being conducted by NASA/JPL for all-weather imaging with full-polarization radars. In addition, Japan has PALSAR (Phased Array type L-band Synthetic Aperture Radar) with polarimetric capabilities, and Canada continues to operate the RADARSAR-2 with full-polarization capabilities. NAVY/P-3 has the FOPEN (FOliage PENetration) radar with full polarization capabilities on board for better detection of objects. Furthermore, a ground-based space fence prototype PPAR has been developed by the Lockheed Martin Corporation to detect debris from space (http://www.lockheedmartin.com/us/products/space-fence.html).

For ground-based weather surveillance, however, it is difficult to use PPAR technology because of the requirement of highly accurate polarimetric radar measurements, given that there are limited resources for the development of this technology. The technical challenges include (i) 2D wide-angle scan (vs. 1D narrow angle scan in military and NASA missions) and (ii) high accuracy in polarimetric radar measurements (e.g., Z_{DR} error <0.2 dB and ρ_{hv} error <0.01), which exceeds the requirements for the detection of traditional airborne targets. These accuracies have been achieved with dish antenna radar systems through a few decades of research and development, but it is difficult for a PPAR to realize these accuracies.

A planar PPAR (PPPAR) has inherent limitations in making accurate polarimetric measurements. For a PPPAR, four faces are normally used to cover the 360° in the azimuth. Because the antenna faces and their broadside directions are fixed, the beam and polarization characteristics change depending on the electronic beam direction, causing geometrically induced cross-polarization coupling (Doviak et al. 2011; Lei et al. 2013; Zhang et al. 2009; Zrnić et al. 2011). The scan-dependent beam characteristics and strong cross-polarization on the beam axis are not desirable features for polarimetric weather measurements.

As shown by Zhang et al. (2011c), there are sensitivity losses and measurement bias errors when the PPPAR beam points off the broadside. The cross-polarization coupling causes biases in polarimetric measurements of weather. Figure 8.3 shows the maximal bias of differential reflectivity due to the cross-polar coupling between the dual-polarization channels. To have a Z_{DR} bias of less than 0.2 dB, the cross-polarization isolation needs to be better than 20 dB for ATSR mode and 40 dB for

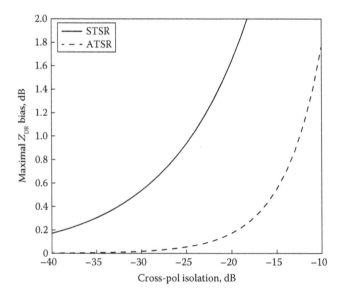

FIGURE 8.3 Dependence of Z_{DR} bias on cross-polarization coupling.

the STSR mode. This is difficult to achieve, especially for the STSR mode. The Z_{DR} bias is correctable (Zhang et al. 2009) through calibration to the scattering matrix or to the radar variables, but calibration over thousands of beams is extremely challenging from an operational perspective.

To avoid geometrically induced cross-polarization coupling, a slot-dipole array was proposed (Crain and Staiman 2007). An array of patch-dipole radiating elements would also serve for this purpose (Lei et al. 2013). If the idealized dipoles were used for the vertical polarization and the slots/patches were used for the horizontal polarization, there would be no cross-pol coupling in all directions. Theoretically, the slot-dipole or patch-dipole arrays can avoid geometrically induced cross-polarization coupling, but they may not be as feasible as in theory, need to be tested, and will be expensive. Although the planar PPAR is feasible, it is not necessarily the best choice for surveillance over the 360° in the azimuth. Other possible array configurations continue to be considered.

Besides planar arrays, other notional antenna configurations for a PPAR include linear arrays, circular/cylindrical arrays, and spherical arrays. The linear array needs one axis of mechanical rotation for weather surveillance like the rapid DOW (see Wurman 2003) and the proposed instrument design created by the Center for Collaborative Adaptive Sensing of the Atmosphere (Hopf et al. 2009; Knapp et al. 2011). Mechanical rotation is not desirable for PPAR, as it places significant constraints on the mission capabilities. For satellite communication applications, the spherical array is optimal and flexible in its use of the antenna aperture size and in its symmetry (Tomasic et al. 2002). For weather surveillance, however, the spherical array cannot provide the high cross-polar isolation required to accurately measure precipitation. A circular or cylindrical configuration has been used for direction finding and communications (Raffaelli and Johansson 2003; Royer 1966). A cylindrical array radar, LSTAR, developed by the Syracuse Research Center (SRC Inc., http://srcinc.com/what-we-do/srctec_product.aspx?id=1087) has been used for gap filling target detection by the FAA and DHS. However, this system uses single polarization and has other limitations. Cylindrical PPAR (CPPAR) was introduced to overcome the deficiencies encountered with PPPAR (Zhang et al. 2011c).

In Section 8.2, we formulate the theory behind PPPAR and provide the calibration formulas. In Section 8.3, we discuss CPPAR.

8.2 FORMULATION FOR PLANAR POLARIMETRIC PHASED ARRAY RADAR

A phased array antenna is composed of elements whereby each is treated as a pair of crossed dipoles. In this section a theory relating the electric fields at the PPAR's antenna elements and at the hydrometeors is formulated through radiation, scattering, and propagation.

8.2.1 DIPOLE RADIATION

Figure 8.4a shows a conceptual sketch of radiated electric fields by a pair of Hertzian dipoles located at the center/origin. The red lines represent the electric field by the

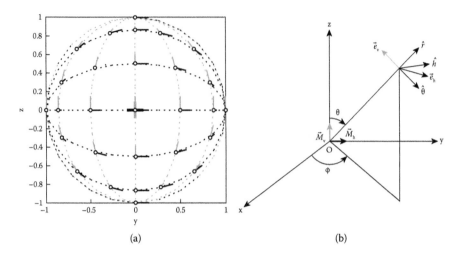

(a)   (b)

FIGURE 8.4 Dipole radiation: (a) a sketch of electric fields radiating from a pair of dipoles and (b) the spherical coordinate system to represent electric fields from dipoles that have moments $\vec{M}_h$, $\vec{M}_v$.

horizontal dipole, and the gray lines indicate that by the vertical dipole. It is evident that the electric fields are in longitudinal directions. The electric field generated by the horizontal dipole (black lines) is not orthogonal to that by the vertical dipole, except for in the principal planes. A coordinate system (Figure 8.4b) is chosen with the PPAR array face in the y, z plane.

Similar to the scattered wave field expression (Equations 3.56 and 3.57) provided in Chapter 3, the electric field radiated by a dipole of moment $\vec{M}$ is as follows (Ishimaru 1978, 1997):

$$\vec{E}_q(\vec{r}) = -\frac{k^2 e^{-jkr}}{4\pi\varepsilon r}\left[\hat{r}\times\left(\hat{r}\times\vec{M}_q\right)\right], \tag{8.1}$$

where $k = 2\pi/\lambda$, λ is the radar wavelength, ε is the permittivity for an assumed uniform precipitation-free atmosphere, and the dipole moment

$$\vec{M}_q \equiv A_q\exp(-j\phi_q)\hat{e}_q \tag{8.2}$$

has amplitude A_q, phase ϕ_q, and direction $\hat{e}_q$, the unit vector along which the dipole moment is directed. The subscript q (h or v) denotes the qth dipole.

Writing Equation 8.1 in terms of spherical components

$$\vec{E}_q(\vec{r}) = E_{\theta q}\hat{\theta} + E_{\phi q}\hat{\phi}, \tag{8.3}$$

where $E_{\theta q}$ (i.e., the V-pol wave) and $E_{\phi q}$ (i.e., the H-pol wave) are the two spherical components of the wave radiated by the qth dipole.

The horizontal dipole $\left(\vec{M}_h = M_h \hat{y}\right)$ is aligned with the y-axis, and the vertical dipole $\left(\vec{M}_v = M_v \hat{z}\right)$ is aligned with the z-axis. After applying a vector identity to the cross products in Equation 8.1, the electric field at $\vec{r}$ from $\vec{M}_h$ and that from $\vec{M}_v$ are, respectively

$$\vec{E}_h = E_{th}^{(p)} \left[y\hat{y} - \left(\sin\theta\cos\phi\hat{x} + \sin\theta\sin\phi\hat{x} + \cos\theta\hat{y}\right)\sin\theta\sin\phi \right] = E_{th}^{(p)}\vec{e}_h \quad (8.4)$$

$$\vec{E}_v = E_{tv} \left[\sin^2\theta\hat{z} - \left(\cos\phi\hat{x} + \sin\phi\hat{y}\right)\sin\theta\cos\theta \right] = E_{tv}^{(p)}\vec{e}_v, \quad (8.5)$$

where

$$E_{tq}^{(p)} = \frac{k^2}{4\pi\varepsilon r} M_q \quad (8.6)$$

is the magnitude of the electric field transmitted along the normal to the array broadside. It is noted that the phase term e^{-jkr} is omitted because it can be represented by the transmission matrix $\overline{\overline{T}}$, which has been included in the propagation-included scattering matrix $\overline{\overline{S}}' = \overline{\overline{T}}\,\overline{\overline{S}}\,\overline{\overline{T}}$. Vector $\vec{e}_h$ gives the direction of the electric field of $\vec{M}_h$ propagating along the (θ, ϕ) direction, and its magnitude, $|\vec{e}_h|$, is the fraction of the corresponding electric field propagating along the (θ, ϕ) direction, compared that normal to the array.

Projections of $\vec{e}_h, \vec{e}_v$ onto the local horizontal $\left(\hat{h} = \hat{\phi}\right)$ and vertical $\left(\hat{v} \equiv -\hat{\theta}\right)$ directions yield

$$\hat{h} \cdot \vec{e}_h = \hat{\phi} \cdot \vec{e}_h = \cos\phi \quad (8.7)$$

$$\hat{v} \cdot \vec{e}_h = -\hat{\theta} \cdot \vec{e}_h = -\cos\theta\sin\phi \quad (8.8)$$

$$\hat{h} \cdot \vec{e}_v = \hat{\phi} \cdot \vec{e}_v = 0 \quad (8.9)$$

$$\hat{v} \cdot \vec{e}_v = -\hat{\theta} \cdot \vec{e}_v = \sin\theta. \quad (8.10)$$

Note that (i) the intensities of the horizontally (H-pol, from Equation 8.7) and "vertically" (V-pol, from Equations 8.8 and 8.10) polarized waves are functions of beam direction (θ, ϕ) and (ii) if the beam is directed away from the intersection of the equatorial planes of the respective dipoles, the horizontal dipole $\vec{M}_h$ produces a vertical component given by Equation 8.8. This is the reason why the PPAR transmits cross-polar components, increasing in intensity with direction from broadside, generating polarization biases that are not significant when using a mechanically steered beam. Thus the transmitted electric fields $\vec{E}_{th}^{(p)}$ and $\vec{E}_{tv}^{(p)}$ generated by the dipoles h and v and projected onto the local H and V directions give the transmitted H, V electric fields

$$\vec{E}_t = \begin{bmatrix} E_{th} \\ E_{tv} \end{bmatrix} = \overline{\overline{P}} \begin{bmatrix} E_{th}^{(p)} \\ E_{tv}^{(p)} \end{bmatrix} = \overline{\overline{P}}\vec{E}_t^{(p)}, \quad (8.11)$$

where $\overline{\overline{P}}$ is the matrix

$$\overline{\overline{P}} \equiv \begin{bmatrix} \hat{h} \cdot \vec{e}_h & \hat{h} \cdot \vec{e}_v \\ \hat{v} \cdot \vec{e}_h & \hat{v} \cdot \vec{e}_v \end{bmatrix} = \begin{bmatrix} \cos\phi & 0 \\ -\cos\theta\sin\phi & \sin\theta \end{bmatrix} \qquad (8.12)$$

that projects the oblique $\vec{E}_h$ and $\vec{E}_v$ onto the local H and V coordinates. Because there is no conversion of one polarized field into the other in a precipitation-free atmosphere, the projection matrix can be applied at any range. However, in precipitation-filled atmospheres, there is conversion and it is important that the projection matrix be applied before the wave enters the precipitating medium. As we will see later, "local" will imply "at the beginning of the precipitation."

8.2.2   Backscattering Matrix

As in Equation 4.122 the received electric field in the backscattering direction, $\vec{E}_r$, at the receiving array element (assumed to be the same as the transmitting array element) can be expressed as

$$\vec{E}_r \equiv \begin{bmatrix} E_{rh} \\ E_{rv} \end{bmatrix} = \frac{1}{r}\begin{bmatrix} s'_{hh}(\pi) & s'_{hv}(\pi) \\ s'_{vh}(\pi) & s'_{vv}(\pi) \end{bmatrix}\begin{bmatrix} E_{th} \\ E_{tv} \end{bmatrix} \equiv \frac{1}{r}\overline{\overline{S'}}\vec{E}_t, \qquad (8.13)$$

where $\overline{\overline{S'}}$ is the propagation-included backscattering matrix of a hydrometeor, including propagation effects.

Although Equation 8.13 gives the H, V electric fields at the receiving array element, we need to determine the fields parallel to the respective dipole axes because these fields enter the H, V channels of the receiver. The fields parallel to the dipole axes are obtained by projecting $\vec{E}_r$ onto the respective dipole directions, and these projections are given by

$$\vec{E}_r^{(p)} \equiv \begin{bmatrix} E_{rh}^{(p)} \\ E_{rv}^{(p)} \end{bmatrix} = \begin{bmatrix} \cos\phi & -\cos\theta\sin\phi \\ 0 & \sin\theta \end{bmatrix}\begin{bmatrix} E_{rh} \\ E_{rv} \end{bmatrix} = \overline{\overline{P}}^t\vec{E}_r, \qquad (8.14)$$

where $\overline{\overline{P}}^t$ is the transpose of $\overline{\overline{P}}$. By combining Equations 8.11 and 8.12 with Equation 8.14, $\vec{E}_r^{(p)}$ can be expressed as

$$\vec{E}_r^{(p)} = \overline{\overline{P}}^t\vec{E}_r = \frac{1}{r}\overline{\overline{P}}^t\overline{\overline{S'}}\overline{\overline{P}}\vec{E}_t^{(p)} \equiv \frac{1}{r}\overline{\overline{S}}^{(p)}\vec{E}_t^{(p)}, \qquad (8.15)$$

where $\overline{\overline{S}}^{(p)} = \overline{\overline{P}}^t\overline{\overline{S'}}\overline{\overline{P}}$ is the PPAR scattering matrix, including polarimetric effects

$$\overline{\overline{S}}^{(p)} = \begin{bmatrix} \begin{matrix} s'_{hh}\cos^2\phi - (s'_{hv} + s'_{vh})\cos\theta\sin\phi\cos\phi \\ + s'_{vv}\cos^2\theta\sin^2\phi \end{matrix} & \begin{matrix} s'_{hv}\sin\theta\cos\phi \\ - s'_{vv}\sin\theta\cos\theta\sin\phi \end{matrix} \\ s'_{vh}\sin\theta\cos\phi - s'_{vv}\sin\theta\cos\theta\sin\phi & s'_{vv}\sin^2\theta \end{bmatrix}, \qquad (8.16)$$

is the backscattering matrix for the H and V channels of the PPAR; $\overline{\overline{S}}^{(p)}$ now includes cross coupling of the fields during propagation as well as in backscatter.

Further analysis of Equation 8.16 indicates that the PPAR not only causes bias in co-polar measurements (diagonal terms), but also generates extra cross-polarization terms (i.e., the off-diagonal terms). As a simple example, consider the case $s'_{hv} = s'_{vh} = 0$, whereby hydrometeors do not introduce cross coupling, either along the propagation path or in backscatter. In this case the cross term $s^{(p)}_{hv} = s^{(p)}_{vh} = -s'_{vv} \sin\theta \cos\theta \sin\phi$, coupling backscattered power from horizontal dipole to vertical dipole, can be comparable to the co-polar terms, especially if the beam is directed far from the intersection (i.e., $\theta = \pi/2$, $\phi = 0$, π) of the dipole's equatorial planes. Corrections are needed, and they can be implemented to the scattering matrix or to the radar variables, as discussed below.

8.2.3 SCATTERING MATRIX CORRECTIONS

There are two polarimetric transmitting/receiving modes currently in use by the weather radar community. They are (i) the ATSR mode and (ii) the STSR mode. Using these modes as examples, we demonstrate bias correction to the scattering matrix measured with a PPAR.

8.2.3.1 Alternate Transmission

In this mode all four elements of $\overline{\overline{S}}^{(p)}$ can be calculated using Equation 8.16 and measurements of the transmitted and received fields. For a mechanically steered beam, $\overline{\overline{P}}$ is a unit matrix and $\overline{\overline{S}}' = \overline{\overline{S}}^{(p)}$ is directly recovered. If a PPAR is used, $\overline{\overline{S}}'$ can still be recovered, but in this case it is necessary to multiply the calculated $\overline{\overline{S}}^{(p)}$ with the inverse of the projection matrices. Specifically,

$$\overline{\overline{S}}' = \left(\overline{\overline{P}}^{t}\right)^{-1} \overline{\overline{S}}^{(p)} \, \overline{\overline{P}}^{-1}. \tag{8.17}$$

Equation 8.17 shows that $\overline{\overline{S}}'$, obtained with a PPAR, is mathematically equivalent to a mechanically steered beam. Such equivalency allows us to use results obtained for mechanically steered beams to calculate $\overline{\overline{S}}$, the matrix that is of prime interest. Write Equation 8.17 as

$$\overline{\overline{S}}' = \overline{\overline{C}}^{t} \overline{\overline{S}}^{(p)} \overline{\overline{C}}, \tag{8.18}$$

where we have defined a correction matrix

$$\overline{\overline{C}} = \overline{\overline{P}}^{-1} = \begin{bmatrix} \dfrac{1}{\cos\phi} & 0 \\[2ex] \dfrac{\cos\theta\sin\phi}{\sin\theta\cos\phi} & \dfrac{1}{\sin\theta} \end{bmatrix}. \tag{8.19}$$

Substituting Equation 8.19 into Equation 8.18 yields $\bar{\bar{S}}'$, indicating that the polarization biases incurred by PPAR are fully correctable using the matrix $\bar{\bar{C}}$. Because $\left(s_{hh}^{(p)}, s_{vh}^{(p)}\right)$ and $\left(s_{hv}^{(p)}, s_{vv}^{(p)}\right)$ could be measured at different pulses, either the Doppler effects need to be taken into account or a covariance method should be used for the bias correction, as in Zrnić et al. (2011).

8.2.3.2   Simultaneous Transmission

Because we measure both components of $\vec{E}_r$ when both components of $\vec{E}_t$ are simultaneously transmitted, we cannot determine, using STSR, all four elements of $\bar{\bar{S}}^{(p)}$. Nevertheless, if there is negligible coupling between the H and V waves as is common, then we can determine the two main diagonal components of $\bar{\bar{S}}'$. Writing in matrix form, we have

$$
\begin{bmatrix} E_{rh}^{(p)} \\ E_{rv}^{(p)} \end{bmatrix} = \frac{1}{r} \begin{bmatrix} \cos\phi & -\cos\theta\sin\phi \\ 0 & \sin\theta \end{bmatrix} \begin{bmatrix} s_{hh}' & 0 \\ 0 & s_{vv}' \end{bmatrix} \begin{bmatrix} \cos\phi & 0 \\ -\cos\theta\sin\phi & \sin\theta \end{bmatrix} \begin{bmatrix} E_{th}^{(p)} \\ E_{tv}^{(p)} \end{bmatrix} \tag{8.20}
$$

or

$$
\begin{bmatrix} E_{rh}^{(p)} \\ E_{rv}^{(p)} \end{bmatrix} = \frac{1}{r} \begin{bmatrix} E_{th}^{(p)}\left(s_{hh}'\cos^2\phi + s_{vv}'\cos^2\theta\sin^2\phi\right) & -E_{tv}^{(p)}s_{vv}'\sin\theta\cos\theta\sin\phi \\ -E_{th}^{(p)}s_{vv}'\sin\theta\cos\theta\sin\phi & E_{tv}^{(p)}s_{vv}'\sin^2\theta \end{bmatrix} \tag{8.21}
$$

Because both $\vec{E}_t$ and $\vec{E}_r$ are known through measurement and calibration, the two equations in Equation 8.21 are used to solve s_{hh}' and s_{vv}'.

It has been shown that the polarization bias can be corrected by rectifying the scattering matrix in either the ATSR or STSR mode. Although it is not necessary to force the transmitted H and V waves to be equal, they do need to be known through measurement and calibration. That is, corrections can be made if the transmitted field magnitudes and phases are known and received fields are accurately measured along with the electronic beam position.

8.2.4   CORRECTION TO POLARIMETRIC VARIABLES

To obtain a quantitative assessment of the effects that an electronically steered beam has on weather radar measurements, we compare the polarimetric radar variables measured by PPAR with those by radar using a mechanically steered beam. In the following sections the intrinsic scattering matrix is assumed to be diagonal, a valid assumption for most meteorological conditions. However, the developed formulation in Sections 8.2.2 and 8.2.3 describes the more general case.

8.2.4.1 Reflectivity Factor

Following the same definitions as in Equations 4.59 through 4.61, the reflectivity factors for the PPAR are

$$
Z_h^{(p)} = \frac{4\lambda^4}{\pi^4 |K_w|^2} \left\langle n \left| s_{hh}^{(p)} \right|^2 \right\rangle = \frac{4\lambda^4}{\pi^4 |K_w|^2} \left\langle n \left| s_{hh}' \cos^2 \phi + s_{vv}' \cos^2 \theta \sin^2 \phi \right|^2 \right\rangle ,
$$

$$
= Z_h' \cos^4 \phi + Z_v' \cos^4 \theta \sin^4 \phi + \tfrac{1}{2} \sqrt{Z_h' Z_v'} \operatorname{Re}\left[\tilde{\rho}_{hv}' \right] \cos^2 \theta \sin^2 2\phi
$$
(8.22)

for horizontal polarization ($\tilde{\rho}_{hv}'$ is defined in Section 4.3) and

$$
Z_v^{(p)} = \frac{4\lambda^4}{\pi^4 |K_w|^2} \left\langle n \left| s_{vv}^{(p)} \right|^2 \right\rangle = \frac{4\lambda^4 \sin^4 \theta}{\pi^4 |K_w|^2} \left\langle n \left| s_{vv}' \right|^2 \right\rangle = Z_v' \sin^4 \theta
$$
(8.23)

for vertical polarization. It is evident that reflectivity factors are biased low compared to those for a mechanically steered beam. This is due to the fact that scatterers located along directions other than the direction of the intersection of the dipoles' equatorial planes experience weaker fields. The bias correction for $Z_v^{(p)}$ is simpler than that for $Z_h^{(p)}$, which depends on both Z_h' and Z_v' as well as ρ_{hv}'.

8.2.4.2 Differential Reflectivity

The differential reflectivity measured with a mechanically steered beam is the intrinsic differential reflectivity minus the two-way differential PIA. The differential reflectivity measured with PPAR is

$$
Z_{DR}^{(p)} = 10 \log \left(\frac{\left\langle n \left| s_{hh}^{(p)} \right|^2 \right\rangle}{\left\langle n \left| s_{vv}^{(p)} \right|^2 \right\rangle} \right).
$$
(8.24)

In the case of ATSR mode, the PPAR-measured differential reflectivity $Z_{DR}(\text{ATSR})$ is related to the differential reflectivity Z_{DR}' measured with a mechanically steered beam as

$$
Z_{DR}(\text{ATSR}) = Z_{DR}' + 10 \log \frac{a^2 + b^2 Z_{dr}'^{-1} + 2ab Z_{dr}'^{-1/2} \operatorname{Re}(\rho_{hv}')}{c^2},
$$
(8.25)

$$
Z_{DR}(\text{ATSR}) = Z_{DR}' + Z_{DR} \operatorname{Bias}(\text{STSR})
$$
(8.26)

where $a = \cos^2\phi$, $b = \cos^2\theta \sin^2\phi$, and $c = \sin^2\theta$. Thus biases incurred with PPAR measurements using ATSR mode can be corrected by subtracting Z_{DR} Bias(ATSR) from the measured $Z_{DR}(\text{ATSR})$, yielding the conventional (i.e., that measured with a mechanically steered beam) differential reflectivity Z'_{DR}. However, Z'_{DR} is coupled with ρ'_{hv} in Equations 8.25 and 8.26, and they need to be solved jointly. This is mathematically more complicated than directly correcting the scattering matrix, as presented in Section 8.2.3.

Figure 8.5 shows Z_{DR} biases for the ATSR (Figure 8.5a) and STSR (Figure 8.5b through d) modes. The biases are plotted as a function of the azimuth angle ϕ using θ as a parameter. Z_{DR} Bias(ATSR) increases for $|\phi| > 0$, and the bias can be positive for small azimuth and zenith angles. The biases are due to the fact that the H, V radiations from an array element change differently as the beam moves away from equatorial planes. At $\phi = 0°$ and $\theta < 90°$, the field generated by the vertical dipole is weaker than that of the horizontal dipole, and thus Z_{DR} Bias(ATSR) is positive. On the other hand, at large azimuth angles (e.g., $\phi = 45°$), the horizontally polarized wave is weaker than the vertically polarized one, leading to large negative biases.

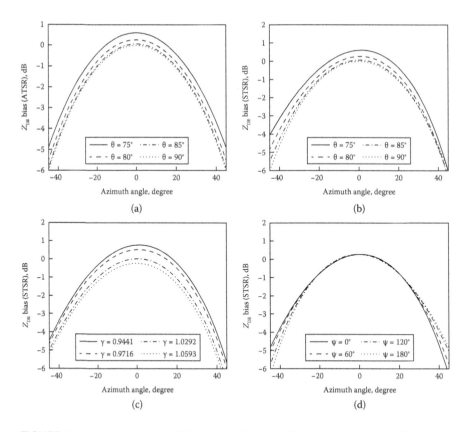

FIGURE 8.5 Dependence of differential reflectivity bias on the electronically steered beam direction. (a) Z_{DR} Bias(ATSR); (b) Z_{DR} Bias(STSR) for various θ, but $\gamma = 1$, $\psi = 0°$; and (c) Z_{DR} Bias(STSR) for various γ but $\theta = 80°$ and $\psi = 0°$; and (d) Z_{DR} Bias(STSR) for various phase differences ψ, but $\theta = 80°$, $\gamma = 1.0$ ($Z'_{dr} = 1.0$, $\rho_{hv} = 0.9$ in all cases).

If the STSR mode is used, the biases are different and more complicated than those for ATSR mode. The measured differential reflectivity can be obtained by using Equation 8.21 with the transmitted wave fields related by $E_{tv}^{(p)} = \gamma E_{th}^{(p)} e^{i\psi}$, where γ is the amplitude ratio of the transmitted electric fields and ψ is their relative phase. Then

$$Z_{DR}(STSR) = 10\log\left[\frac{\left\langle\left|\left(s'_{hh}\cos^2\phi + s'_{vv}\cos^2\theta\sin^2\phi\right)E_{t1} - s'_{vv}\sin\theta\cos\theta\sin\phi E_{t2}\right|^2\right\rangle}{\left\langle\left|s'_{vv}\sin\theta\cos\theta\sin\phi E_{t1} + s'_{vv}\sin^2\theta E_{t2}\right|^2\right\rangle}\right]$$

$$= Z'_{DR} + 10\log\frac{\left[a^2 + |b_s|^2 Z'^{-1}_{dr} + 2a\,\mathrm{Re}(b_s\rho'_{hv})Z'^{-1/2}_{dr}\right]}{|c_s|^2} \tag{8.27}$$

$$= Z'_{DR} + Z_{DR}\,\mathrm{Bias(STSR)} \tag{8.28}$$

where $b_s = b - \gamma e^{i\psi}d$, $c_s = d + \gamma e^{i\psi}\sin^2\theta$, and $d = \sin\theta\cos\theta\sin\phi$. Z_{DR} Bias(STSR) depends not only on the beam direction, but also on the amplitude ratio and the relative phase of transmitted waves, as well as the hydrometeor characteristics.

Figure 8.5b through d shows Z_{DR} Bias(STSR). The results for transmitting equal H and V amplitudes (i.e., $\gamma = 1$) are shown in Figure 8.6b. It is evident that Z_{DR} Bias(STSR) is not symmetric about the x–z plane because the V dipole contributes to the $E_{rh}^{(p)}$ field, in phase with that of the H dipole for $-\pi/2 < \phi < 0$, whereas the contribution is out of phase with the H dipole's contribution for $0 < \phi < \pi/2$, as shown by Equations 8.21, 8.27, and 8.28. However, the projection of the V dipole field in the $E_{rv}^{(p)}$ direction has the same magnitude for symmetrical azimuths. Thus Z_{DR}, proportional to the ratio of $E_{rh}^{(p)}$ to $E_{rv}^{(p)}$, is larger for negative ϕ than for equal positive ϕ. The results for unequal transmitted H and V fields are shown in Figure 8.6c and d for various amplitude ratios γ and phase differences ψ. The amplitude ratios of 0.9441, 0.9716, 1.0292, and 1.0593 correspond to −0.5, −0.25, 0.25, and 0.5 dB differences in power, respectively. The power imbalance causes Z_{DR} Bias(STSR) in a similar way as that of the elevation angle, because they both yield relative differences in the projection of the transmitted fields to the local polarization directions. It is interesting to note that Z_{DR} Bias(STSR) also depends on the relative phase of the transmitted fields and it becomes antisymmetric when $\psi = 180°$. This is because the polarization of the transmitted fields changes depending on the relative phase. The differential reflectivity can also be corrected through the PPAR estimated value $Z_{DR}^{(p)}$ by inverting Equations 8.27 and 8.28 to solve for Z'_{DR} if ρ'_{hv} is known, but the power imbalance needs to be known within 0.1 dB and the relative phase within a few degrees, which is challenging.

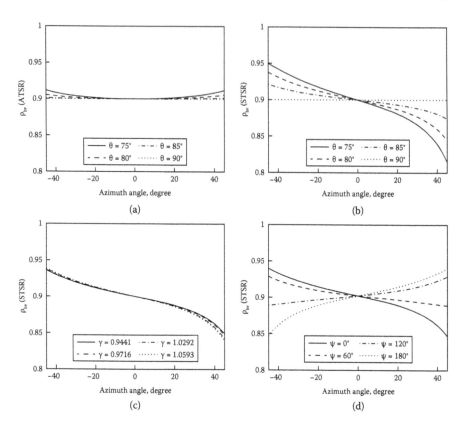

FIGURE 8.6   Dependence of $\rho_{hv}^{(p)}$ on the electronically steered beam direction: (a) ρ_{hv}(ATSR); (b) ρ_{hv}(STSR) for balanced transmission (i.e., $\gamma = 1$, $\psi = 0$), (c) ρ_{hv}(STSR) for γ as a parameter, but $\psi = 0$ and $\theta = 80°$; and (d) ρ_{hv}(STSR) for ψ as a parameter, but $\gamma = 1.0$ and $\theta = 80°$. $Z'_{dr} = 1.0$ and $\rho_{hv} = 0.9$ for all cases.

8.2.4.3 Correlation Coefficient

In the case of an ATSR PPAR, the computation of the correlation coefficient produces

$$\rho_{hv}^{(p)}(\text{ATSR}) = \frac{\left| \left\langle n s_{hh}^{*(p)} s_{vv}^{(p)} \right\rangle \right|}{\sqrt{\left\langle n \left| s_{hh}^{(p)} \right|^2 \right\rangle \left\langle n \left| s_{vv}^{(p)} \right|^2 \right\rangle}}$$

$$= \frac{\left| \left\langle n \left(s'_{hh} \cos^2 \phi + s'_{vv} \cos^2 \theta \sin^2 \phi \right)^* s'_{vv} \sin^2 \theta \right\rangle \right|}{\sqrt{\left\langle n \left| s'_{hh} \cos^2 \phi + s'_{vv} \cos^2 \theta \sin^2 \phi \right|^2 \right\rangle \left\langle n \left| s'_{vv} \sin^2 \theta \right|^2 \right\rangle}} \qquad (8.29)$$

$$= \frac{\left| a \tilde{\rho}'_{hv} + b (Z'_{dr})^{-1/2} \right|}{\sqrt{a^2 + b^2 (Z'_{dr})^{-1} + 2ab (Z'_{dr})^{-1/2} \operatorname{Re}(\rho'_{hv})}},$$

where a and b are defined in the paragraph after Equations 8.25 and 8.26. Figure 8.6a shows ρ_{hv}(ATSR) dependence on the electronically steered beam direction. This figure shows that the bias for an intrinsic $\rho_{hv} = 0.9$ is smaller than 0.02; calculations (not given here) show that the bias is even smaller if ρ_{hv} is larger, as is the case for most precipitation.

If the STSR mode is used, the correlation coefficient is

$$\rho_{hv}^{(p)}(\text{STSR}) = \frac{\left|\left\langle n[as'_{hh} + b_s s'_{vv}]^*[c_s s'_{vv}]\right\rangle\right|}{\sqrt{\left\langle n|as'_{hh} + b_s s'_{vv}|^2\right\rangle\left\langle n|c_s s'_{vv}|^2\right\rangle}}$$

$$= \frac{a\rho'_{hv} + b_s(Z'_{dr})^{-1/2}}{\sqrt{a^2 + |b_s|^2(Z'_{dr})^{-1} + 2a(Z'_{dr})^{-1/2}\,\text{Re}(b_s\rho'_{hv})}},$$

(8.30)

where a, b_s, and c_s are given after Equations 8.27 and 8.28. It is evident that the bias of ρ_{hv}(STSR) depends on the beam direction, relative amplitude, and phase of the transmitted H, V fields and on the scattering properties of the hydrometeors.

Figure 8.6b through d shows ρ_{hv}(STSR) as a function of beam direction as well as amplitude and phase imbalance. The ρ_{hv}(STSR) measurement can be substantially biased, as can Z_{hv}(STSR). The bias can be either positive or negative depending on the beam position, balanced factors, and hydrometeor properties; this is due to the change in the relative strengths of the horizontal and vertical wave fields. The bias, however, can be corrected through either the scattering matrix or polarimetric variables. The bias correction for the polarimetric variables is done by jointly solving Equations 8.27 and 8.28 and Equation 8.30 for Z'_{dr} and ρ'_{hv}, respectively.

8.2.4.4 Linear Depolarization Ratio

The LDR cannot be measured if the STSR mode is used. However, if observations are made with the ATSR mode, we can calculate LDR and relate it to the backscattering matrix elements $s_{hv}^{(b)}, s_{vh}^{(b)}$. By assuming $s_{hv}^{(b)} = s_{vh}^{(b)} = 0$, the measured LDR is strictly the LDR bias for PPAR. Thus it can be shown that Bias(LDR) is

$$\text{Bias}(LDR)_h \equiv 10\log\left(\frac{\left\langle n|s_{vh}^{(p)}|^2\right\rangle}{\left\langle n|s_{hh}^{(p)}|^2\right\rangle}\right) = 10\log\left(\frac{\left\langle n|s'_{vv}\sin\theta\cos\theta\sin\phi|^2\right\rangle}{\left\langle n|s'_{hh}\cos^2\phi + s'_{vv}\cos^2\theta\sin^2\phi|^2\right\rangle}\right)$$

(8.31)

$$= 10\log\left(\frac{d^2}{a^2 Z'_{dr} + b^2 + 2ab\sqrt{Z'_{dr}}\,\text{Re}(\tilde{\rho}'_{hv})}\right)$$

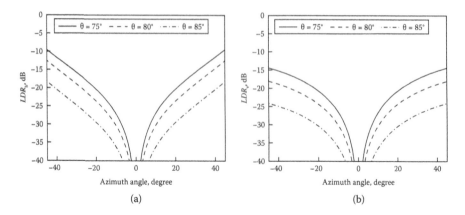

FIGURE 8.7 Bias of the linear depolarization ratio versus the direction of the electronically steered beam. (a) LDR_h and (b) LDR_v ($Z'_{dr} = 1.0$, $\rho_{hv} = 0.9$ for all cases).

$$
\text{Bias}(LDR)_v \equiv 10\log\left(\frac{\left\langle n\left|s_{hv}^{(p)}\right|^2\right\rangle}{\left\langle n\left|s_{vv}^{(p)}\right|^2\right\rangle}\right) = 10\log\left(\frac{\left\langle n\left|s'_{vv}\sin\theta\cos\theta\sin\phi\right|^2\right\rangle}{\left\langle n\left|s'_{vv}\sin^2\theta\right|^2\right\rangle}\right)
$$
$$
= 10\log\left(\frac{d^2}{c^2}\right). \tag{8.32}
$$

Bias$(LDR)_{h,v}$ increases as the beam points away from broadside in Figure 8.7. Bias$(LDR)_h$ is a few decibels larger than Bias$(LDR)_v$ because the co-polar power $\left(Z_h^{(p)}\right)$ for horizontal polarization is lower than that for vertical polarization in the directions of a large $|\phi|$. The system (LDR) can be -15 to -10 dB for the $25°$ elevation and $45°$ azimuth span needed for weather observations. This is too large to make meaningful meteorological observations (i.e., typically hydrometeor $LDRs < -20$ dB). Fortunately, this is amendable by calibrating the scattering matrix.

The polarization bias can also be corrected for in the scattering matrix or directly in the radar measurements of polarimetric variables, as formulated for Z_{DR}, ρ_{hv}, and LDR. However, these corrections are dependent on the scan beam direction. Calibrating over thousands of beams and knowing the power balance within 0.1 dB and the relative phase within a few degrees is challenging. The feasibility and performance on a real PPAR system remains to be examined with known PPAR characteristics.

8.3 CYLINDRICAL POLARIMETRIC PHASED ARRAY RADAR

Realizing the issue of geometrically induced cross-pol coupling with a PPAR and to avoid calibrating thousands of beams, CPPAR was proposed for future weather measurements and multiple missions (Zhang et al. 2011c). We discuss its concept and pattern characteristics next.

8.3.1 CPPAR CONCEPT AND FORMULATION

Figure 8.8 shows the CPPAR concept. As sketched in Figure 8.8a, there are $M \times N$ dual-polarized radiating elements arranged azimuthally (M) and axially (N) on the surface of a cylinder. Multiple simultaneous beams are formed, with each beam generated from a sector of the cylindrical surface and the broadside direction along the bisector of the radiating/illuminated sector. Because the beam axis is always at the bisector on which there is symmetry for the contributing radiation elements, polarization orthogonality is preserved in all directions. As shown in Figure 8.8b, the cross-pol components on both sides of the bisector cancel each other, yielding the orthogonal dual-pol radiation from the sector. On the other hand, PPPAR does not have such symmetry and orthogonality when its beams point away from the principal planes, as shown in Figure 8.8c.

To study the radiation characteristics of a CPPAR, we choose a coordinate system with its z direction along the cylinder's axis (Figure 8.9). An array element (mn: mth row, nth column), comprised of crossed h and v dipoles is located at ϕ_n, z_m on the cylindrical surface at $\vec{r}_{mn} = R\cos\phi_n\hat{x} + R\sin\phi_n\hat{y} + z_m\hat{z}$, where R is the cylinder's radius and the row height z_m ranges from $-D/2$ to $+D/2$, where D is the axial length of the cylindrical array (equal to the diameter D of the WSR-88D). The azimuth location ϕ_n is measured relative to the x-axis and is $\phi_n = n\Delta\phi$, $n = 1,2,3...$.

The electric field at $\vec{r} = r\sin\theta\cos\phi\hat{x} + r\sin\theta\sin\phi\hat{y} + r\cos\theta\hat{z}$, transmitted by the mnth q (i.e., $q = h$ or v) dipole, is similar to Equation 8.1:

$$\vec{E}_{q,mn}(\vec{r}) = -\frac{k^2 e^{-jk|\vec{r}-\vec{r}_{mn}|}}{4\pi\varepsilon|\vec{r}-\vec{r}_{mn}|}\{\hat{r}\times[\hat{r}\times\vec{M}_{q,mn}),\tag{8.33}$$

where $\vec{M}_{q,mn}$ is the moment of dipole q at location mn, and $\hat{r}$ is the unit vector along $\vec{r}$.

Using the far-field approximation, we have the electric field at $\vec{r}$ radiated by the mnth q dipole

$$\vec{E}_{\text{th},mn} = E_{\text{th},mn}^{(p)}\vec{e}_{h,n} \quad \text{and} \quad \vec{E}_{\text{tv},mn} = E_{\text{tv},mn}^{(p)}\vec{e}_{v,n},\tag{8.34}$$

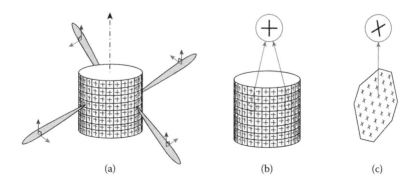

(a)　　　　　　　　(b)　　　　　　　　(c)

FIGURE 8.8 A sketch of the cylindrical polarimetric phased array radar with dual-pol radiation elements: (a) CPPAR conceptual sketch, (b) orthogonal dual-polarizations preserved with the symmetry in a CPPAR, and (c) cross-pol coupling in a PPPAR.

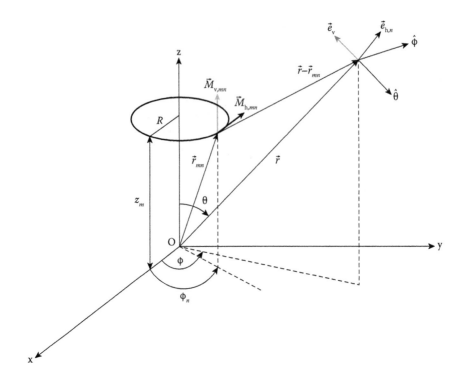

FIGURE 8.9 Coordinate system for CPPAR element radiation.

where $E^{(p)}_{\text{th, }mn}$ and $E^{(p)}_{\text{tv, }mn}$ are the fields transmitted by the h and v dipoles, respectively, along the normal to the plane of the dipoles (i.e., the crossed dipole's broadside direction) located at ϕ_n, z_m, expressed as

$$\begin{bmatrix} E^{(p)}_{\text{th, }mn} \\ E^{(p)}_{\text{tv, }mn} \end{bmatrix} \approx \frac{k^2}{4\pi\varepsilon r} e^{jk[z_m \cos\theta + R\sin\theta\cos(\phi-\phi_n)]} \begin{bmatrix} M_{\text{h, }mn} \\ M_{\text{v, }mn} \end{bmatrix}, \tag{8.35}$$

with

$$\vec{e}_{\text{h, }n} = \hat{y}' - \left[\hat{x}'\sin\theta\cos(\phi-\phi_n) + \hat{y}'\sin\theta\sin(\phi-\phi_n) + \hat{z}\cos\theta \right]\sin\theta\sin(\phi-\phi_n) \tag{8.36}$$

representing wave field polarization and magnitude for the ϕ_n angular rotation about z of the coordinate x- and y-axes, to x', y' for the mnth element. The value of $\vec{e}_{\text{v, }n}$ is identical to that given in Eq. 5b of Zhang et al. (2009).

To form a beam pointing in the (θ_0, ϕ_0) direction, a phase shift

$$\psi_{mn} = -k[z_m \cos\theta_0 + R\sin\theta_0 \cos(\phi_0 - \phi_n)] \tag{8.37}$$

is applied to each of the mn elements that are used to form the beam. The phase shifts given by Equations 8.34, 8.35, and 8.37 produce a beam in the (θ_0, ϕ_0) direction.

The transmitted horizontal and so-called vertical* fields $E_{th,mn}$ and $E_{tv,mn}$ in the plane of polarization are given by

$$\begin{bmatrix} E_{th\,mn} \\ E_{tv\,mn} \end{bmatrix} = \frac{k^2}{4\pi\varepsilon r} \overline{\overline{P}}_{mn} \begin{bmatrix} M_{h,\,mn} \\ M_{v,\,mn} \end{bmatrix} \exp\left(j\psi_{mn}^{(0)}\right), \tag{8.38}$$

where

$$\psi_{mn}^{(0)} = k\left\{ z_m \left[\cos\theta - \cos\theta_0\right] + R\left[\sin\theta\cos(\phi - \phi_n) - \sin\theta_0\cos(\phi_0 - \phi_n)\right]\right\},$$

and

$$\overline{\overline{P}}_{mn} = \begin{bmatrix} \cos(\phi - \phi_n) & 0 \\ -\cos\theta\sin(\phi - \phi_n) & \sin\theta \end{bmatrix} \tag{8.39}$$

is a matrix that projects the elements' broadside electric field to the plane of polarization at $\hat{r}$ and accounts for h dipole orientation at ϕ_n.

Radiation patterns with specified sidelobe levels and beamwidths can be achieved with a proper weight $\left(w_{mn}^{(q)}\right)$ applied to each element. Hence, the total transmitted field at $\hat{r}$ is the weighted contributions from all the active elements used to form the beam at (θ_0, ϕ_0). This field can be expressed as

$$\begin{bmatrix} E_{th} \\ E_{tv} \end{bmatrix} = \frac{k^2}{4\pi\varepsilon r} \sum_{m,\,n} \overline{\overline{P}}_{mn} \overline{\overline{W}}_{mn} \exp\left(j\psi_{mn}^{(0)}\right), \tag{8.40}$$

where the weighting matrix is applied to each element, and the angular dependence of the broadside field generated by the mnth H and V dipole moments is incorporated into $\overline{\overline{W}}_{mn}$, which is

$$W_{mn} = \begin{bmatrix} \dfrac{1}{\cos(\phi_0 - \phi_n)} & 0 \\ 0 & \dfrac{1}{\sin\theta_0} \end{bmatrix} w_{mn}^{(i)}, \tag{8.41}$$

where the upper-left matrix element $\dfrac{1}{\cos(\phi_0 - \phi_n)}$ compensates for the projection loss of the H-dipole radiated field onto the horizontal polarization direction along the beam's boresight. The boresight always lies in the plane containing the bisector of the angle that encompasses the azimuth sector containing the elements that form the beam; in effect the boresight of the CPPAR is always in the broadside direction. Alternatively

$$\phi_n = n\Delta\phi = \phi_0 \pm n'\Delta\phi \ (n_0 \pm n')\Delta\phi, \ [n' = 0, 1, 2, \ldots N_a], \tag{8.42}$$

---

* The vertical field lies in the vertical plane but is only vertical at the 90° zenith angle.

is the location of the active dipoles in an angular sector (e.g., 120° for a three-beam CPPAR) centered on ϕ_0 with $(2N_a+1)$ active array elements in the azimuthal span of $[n_0 - N_a, n_0 + N_a]$. Similarly, the lower-right matrix element $\dfrac{1}{\sin\theta_0}$ compensates for the projection loss of the V-dipole radiated field onto the vertical direction; this correction is normally close to unity because the elevation angle $(\pi/2 - \theta_0)$ for weather measurements is typically small. Cross-pol coupling can be corrected by including a term of $\dfrac{\cos\theta_0 \sin(\phi_0 - \phi_n)}{\sin\theta_0 \cos(\phi_0 - \phi_n)}$ in lower-left, as in the correction (Equation 8.19).

The scalar weight $w_{mn}^{(i)}$ is for isotropic radiators; these weights are selected to control the sidelobe levels. The WSR-88D antenna pattern is mimicked by selecting

$$w_{mn}^{(i)} = \left(\frac{\left\{ 1 - 4\left[R^2 \sin^2(\phi_0 - \phi_n) + z_m^2 \right]/D^2 \right\} + b}{1 + b} \right) \cos(\phi_0 - \phi_n). \qquad (8.43)$$

The term in the big parentheses is equivalent to the WSR-88D illumination taper but applied to those mnth dipoles whose projection onto the vertical plane bisecting the cylinder lies within the $\pi D^2/4$ area, where D is the diameter of the WSR-88D dish antenna (dipoles outside this circular area, but lying within the angular sector of elements forming the beam, have zero weight); the term $\cos(\phi - \phi_0)$ accounts for the change of the density of the array elements projected onto the vertical plane and the term $b = 0.16$ accounts for edge illumination of the WSR-88D reflector (Doviak et al. 1998). Although $w_{mn}^{(i)}$ mimics the illumination taper on the WSR-88D antenna for the boresight direction, the analogy no longer exists for azimuths in off-boresight directions. This is because the active elements on the cylinder have a density that lacks the symmetry of the dish antenna about the vertical bisector of the circular area.

On the beam's boresight (i.e., $\theta = \theta_0$, $\phi = \phi_0$), the radiated fields from all the elements are in phase so the phase term in Equation 8.40 disappears. Because the active elements and the weighting factor $w_{mn}^{(i)}$ are symmetric about ϕ_0 and $z_m = 0$, there is no on-axis cross-polar radiation. That is, the vertically polarized wave field caused by the horizontal dipole at $\phi_0 - n'\Delta\phi$ cancels that field from the dipole at the opposite azimuth, $\phi_0 + n'\Delta\phi$. This cross-polar null on-axis is important for accurate polarimetric radar measurement of weather (Wang and Chandrasekar 2006; Zrnić et al. 2010a). This is one of the main reasons for using a CPPAR commutating scan in which the beam direction changes in azimuth by shifting a column of active elements and maintaining the weight symmetry about the beam center. This way, the beam characteristics of the CPPAR are scan invariant; this is not so for the PPPAR.

For the received wave field, the scattered wave fields are projected onto the respective dipole directions, and with the proper weighting and phase shifts. In this case the total received wave field is expressed as

$$\vec{E}_r^{(p)} = \frac{k^2}{4\pi\varepsilon r^2} \sum_{m,n} \overline{\overline{W}}_{mn}^{t} \overline{\overline{P}}_{mn}^{t} \overline{\overline{S}}' \overline{\overline{P}}_{mn} \overline{\overline{W}}_{mn} \vec{M}. \qquad (8.44)$$

8.3.2 Sample Calculation of CPPAR Patterns

The operational WSR-88D radar performs well for meteorological observations: it has a dish antenna with a diameter of 8.54 m, a beamwidth of about 1°, and the first sidelobe below −26 dB. It is desirable for the MPAR/CPPAR to have similar or better performance. Considering the tradeoff for maximizing the effective aperture and the number of beams, it is efficient to use four simultaneous beams for a CPPAR (Zhang et al. 2013, 2015). The simulated 3D power patterns are shown in Figure 8.10. Co-polar patterns are on the left, and cross-polar patterns on the right. The top row is that the

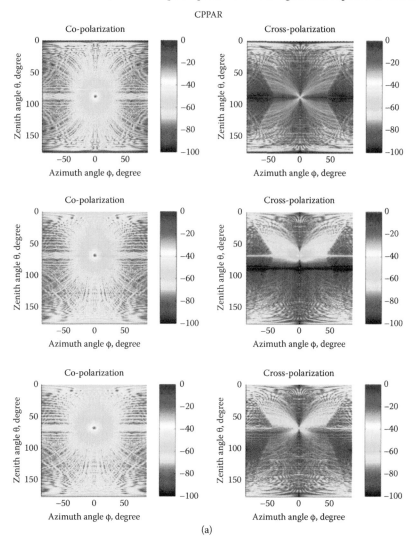

(a)

FIGURE 8.10 The simulated 3D power patterns. (a) CPPAR co-polar (left column) and cross-polar (right column) power patterns at elevation of 0° (top row), 20° without calibration (middle row), and 20° with calibration (bottom row). *(Continued)*

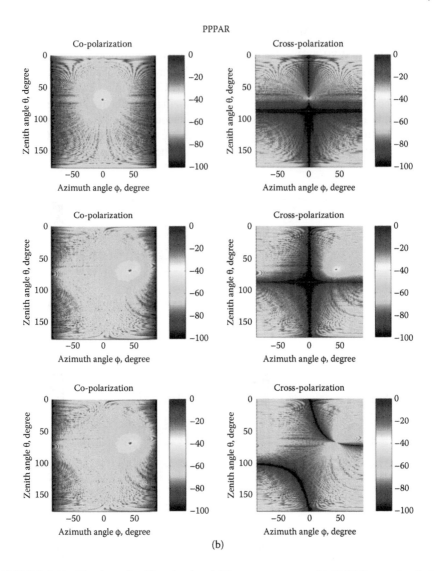

FIGURE 8.10 (Continued)   The simulated 3D power patterns. (b) PPPAR co-polar (left column) and cross-polar (right column) power patterns at azimuth of 0° (top row), 45° without calibration (middle row), and 45° with calibration (bottom row).

beam points to the broadside direction (0° elevation). The cross-polar radiation is at least 60 dB below the co-polar peak throughout, indicating the CPPAR performs well for preserving polarization purity. The middle row is for a beam direction of 20° elevation, without applying calibration. The cross-pol level increases but is still below –20 dB and there is still a null at the vertical plane. The bottom row is for 20° elevation and, with calibration, has a cross-pol level below –50 dB.

For comparison, the PPPAR patterns are shown in Figure 8.10(b). As expected, cross-polarization level is low (<–55 dB) when its beam points on the principal planes (top row). When the beam points off principal planes at a direction of 20° elevation and 45° azimuth, however, the PPPAR has a cross-polar main lobe coaxial with the co-polar main lobe, and it peaks at −9.3 dB from the co-polar peak (middle row). Such high cross-polar level is unacceptable for weather measurements. Although the cross-polarization coupling can be calibrated using the formulation described in Section 8.2, the calibration over thousands of beam is not desirable in real operations. Hence, the CPPAR is better suited for quantitative polarimetric measurements.

8.3.3   CPPAR Development

To test the CPPAR concept, a small-scale CPPAR demonstrator is being built with a joint effort between the Advanced Radar Research Center staff at OU and NSSL engineers. Figure 8.11 displays a picture of the CPPAR demonstrator mounted on a trailer and pattern results. The cylinder has a diameter of 2 m and is 2 m tall. There are 96 columns, but only half of them are populated due to budget constraints. Furthermore, a frequency scan antenna is used for each column to save on costs while maintaining high performance in high isolation, low cross-polarization, and low sidelobe level (Karimkashi and Zhang 2013, 2015). The bottom row shows the isolate and embedded antenna patterns of each column for horizontal polarization (left) and vertical polarization (right). Although the embedded patterns are different, the formed H- and V-pol patterns match with each other well, through pattern synthesis.

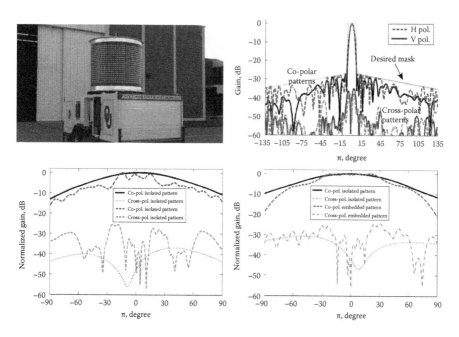

FIGURE 8.11   Preliminary results of the cylindrical polarimetric phased array radar demonstrator (From Karimkashi, S., and G. Zhang, 2015. *IEEE Transactions on Geoscience and Remote Sensing*, 53, 2810–2818.).

While studying both the PPPAR and CPPAR for future weather observation, we have found the advantages of the CPPAR to be as follows:

1. Scan-invariant polarimetric radar measurements with the same beamwidth and polarization characteristics in all azimuth angles for each elevation, allowing for easier calibration and data interpretation.
2. *Polarization purity.* Dual-polarized (H and V) wave fields are orthogonal in all directions, resulting in high-quality polarimetric data. Compensation is only needed for horizontal and vertical polarizations separately, but cross-polarization isolation is maintained.
3. *High efficiency of utilizing radiation power.* Only certain array elements are activated and properly weighted to achieve the desired beams. The elements on the broadside are mostly activated and more heavily weighted. Hence there is less scan loss due to the element radiation pattern.
4. Optimal use of the antenna aperture for fast data update or for multiple functionality with simultaneous multiple beams.
5. Flexibility to choose the number of beams (e.g., two, three, or four) and assign different tasks among the beams. For example, if four beams are generated, two beams can be used for weather surveillance and the other two for aircraft tracking—making it a candidate for the future MAPR. This flexibility can be combined with multiple frequencies used in the currently proposed PPPAR, namely, one band of frequencies for weather functions and another band for aircraft surveillance.
6. No need for face-to-face matching as required for a PPPAR, where each face is an individual radar system and could have different characteristics that need to be matched.

Although the CPPAR is a natural fit for MPAR and has the aforementioned advantages, it is not without problems. These issues include the complexity of the system design and development, difficulty in controlling the sidelobes, and the need to synchronize all the elements to form simultaneous multiple beams. There are other common PPPAR issues such as polarization mode selection, design of radiating elements and array optimization, waveform design, and so forth. Although these issues are challenging, they are solvable. Hence, the challenges can be viewed as good research opportunities for the weather radar community to advance its radar technology with potential for new scientific findings and better weather service.

APPENDIX 8A: PPAR FORMULATION FOR APERTURE AND PATCH ELEMENTS

In the text, we discussed the PPAR biases and their correction for the fixed antenna arrays with the radiating elements of a pair of dipoles. Other dual-polarization radiation elements such as horn aperture and microstrip patch antennas are also commonly used in PAR systems. They have different radiation patterns and hence cause different biases than dipoles. They are discussed in the following section.

A.1 APERTURE

The aperture antenna is an open-ended waveguide. By using the field equivalence principle, the actual sources can be replaced by the equivalent electric and magnetic current sources. For a horizontally polarized rectangular aperture, the longer side lies along the z-axis (Figure 8A.1a). A TE_{10} mode is assumed to propagate inside the waveguide feeding the aperture. After omitting the co-factor $\dfrac{jke^{-jkr}}{4r}abE_0$, the electric field radiated from a horizontally polarized aperture (Balanis 2005; Lei et al. 2013) is as follows:

$$E_\phi^{(h)} = \sin\theta \cdot f^{(h)}(\theta, \phi), \tag{8A.1}$$

$$E_\theta^{(h)} = 0, \tag{8A.2}$$

where

$$f^{(h)}(\theta, \phi) = \frac{\left(\dfrac{\pi}{2}\right)^2 \cos\left(\dfrac{k_0 a}{2}\cos\theta\right)}{\left(\dfrac{\pi}{2}\right)^2 - \left(\dfrac{k_0 a}{2}\cos\theta\right)^2} \times \frac{\sin\left(\dfrac{k_0 b}{2}\sin\theta\sin\phi\right)}{\dfrac{k_0 b}{2}\sin\theta\sin\phi}. \tag{8A.3}$$

The superscript (h) denotes that the radiated field originates from the horizontally polarized aperture, and $k_0 = 2\pi/\lambda_0$ is the free space wave number.

Similarly, for the vertically polarized aperture, the longer side lies along the y-axis (Figure 8A.1b). Similarly, a TE_{10} mode and infinite ground plane are assumed, the radiated field for which is as follows:

$$E_\phi^{(v)} = \cos\theta\sin\phi \cdot f^{(v)}(\theta, \phi), \tag{8A.4}$$

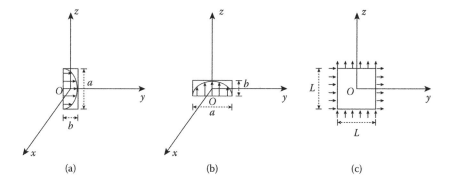

FIGURE 8A.1 Sketches of aperture and patch antenna radiation: (a) horizontally polarized aperture, (b) vertically polarized aperture, and (c) dual-pol patch.

$$E_\theta^{(v)} = -\cos\phi \cdot f^{(v)}(\theta, \phi), \tag{8A.5}$$

where

$$f^{(v)}(\theta, \phi) = \frac{\left(\dfrac{\pi}{2}\right)^2 \cos\left(\dfrac{k_0 a}{2}\sin\theta\sin\phi\right)}{\left(\dfrac{\pi}{2}\right)^2 - \left(\dfrac{k_0 a}{2}\sin\theta\sin\phi\right)^2} \cdot \frac{\sin\left(\dfrac{k_0 b}{2}\cos\theta\right)}{\dfrac{k_0 b}{2}\cos\theta}. \tag{8A.6}$$

Following the procedure to obtain Equations 8.11 and 8.12 in Section 8.2.1, we obtain the projection matrix for the aperture as follows:

$$\overline{\overline{P}} = \begin{bmatrix} \sin\theta \cdot f^{(h)}(\theta, \phi) & \cos\theta\sin\phi \cdot f^{(v)}(\theta, \phi) \\ 0 & \cos\phi \cdot f^{(v)}(\theta, \phi) \end{bmatrix}. \tag{8A.7}$$

The $f^{(h)}(\theta, \phi)$ and $f^{(v)}(\theta, \phi)$ given by Equations 8A.3 and 8A.6 in the above matrix are due to the finite size of the aperture. The terms other than $f^{(h)}(\theta, \phi)$ and $f^{(v)}(\theta, \phi)$ have similar forms as the P-matrix of dipoles (Zhang et al. 2009). The slightly different form is due to the complementary characteristics (Balanis 2005, chapter 9-2) of the dipole and the rectangular aperture in an infinite ground plane. Whereas a vertically oriented electric dipole generates fields that are isotropic in the x–y plane (i.e., the H-plane of the dipole), a vertically oriented narrow slot, having horizontally polarized fields, is well represented by a vertically oriented magnetic dipole generating fields that are isotropic in the E-plane, as well as the x–y plane. That is, the H-plane of the vertically oriented electric dipole and the E-plane of the vertically oriented narrow slot are both on the horizontal plane at $z = 0$.

A.2 PATCH

A microstrip patch antenna consists of an electrically conducting ground plane, a substrate, and an electrically conducting patch on top forming an open-ended cavity. If the substrate thickness is much thinner than the free space wavelength and the patch and ground plane are perfectly conducting, the four sides of this open-ended cavity can be modeled as perfect magnetic walls. Then the radiation field of a patch element can be calculated by assuming the space below the patch is a resonant cavity bounded on two sides by walls (i.e., the patch and the ground plane) that perfectly conduct electric currents and on the other four sides by walls that perfectly conduct magnetic currents.

For the square patch, the TM_{010} and TM_{100} modes (e.g. TM_{010} is the horizontal polarization; TM_{100} is the vertical polarization) have the same resonant frequency. Both modes can be excited and coexist independently inside the patch. For small value of substrate thickness, the electric fields radiated from the horizontally polarized square patch are given by Balanis (2005, Eqs. 14–44). In order to directly constitute the projection matrix in the next section, the cofactor $j\dfrac{2V_0e^{-jk_0r}}{\pi r}\cdot\dfrac{k_0L}{2}$ is extracted out and normalized co-polar electric field is

$$E_\phi^{(h)}\approx\sin\theta\cdot g^{(h)}(\theta,\phi),\qquad(8A.8)$$

and the cross-polar field is

$$E_\theta^{(h)}\approx 0,\qquad(8A.9)$$

where

$$g^{(h)}(\theta,\phi)=\frac{\sin\left(\dfrac{k_0L}{2}\cos\theta\right)}{\dfrac{k_0L}{2}\cos\theta}\cos\left(\dfrac{k_0L_e}{2}\sin\theta\sin\phi\right),\qquad(8A.10)$$

V_0 is the voltage across the open edges of the patch and the fringing fields at the edges account for most of the radiation.

Similarly, the normalized E fields of the vertically polarized patch are

$$E_\phi^{(v)}\approx\cos\theta\sin\phi\cdot g^{(v)}(\theta,\phi),\qquad(8A.11)$$

$$E_\theta^{(v)}\approx-\cos\phi\cdot g^{(v)}(\theta,\phi),\qquad(8A.12)$$

where

$$g^{(v)}(\theta,\phi)=\frac{\sin\left(\dfrac{k_0L}{2}\sin\theta\sin\phi\right)}{\dfrac{k_0L}{2}\sin\theta\sin\phi}\cos\left(\dfrac{k_0L_e}{2}\cos\theta\right).\qquad(8A.13)$$

L is the physical length of the square patch and is determined by the permittivity of the material in the cavity (typically $\lambda_o/3 < L < \lambda_o/2$). For an air substrate L is $\lambda_o/2$; for high permittivity substrate L approaches $\lambda_o/3$. The co- and cross-polarization patterns are not simple versions of one another simply rotated by 90° when the patch excitation is rotated by 90°. This is so because the coordinated system that defines the co- and cross-polar fields does not rotate [see definition 2 in figure 1 of Ludwig (1973)].

Similarly, the projection matrix for the patch is

$$\bar{\bar{P}}=\begin{bmatrix}\sin\theta\cdot g^{(h)}(\theta,\phi) & \cos\theta\sin\phi\cdot g^{(v)}(\theta,\phi)\\ 0 & \cos\phi\cdot g^{(v)}(\theta,\phi)\end{bmatrix},\qquad(8A.14)$$

where the sine function in $g^{(h)}$ and $g^{(v)}$ given by Equations 8A.10 and 8A.13 is due to the finite size of antenna, and the cosine function in them is the array factor for the two radiating slots on opposite sides of the patch. By comparing the P-matrix of the aperture and patch elements with the P-matrix of the crossed dipole element in Equation 8.9, and assuming that f and g in Equations 8A.7 and 8A.14 are equal to one, the P-matrices have the following relationships:

$$\left(\mathbf{P}_{dipole}^{t}\right)^{-1} \propto \mathbf{P}_{aperture}, \left(\mathbf{P}_{dipole}^{t}\right)^{-1} \propto \mathbf{P}_{patch}, \left(\mathbf{P}_{aperture}^{t}\right)^{-1} \propto \mathbf{P}_{dipole},$$

and   (8A.15)

$$\left(\mathbf{P}_{patch}^{t}\right)^{-1} \propto \mathbf{P}_{dipole}.$$

This shows that the radiation patterns of aperture and patch array antennas are complementary to that of dipoles.

Problems

8.1 What are the challenges to developing a PPAR for weather measurements?

8.2 Describe and compare the advantages and disadvantages of a PPPAR and a CPPAR for multiple missions. Discuss why a spherical array, like the geodesic dome, is not recommended for MPAR.

8.3 Discuss the difference in geometrically-induced cross-pol coupling for planar arrays composed of pairs of dipoles and slot-dipoles or dual-pol slots and dual-pol patches, respectively.

8.4 Can the geometrically induced cross-pol coupling be calibrated/corrected? What is the challenge in developing a PPPAR?

8.5 How do you weather radars in the future? What are the most important functionalities that you'd like to add to the current WSR-88DP radars?

References

Andrić, J., M. R. Kumjian, D. S. Zrnić, J. M. Straka, and V. M. Melnikov, 2013. Polarimetric signatures above the melting layer in winter storms: An observational and modeling study. *Journal of Applied Meteorology and Climatology*, **52**, 682–700.

Andsager, K., K. V. Beard, and N. F. Laird, 1999. Laboratory measurements of axis ratios for large raindrops. *Journal of the Atmospheric Sciences*, **56**, 2673–2683.

Atlas, D., 1964. *Advances in radar meteorology.* Vol. 10. Academic Press, pp. 317–478.

Atlas, D., 1990. *Radar in meteorology.* Vol. 806. American Meterological Society.

Atlas, D., R. C. Srivastava, and R. S. Sekhon, 1973. Doppler radar characteristics of precipitation at vertical incidence. *Reviews of Geophysics*, **11**, 1–35.

Atlas, D., and C. W. Ulbrich, 1977. Path- and area-integrated rainfall measurement by microwave attenuation in the 1–3 cm band. *Journal of Applied Meteorology*, **16**, 1322–1331.

Aydin, K., T. Seliga, and V. Balaji, 1986. Remote sensing of hail with a dual linear polarization radar. *Journal of Climate and Applied Meteorology*, **25**, 1475–1484.

Balakrishnan, N., and D. S. Zrnić, 1990. Estimation of rain and hail rates in mixed-phase precipitation. *Journal of the Atmospheric Sciences*, **47**, 565–583.

Balakrishnan, N., D. S. Zrnic, J. Goldhirsh, and J. Rowland, 1989. Comparison of simulated rain rates from disdrometer data employing polarimetric radar algorithms. *Journal of Atmospheric and Oceanic Technology*, **6**, 476–486.

Balanis, C. A., 2005. *Antenna theory: Analysis and design.* Vol. 1. Wiley.

Barber, P., and C. Yeh, 1975. Scattering of electromagnetic waves by arbitrarily shaped dielectric bodies. *Applied Optics*, **14**, 2864–2872.

Battaglia, A., S. Tanelli, G. M. Heymsfield, and L. Tian, 2014. The dual wavelength ratio knee: A signature of multiple scattering in airborne ku–ka observations. *Journal of Applied Meteorology and Climatology*, **53**, 1790–1808.

Battan, L., 1959. *Radar meteorology.* University of Chicago Press, 161 pp.

Battan, L., 1973. *Radar observation of the atmosphere.* University of Chicago Press, 323 pp.

Battan, L. J., 1953. Observation on the formation of precipitation in convective clouds. *Journal of Meteorology*, **10**, 311–324.

Beard, K. V., and C. Chuang, 1987. A new model for the equilibrium shape of raindrops. *Journal of the Atmospheric Sciences*, **44**, 1509–1524.

Beard, K. V., and A. R. Jameson, 1983. Raindrop canting. *Journal of the Atmospheric Sciences*, **40**, 448–454.

Beard, K. V., D. B. Johnson, and A. R. Jameson, 1983. Collisional forcing of raindrop oscillations. *Journal of the Atmospheric Sciences*, **40**, 455–462.

Beard, K. V., and R. J. Kubesh, 1991. Laboratory measurements of small raindrop distortion. Part 2: Oscillation frequencies and modes. *Journal of the Atmospheric Sciences*, **48**, 2245–2264.

Beard, K. V., and A. Tokay, 1991. A field study of small raindrop oscillations. *Geophysical Research Letters*, **18**, 2257–2260.

Bent, A. E., Massachusetts Institute of Technology, and Radiation Laboratory, 1943. *Radar echoes from atmospheric phenomena.* Radiation Laboratory, Massachusetts Institute of Technology.

Bohren, C. F., and D. R. Huffman, 1983. *Absorption and scattering of light by small particles.* Wiley, 530 pp.

Borgeaud, M., R. T. Shin, and J. A. Kong, 1987. Theoretical models for polarimetric radar clutter. *Journal of Electromagnetic Waves and Applications*, **1**, 73–89.

Born, M., and E. Wolf, 1999. *Principles of optics: Electromagnetic theory of propagation, interference and diffraction of light.* Cambridge University Press.

Brandes, E. A., and K. Ikeda, 2004. Freezing-level estimation with polarimetric radar. *Journal of Applied Meteorology*, **43**, 1541–1553.

Brandes, E. A., G. Zhang, and J. Vivekanandan, 2002. Experiments in rainfall estimation with a polarimetric radar in a subtropical environment. *Journal of Applied Meteorology*, **41**, 674–685.

Brandes, E. A., G. Zhang, and J. Vivekanandan, 2003. An evaluation of a drop distribution-based polarimetric radar rainfall estimator. *Journal of Applied Meteorology*, **42**, 652–660.

Brandes, E. A., G. Zhang, and J. Vivekanandan, 2004a. Comparison of polarimetric radar drop size distribution retrieval algorithms. *Journal of Atmospheric and Oceanic Technology*, **21**, 584–598.

Brandes, E. A., G. Zhang, and J. Vivekanandan, 2004b. Drop size distribution retrieval with polarimetric radar: Model and application. *Journal of Applied Meteorology*, **43**, 461–475.

Brandes, E. A., G. Zhang, and J. Vivekanandan, 2005. Corrigendum. *Journal of Applied Meteorology*, **44**, 186.

Brandes, E. A., K. Ikeda, G. Zhang, M. Schönhuber, and R. M. Rasmussen, 2007. A statistical and physical description of hydrometeor distributions in Colorado snowstorms using a video disdrometer. *Journal of Applied Meteorology and Climatology*, **46**, 634–650.

Bringi, V., and V. Chandrasekar, 2001. *Polarimetric Doppler weather radar: Principles and applications.* Cambridge University Press.

Bringi, V. N., V. Chandrasekar, N. Balakrishnan, and D. S. Zrnić, 1990. An examination of propagation effects in rainfall on radar measurements at microwave frequencies. *Journal of Atmospheric and Oceanic Technology*, **7**, 829–840.

Bringi, V. N., V. Chandrasekar, and R. Xiao, 1998. Raindrop axis ratios and size distributions in Florida rainshafts: An assessment of multiparameter radar algorithms. *IEEE Transactions on Geoscience and Remote Sensing*, **36**, 703–715.

Bringi, V. N., and A. Hendry, 1990. Technology of polarization diversity radars for meteorology. In *Radar in meteorology*, D. Atlas, Ed., American Meteorological Society, pp. 153–190.

Bringi, V. N., G.-J. Huang, V. Chandrasekar, and E. Gorgucci, 2002. A methodology for estimating the parameters of a gamma raindrop size distribution model from polarimetric radar data: Application to a squall-line event from the TRMM/Brazil campaign. *Journal of Atmospheric and Oceanic Technology*, **19**, 633–645.

Bringi, V. N., T. Keenan, and V. Chandrasekar, 2001. Correcting C-band radar reflectivity and differential reflectivity data for rain attenuation: A self-consistent method with constraints. *IEEE Transactions on Geoscience and Remote Sensing*, **39**, 1906–1915.

Bringi, V. N., R. M. Rasmussen, and J. Vivekanandan, 1986a. Multiparameter radar measurements in Colorado convective storms. Part I: Graupel melting studies. *Journal of the Atmospheric Sciences*, **43**, 2545–2563.

Bringi, V. N., T. A. Seliga, and K. Aydin, 1984. Hail detection with a differential reflectivity radar. *Science*, **225**, 1145–1147.

Bringi, V. N., T. A. Seliga, and S. M. Cherry, 1983. Statistical properties of the dual-polarization differential reflectivity (ZDR) radar signal. *IEEE Transactions on Geoscience and Remote Sensing*, **GE–21**, 215–220.

Bringi, V. N., J. Vivekanandan, and J. D. Tuttle, 1986b. Multiparameter radar measurements in Colorado convective storms. Part II: Hail detection studies. *Journal of the Atmospheric Sciences*, **43**, 2564–2577.

Brookner, E., 2008. Now: Phased-array radars: Past, astounding breakthroughs and future trends (January 2008). *Microwave Journal*, **51**, 31–18.

Bukovčić, P., D. Zrnić, and G. Zhang, 2015. Convective–stratiform separation using video disdrometer observations in central Oklahoma–the Bayesian approach. *Atmospheric Research*, **155**, 176–191.

Cao, Q., M. B. Yeary, and G. Zhang, 2012a. Efficient ways to learn weather radar polarimetry. *IEEE Transactions on Education*, **55**, 58–68.

Cao, Q., and G. Zhang, 2009. Errors in estimating raindrop size distribution parameters employing disdrometer and simulated raindrop spectra. *Journal of Applied Meteorology and Climatology*, **48**, 406–425.

Cao, Q., G. Zhang, E. Brandes, T. Schuur, A. Ryzhkov, and K. Ikeda, 2008. Analysis of video disdrometer and polarimetric radar data to characterize rain microphysics in Oklahoma. *Journal of Applied Meteorology and Climatology*, **47**, 2238–2255.

Cao, Q., G. Zhang, E. A. Brandes, and T. J. Schuur, 2010. Polarimetric radar rain estimation through retrieval of drop size distribution using a Bayesian approach. *Journal of Applied Meteorology and Climatology*, **49**, 973–990.

Cao, Q., G. Zhang, R. D. Palmer, M. Knight, R. May, and R. J. Stafford, 2012b. Spectrum-Time Estimation and Processing (STEP) for improving weather radar data quality. *IEEE Transactions on Geoscience and Remote Sensing*, **50**, 4670–4683.

Cao, Q., G. Zhang, and M. Xue, 2013. A variational approach for retrieving raindrop size distribution from polarimetric radar measurements in the presence of attenuation. *Journal of Applied Meteorology and Climatology*, **52**, 169–185.

Chandrasekar, V., and V. N. Bringi, 1988. Error structure of multiparameter radar and surface measurements of rainfall. Part I: Differential reflectivity. *Journal of Atmospheric and Oceanic Technology*, **5**, 783–795.

Chandrasekar, V., W. A. Cooper, and V. N. Bringi, 1988. Axis ratios and oscillations of raindrops. *Journal of the Atmospheric Sciences*, **45**, 1323–1333.

Cheng, L., and M. English, 1983. A relationship between hailstone concentration and size. *Journal of the Atmospheric Sciences*, **40**, 204–213.

Cheong, B., J. Kurdzo, G. Zhang, and R. Palmer, 2013. The impacts of multi-lag moment processor on a solid-state polarimetric weather radar. *AMS 36th Conference on Radar Meteorology*.

Chuang, C. C., and K. V. Beard, 1990. A numerical model for the equilibrium shape of electrified raindrops. *Journal of the Atmospheric Sciences*, **47**, 1374–1389.

Cole, K. S., and R. H. Cole, 1941. Dispersion and absorption in dielectrics I. Alternating current characteristics. *The Journal of Chemical Physics*, **9**, 341–351.

Crain, G., and D. Staiman, 2007. Polarization selection for phased array weather radar. *Proceedings of the AMS Annual Meeting: 23rd Conference on IIPS*, San Antonio, TX.

Crane, R. K., 1996. *Electromagnetic wave propagation through rain*. Wiley-Interscience.

Debye, P. J. W., 1929. *Polar molecules*. Chemical Catalog Company.

Doviak, R., D. Zrnic, J. Carter, A. Ryzhkov, S. Torres, and A. Zahrai, 1998. *Polarimetric upgrades to improve rainfall measurements*. NSSL report, 110 pp. Available at https://www.nssl.noaa.gov/publications/wsr88d_reports/2pol_upgrades.pdf

Doviak, R. J., V. Bringi, A. Ryzhkov, A. Zahrai, and D. Zrnić, 2000. Considerations for polarimetric upgrades to operational WSR-88D radars. *Journal of Atmospheric and Oceanic Technology*, **17**, 257–278.

Doviak, R. J., L. Lei, G. Zhang, J. Meier, and C. Curtis, 2011. Comparing theory and measurements of cross-polar fields of a phased-array weather radar. *IEEE on Geoscience and Remote Sensing Letters*, **8**, 1002–1006.

Doviak, R. J., and D. Sirmans, 1973. Doppler radar with polarization diversity. *Journal of the Atmospheric Sciences*, **30**, 737–738.

Doviak, R. J., and D. S. Zrnić, 1984, 1993, 2006. *Doppler radar and weather observations*. Academic Press, 562 pp.

Fabry, F., and W. Szyrmer, 1999. Modeling of the melting layer. Part II: Electromagnetic. *Journal of the Atmospheric Sciences*, **56**, 3593–3600.

Fradin, A. Z., 1961. *Microwave antennas*. Pergamon Press.

Fulton, R. A., J. P. Breidenbach, D.-J. Seo, D. A. Miller, and T. O'Bannon, 1998. The WSR-88D rainfall algorithm. *Weather and Forecasting*, **13**, 377–395.

Gans, R., 1912. Über die form ultramikroskopischer goldteilchen. *Annalen der Physik*, **342**, 881–900.

Gao, J., M. Xue, K. Brewster, and K. K. Droegemeier, 2004. A three-dimensional variational data analysis method with recursive filter for Doppler radars. *Journal of Atmospheric and Oceanic Technology*, **21**, 457–469.

Giangrande, S. E., and A. V. Ryzhkov, 2008. Estimation of rainfall based on the results of polarimetric echo classification. *Journal of Applied Meteorology and Climatology*, **47**, 2445–2462.

Goodman, J., B. T. Draine, and P. J. Flatau, 1991. Application of fast-Fourier-transform techniques to the discrete-dipole approximation. *Optics Letters*, **16**, 1198–1200.

Gossard, E., D. Wolfe, K. Moran, R. Paulus, K. Anderson, and L. Rogers, 1998. Measurement of clear-air gradients and turbulence properties with radar wind profilers. *Journal of Atmospheric and Oceanic Technology*, **15**, 321–342.

Gradshteyn, I., and I. Ryzhik, 1994. *Table of integrals, series and products*. 5th ed. Academic Press, 1204 pp.

Green, A. W., 1975. An approximation for the shapes of large raindrops. *Journal of Applied Meteorology*, **14**, 1578–1583.

Groginsky, H. L., and K. M. Glover, 1980. Weather radar canceller design. *19th Conference on Radar Meteorology*, pp. 192–198.

Gunn, R., and G. D. Kinzer, 1949. The terminal velocity of fall for water droplets in stagnant air. *Journal of Meteorology*, **6**, 243–248.

Hall, M., S. Cherry, J. Goddard, and G. Kennedy, 1980. Rain drop sizes and rainfall rate measured by dual-polarization radar. *Nature* **285**, 195–198.

Han, J., M. Kamber, and J. Pei, 2011, *Data mining: Concepts and techniques*. 3rd ed. Morgan Kaufmann, 744 pp.

Handbook, F., 2006. *Federal meteorological handbook No. 11, Doppler radar meteorological observations part C*. 390 pp.

Harrington, R. F., 1968. *Field computation by moment methods*. Macmillan.

Harris, B., and G. Kelly, 2001. A satellite radiance-bias correction scheme for data assimilation. *Quarterly Journal of the Royal Meteorological Society*, **127**, 1453–1468.

Heinselman, P., D. Priegnitz, K. Manross, and R. Adams, 2006. Comparison of storm evolution characteristics: The NWRT and WSR-88D. *23rd Conference on Severe Local Storms*, American Meteorological Society.

Heinselman, P. L., D. L. Priegnitz, K. L. Manross, T. M. Smith, and R. W. Adams, 2008. Rapid sampling of severe storms by the national weather radar testbed phased array radar. *Weather and Forecasting*, **23**, 808–824.

Hildebrand, P. H., and R. Sekhon, 1974. Objective determination of the noise level in Doppler spectra. *Journal of Applied Meteorology*, **13**, 808–811.

Hitschfeld, W., and J. Bordan, 1954. Errors inherent in the radar measurement of rainfall at attenuating wavelengths. *Journal of Meteorology*, **11**, 58–67.

Hogan, R. J., 2007. A variational scheme for retrieving rainfall rate and hail reflectivity fraction from polarization radar. *Journal of Applied Meteorology and Climatology*, **46**, 1544–1564.

Holroyd, E. W., 1972. *The meso- and microscale structure of Great Lakes snowstorm bands: A synthesis of ground measurements, radar data, and satellite observations*. State University of New York, Albany, NY.

Hong, S.-Y., and J.-O. J. Lim, 2006. The WRF single-moment 6-class microphysics scheme (WSM6). *Asia-Pacific Journal of Atmospheric Sciences*, **42**, 129–151.

Hopf, A. P., J. L. Salazar, R. Medina, V. Venkatesh, E. J. Knapp, S. J. Frasier, and D. J. McLaughlin, 2009. CASA phased array radar system description, simulation and products. *2009 IEEE International Geoscience and Remote Sensing Symposium*.

Huang, X.-Y., 2000. Variational analysis using spatial filters. *Monthly Weather Review*, **128**, 2588–2600.

Hubbert, J., M. Dixon, and S. Ellis, 2009b. Weather radar ground clutter. Part II: Real-time identification and filtering. *Journal of Atmospheric and Oceanic Technology*, **26**, 1181–1197.

Hubbert, J., M. Dixon, S. Ellis, and G. Meymaris, 2009a. Weather radar ground clutter. Part I: Identification, modeling, and simulation. *Journal of Atmospheric and Oceanic Technology*, **26**, 1165–1180.

Hubbert, J. C., V. N. Bringi, and D. Brunkow, 2003. Studies of the polarimetric covariance matrix. Part I: Calibration methodology. *Journal of Atmospheric and Oceanic Technology*, **20**, 696–706.

Hubbert, J. C., S. M. Ellis, M. Dixon, and G. Meymaris, 2010a. Modeling, error analysis, and evaluation of dual-polarization variables obtained from simultaneous horizontal and vertical polarization transmit radar. Part I: Modeling and antenna errors. *Journal of Atmospheric and Oceanic Technology*, **27**, 1583–1598.

Hubbert, J. C., S. M. Ellis, M. Dixon, and G. Meymaris, 2010b. Modeling, error analysis, and evaluation of dual-polarization variables obtained from simultaneous horizontal and vertical polarization transmit radar. Part II: Experimental data. *Journal of Atmospheric and Oceanic Technology*, **27**, 1599–1607.

Ice, R. L., R. D. Rhoton, J. C. Krause, D. S. Saxion, O. E. Boydstun, A. K. Heck, J. N. Chrisman, D. S. Berkowitz, W. D. Zittel, and D. A. Warde, 2009. Automatic clutter mitigation in the WSR-88D, design, evaluation, and implementation. *34th Conference on Radar Meteorology*, Williamsburg, VA.

Ivić I. R., 2014. On the use of a radial-based noise power estimation technique to improve estimates of the correlation coefficient on dual-polarization weather radars. *Journal of Atmospheric and Oceanic Technology* **31**, 1867–1880.

Ivić, I. R., C. Curtis, and S. M. Torres, 2013. Radial-based noise power estimation for weather radars. *Journal of Atmospheric and Oceanic Technology*, **30**, 2737–2753.

Ishimaru, A., 1978, 1997. *Wave propagation and scattering in random media*. Vol. 2. Academic Press, New York.

Ishimaru, A., 1991. *Electromagnetic wave propagation, radiation, and scattering*. Vol. 1. Prentice Hall, Englewood Cliffs, NJ.

Jameson, A., 1985. Deducing the microphysical character of precipitation from multiple-parameter radar polarization measurements. *Journal of Climate and Applied Meteorology*, **24**, 1037–1047.

Jameson, A., and A. Kostinski, 2010. Partially coherent backscatter in radar observations of precipitation. *Journal of the Atmospheric Sciences*, **67**, 1928–1946.

Janssen, L. H., and G. Van Der Spek, 1985. The shape of Doppler spectra from precipitation. *IEEE Transactions on Aerospace and Electronic Systems*, pp. 208–219.

Jones, D. M. A., 1959. The shape of raindrops. *Journal of Meteorology*, **16**, 504–510.

Jordan, R. L., B. L. Huneycutt, and M. Werner, 1995. The SIR-C/X-SAR synthetic aperture radar system. *IEEE Transactions on Geoscience and Remote Sensing*, **33**, 829–839.

Joss, J., and A. Waldvogel, 1969. Raindrop size distribution and sampling size errors. *Journal of the Atmospheric Sciences*, **26**, 566–569.

Jung, Y., M. Xue, and G. Zhang, 2010a. Simulations of polarimetric radar signatures of a supercell storm using a two-moment bulk microphysics scheme. *Journal of Applied Meteorology and Climatology*, **49**, 146–163.

Jung, Y., M. Xue, and G. Zhang, 2010b. Simultaneous estimation of microphysical parameters and the atmospheric state using simulated polarimetric radar data and an ensemble Kalman filter in the presence of an observation operator error. *Monthly Weather Review*, **138**, 539–562.

Jung, Y., M. Xue, G. Zhang, and J. M. Straka, 2008b. Assimilation of simulated polarimetric radar data for a convective storm using the ensemble Kalman filter. Part II: Impact of polarimetric data on storm analysis. *Monthly Weather Review*, **136**, 2246–2260.

Jung, Y., G. Zhang, and M. Xue, 2008a. Assimilation of simulated polarimetric radar data for a convective storm using the ensemble Kalman filter. Part I: Observation operators for reflectivity and polarimetric variables. *Monthly Weather Review*, **136**, 2228–2245.

Kalnay, E., 2003. *Atmospheric modeling, data assimilation, and predictability*. Cambridge University Press.

Karimkashi, S., and G. Zhang, 2013. A dual-polarized series-fed microstrip antenna array with very high polarization purity for weather measurements. *IEEE Transactions on Antennas and Propagation*, **61**, 5315–5319.

Karimkashi, S., and G. Zhang, 2015. Optimizing radiation patterns of a cylindrical polarimetric phased-array radar for multimissions. *IEEE Transactions on Geoscience and Remote Sensing*, **53**, 2810–2818.

Kay, S. M., 1998. *Fundamentals of statistical signal processing: Detection theory*. Vol. 2. Prentice Hall, Upper Saddle River, NJ.

Kerker, M., 1969. *The scattering of light*. Academic Press, New York.

Kessinger, C., S. Ellis, and J. Van Andel, 2003. The radar echo classifier: A fuzzy logic algorithm for the WSR-88D. *Preprints-CD, 3rd Conference on Artificial Applications to the Environmental Science*.

Kessler, E., 1969. On the distribution and continuity of water substance in atmospheric Circulations. Meteorological Monograph No. 32, *American Meteorological Society*, 84 p.

Kliche, D. V., P. L. Smith, and R. W. Johnson, 2008. L-moment estimators as applied to gamma drop size distributions. *Journal of Applied Meteorology and Climatology*, **47**, 3117–3130.

Knapp, E. J., J. Salazar, R. H. Medina, A. Krishnamurthy, and R. Tessier, 2011. Phase-tilt radar antenna array. *Microwave Conference (EuMC), 2011 41st European*, IEEE, pp. 1055–1058.

Knight, N. C., 1986. Hailstone shape factor and its relation to radar interpretation of hail. *Journal of Climate and Applied Meteorology*, **25**, 1956–1958.

Kruger, A., and W. F. Krajewski, 2002. Two-dimensional video disdrometer: A description. *Journal of Atmospheric and Oceanic Technology*, **19**, 602–617.

Kumjian, M. R., and A. V. Ryzhkov, 2008. Polarimetric signatures in supercell thunderstorms. *Journal of Applied Meteorology and Climatology*, **47**, 1940–1961.

Lamb, D., and J. Verlinde, 2011. *Physics and chemistry of clouds*. Cambridge University Press.

Lee, J.-S., M. R. Grunes, and G. De Grandi, 1999. Polarimetric SAR speckle filtering and its implication for classification. *IEEE Transactions on Geoscience and Remote Sensing*, **37**, 2363–2373.

Lee, R., 1978. Performance of the poly-pulse-pair Doppler estimator. *Lassen Research Memo*, 78–03, Chapter 20a.

Lee, R., G. Della Bruna, and J. Joss, 1995. Intensity of ground clutter and of echoes of anomalous propagation and its elimination. *27th Conference on Radar Meteorology*, Vail, CO.

Lei, L., 2009. *Simulations and processing of polarimetric radar signals based on numerical weather prediction model output*. University of Oklahoma.

Lei, L., Z. Guifu, and R. J. Doviak, 2013. Bias correction for polarimetric phased-array radar with idealized aperture and patch antenna elements. *IEEE Transactions on Geoscience and Remote Sensing*, **51**, 473–486.

Lei, L., G. Zhang, B. L. Cheong, R. D. Palmer, and M. Xue, 2009a. Simulations of polarimetric radar signals based on numerical weather prediction model output. *25th Conference on International Interactive Information and Processing Systems (IIPS) for Meteorology, Oceanography, and Hydrology*.

Lei, L., G. Zhang, R. J. Doviak, R. Palmer, B. L. Cheong, M. Xue, Q. Cao, and Y. Li, 2012. Multilag correlation estimators for polarimetric radar measurements in the presence of noise. *Journal of Atmospheric and Oceanic Technology*, **29**, 772–795.

Lei, L., G. Zhang, R. D. Palmer, B. L. Cheong, M. Xue, and Q. Cao, 2009b. A multi-lag correlation estimator for polarimetric radar variables in the presence of noise. *Proceedings of 34th Conference on Radar Meteorology*, pp. 5–9.

Li, X., and J. R. Mecikalski, 2012. Impact of the dual-polarization Doppler radar data on two convective storms with a warm-rain radar forward operator. *Monthly Weather Review*, **140**, 2147–2167.

Li, Y., G. Zhang, and R. J. Doviak, 2014. Ground clutter detection using the statistical properties of signals received with a polarimetric radar. *IEEE Transactions on Signal Processing*, **62**, 597–606.

Li, Y., G. Zhang, R. J. Doviak, L. Lei, and Q. Cao, 2013b. A new approach to detect ground clutter mixed with weather signals. *IEEE Transactions on Geoscience and Remote Sensing*, **51**, 2373–2387.

Li, Y., G. Zhang, R. J. Doviak, and D. S. Saxion, 2013a. Scan-to-scan correlation of weather radar signals to identify ground clutter. *IEEE on Geoscience and Remote Sensing Letters*, **10**, 855–859.

Lim, J.-S., 2005. Reservoir properties determination using fuzzy logic and neural networks from well data in offshore Korea. *Journal of Petroleum Science and Engineering*, **49**, 182–192.

Lim, K.-S. S., and S.-Y. Hong, 2010. Development of an effective double-moment cloud microphysics scheme with prognostic cloud condensation nuclei (CCN) for weather and climate models. *Monthly Weather Review*, **138**, 1587–1612.

Lin, Y.-L., R. D. Farley, and H. D. Orville, 1983. Bulk parameterization of the snow field in a cloud model. *Journal of Climate and Applied Meteorology*, **22**, 1065–1092.

Liu, H., and V. Chandrasekar, 2000. Classification of hydrometeors based on polarimetric radar measurements: Development of fuzzy logic and neuro-fuzzy systems, and in situ verification. *Journal of Atmospheric and Oceanic Technology*, **17**, 140–164.

Löffler-Mang, M., and J. Joss, 2000. An optical disdrometer for measuring size and velocity of hydrometeors. *Journal of Atmospheric and Oceanic Technology*, **17**, 130–139.

Ludwig, A., 1973. The definition of cross-polarization. *IEEE Transactions on Antennas and Propagation*, **21**, 116–119.

Mahale, V. N., G. Zhang, and M. Xue, 2014. Fuzzy logic classification of S-band polarimetric radar echoes to identify three-body scattering and improve data quality. *Journal of Applied Meteorology and Climatology*, **53**, 2017–2033.

Marshall, J. S., and W. M. K. Palmer, 1948. The distribution of raindrops with size. *Journal of Meteorology*, **5**, 165–166.

Martner, B. E., R. M. Rauber, M. K. Ramamurthy, R. M. Rasmussen, and E. T. Prater, 1992. Impacts of a destructive and well-observed cross-country winter storm. *Bulletin of the American Meteorological Society*, **73**, 169–172.

Matrosov, S. Y., K. A. Clark, B. E. Martner, and A. Tokay, 2002. X-band polarimetric radar measurements of rainfall. *Journal of Applied Meteorology*, **41**, 941–952.

Matson, R. J., and A. W. Huggins, 1980. The direct measurement of the sizes, shapes and kinematics of falling hailstones. *Journal of the Atmospheric Sciences*, **37**, 1107–1125.

Maxwell, J. C., 1873. *A treatise on electricity and magnetism.* Vol. 1. Clarendon Press.

May, P. T., and R. G. Strauch, 1989. An examination of wind profiler signal processing algorithms. *Journal of Atmospheric and Oceanic Technology*, **6**, 731–735.

McCormick, G., and A. Hendry, 1974. Polarization properties of transmission through precipitation over a communication link. *Journal de Recherches Atmospheriques*, **8**, 175–187.

McCormick, G., and A. Hendry, 1976. Polarization-related parameters for rain: Measurements obtained by radar. *Radio Science*, **11**, 731–740.

Meischner, P., 2004. *Weather radar: Principles and advanced applications.* Springer Science & Business Media.

Melnikov, V. M., and D. S. Zrnić, 2007. Autocorrelation and cross-correlation estimators of polarimetric variables. *Journal of Atmospheric and Oceanic Technology*, **24**, 1337–1350.

Meneghini, R., and L. Liao, 2007. On the equivalence of dual-wavelength and dual-polarization equations for estimation of the raindrop size distribution. *Journal of Atmospheric and Oceanic Technology*, **24**, 806–820.

Mie, G., 1908. Beiträge zur Optik trüber Medien, speziell kolloidaler Metallösungen. *Annalen der Physik*, **330**, 377–445.

Milbrandt, J., and M. Yau, 2005a. A multimoment bulk microphysics parameterization. Part I: Analysis of the role of the spectral shape parameter. *Journal of the Atmospheric Sciences*, **62**, 3051–3064.

Milbrandt, J., and M. Yau, 2005b. A multimoment bulk microphysics parameterization. Part II: A proposed three-moment closure and scheme description. *Journal of the Atmospheric Sciences*, **62**, 3065–3081.

Milbrandt, J. A., and M. K. Yau, 2006. A multimoment bulk microphysics parameterization. Part IV: Sensitivity experiments. *Journal of the Atmospheric Sciences*, **63**, 3137–3159.

Miller, M., and R. Pearce, 1974. A three-dimensional primitive equation model of cumulonimbus convection. *Quarterly Journal of the Royal Meteorological Society*, **100**, 133–154.

Mueller, E., 1984. Calculation procedures for differential propagation phase shift. *22nd Conference on Radar Meteorology*, Zurich, Switzerland, American Meteor Society, 397–399.

Newell, R. E., and S. G. Geotis, 1955. *Meteorological measurements with a radar provided with variable polarization*. MIT Department of Meteorology.

Oguchi, T., 1960. Attenuation of electromagnetic wave due to rain with distorted raindrops. *Journal of the Radio Research Laboratory*, **7**, 467–485.

Oguchi, T., 1964. Attenuation of electromagnetic wave due to rain with distorted raindrops, Part II. *Journal of the Radio Research Laboratory*, **11**, 19–43.

Oguchi, T., 1975. Rain depolarization studies at centimeter and millimeter wavelengths-theory and measurement. *Journal of the Radio Research Laboratory*, **22**, 165–211.

Oguchi, T., 1983. Electromagnetic wave propagation and scattering in rain and other hydro-meteors. *Proceedings of the IEEE*, **71**, 1029–1078.

Olsen, R., 1982. A review of theories of coherent radio wave propagation through precipitation media of randomly oriented scatterers, and the role of multiple scattering. *Radio Science*, **17**, 913–928.

Pan, Y., M. Xue, G. Ge, and Y. Pan, 2016. Incorporating diagnosed intercept parameters and the graupel category within the ARPS cloud analysis system for the initialization of double-moment microphysics: Testing with a squall line over south China. *Monthly Weather Review*, **144**, 371–392.

Papoulis, A., 1991. *Probability, random variables and stochastic processes*. McGraw-Hill.

Park, H. S., A. Ryzhkov, D. Zrnic, and K.-E. Kim, 2009. The hydrometeor classification algorithm for the polarimetric WSR-88D: Description and application to an MCS. *Weather and Forecasting*, **24**, 730–748.

Park, S. G., V. N. Bringi, V. Chandrasekar, M. Maki, and K. Iwanami, 2005. Correction of radar reflectivity and differential reflectivity for rain attenuation at X band. Part I: Theoretical and empirical basis. *Journal of Atmospheric and Oceanic Technology*, **22**, 1621–1632.

Parrish, D. F., and J. C. Derber, 1992. The National Meteorological Center's spectral statistical-interpolation analysis system. *Monthly Weather Review*, **120**, 1747–1763.

Pazmany, A. L., J. B. Mead, H. B. Bluestein, J. C. Snyder, and J. B. Houser, 2013. A mobile rapid-scanning X-band polarimetric (RaXPol) Doppler radar system. *Journal of Atmospheric and Oceanic Technology*, **30**, 1398–1413.

Poincaré, H., 1892. *Théorie mathématique de la lumiere*. Gauthier Villars.

Posselt, D. J., 2015. A Bayesian examination of deep convective squall line sensitivity to changes in cloud microphysical parameters. *Journal of the Atmospheric Sciences*, **73**, 637–665.

Pruppacher, H., and J. Klett, 1996. *Microphysics of clouds and precipitation*. Vol. 18. Springer Science & Business Media.

Pruppacher, H. R., and K. V. Beard, 1970. A wind tunnel investigation of the internal circulation and shape of water drops falling at terminal velocity in air. *Quarterly Journal of the Royal Meteorological Society*, **96**, 247–256.

Pruppacher, H. R., and R. L. Pitter, 1971. A semi-empirical determination of the shape of cloud and rain drops. *Journal of the Atmospheric Sciences*, **28**, 86–94.

Purcell, E. M., and C. R. Pennypacker, 1973. Scattering and absorption of light by nonspherical dielectric grains. *The Astrophysical Journal*, **186**, 705–714.

Putnam, B., M. Xue, G. Zhang, and Y. Jung, 2013. Simulation of polarimetric radar variables from the CAPS Spring Experiment Storm Scale Ensemble Forecasts. *36th Conference on Radar Meteorology*, Breckenridge, CO.

Raffaelli, S., and M. Johansson, 2003. Conformal array antenna demonstrator for WCDMA applications. *Proceedings of Antenna*, **3**, 207–212.

Ray, P. S., 1972. Broadband complex refractive indices of ice and water. *Applied Optics*, **11**, 1836–1844.

Rayleigh, L., 1871. On the scattering of light by small particles. *Philosophical Magazine*, pp. 447–454.

Rinehart, R. E., 1990, 1991, 1997, 2004. *Radar for meteorologists*. Rinehart, 482 pp.

Roebber, P. J., S. L. Bruening, D. M. Schultz, and J. V. Cortinas, 2003. Improving snowfall forecasting by diagnosing snow density. *Weather and Forecasting*, **18**, 264–287.

Rogers, R. R., and M. K. Yau, 1989. *A short course in cloud physics. International series in natural philosophy*. Pergamon Press.

Rosenfeld, D., and C. W. Ulbrich, 2003. Cloud microphysical properties, processes, and rainfall estimation opportunities. In *Radar and atmospheric science: A collection of essays in honor of David Atlas*, R. M. Wakimoto, and R. C. Srivasva, Eds., Springer, pp. 237–258.

Royer, G., 1966. Directive gain and impedance of a ring array of antennas. *IEEE Transactions on Antennas and Propagation*, **14**, 566–573.

Ryde, J., 1941. Echo intensities and attenuation due to clouds, rain, hail, sand and dust storms at centimetre wavelengths. *Report*, **7831**, 22–24.

Ryzhkov, A., and D. Zrnić, 1996. Assessment of rainfall measurement that uses specific differential phase. *Journal of Applied Meteorology*, **35**, 2080–2090.

Ryzhkov, A. V., S. E. Giangrande, and T. J. Schuur, 2005a. Rainfall estimation with a polarimetric prototype of WSR-88D. *Journal of Applied Meteorology*, **44**, 502–515.

Ryzhkov, A. V., T. J. Schuur, D. W. Burgess, P. L. Heinselman, S. E. Giangrande, and D. S. Zrnić, 2005b. The joint polarization experiment: Polarimetric rainfall measurements and hydrometeor classification. *Bulletin of the American Meteorological Society*, **86**, 809–824.

Ryzhkov, A. V., and D. S. Zrnić, 1995. Comparison of dual-polarization radar estimators of rain. *Journal of Atmospheric and Oceanic Technology*, **12**, 249–256.

Ryzhkov, A. V., and D. S. Zrnić, 2007. Depolarization in ice crystals and its effect on radar polarimetric measurements. *Journal of Atmospheric and Oceanic Technology*, **24**, 1256–1267.

Sachidananda, M., and D. S. Zrnić, 1985. ZDR measurement considerations for a fast scan capability radar. *Radio Science*, **20**, 907–922.

Sachidananda, M., and D. S. Zrnić, 1986. Differential propagation phase shift and rainfall rate estimation. *Radio Science*, **21**, 235–247.

Sachidananda, M., and D. S. Zrnić, 1987. Rain rate estimates from differential polarization measurements. *Journal of Atmospheric and Oceanic Technology*, **4**, 588–598.

Sachidananda, M., and D. S. Zrnić, 1989. Efficient processing of alternately polarized radar signals. *Journal of Atmospheric and Oceanic Technology*, **6**, 173–181.

Schaefer, J. T., 1990. The critical success index as an indicator of warning skill. *Weather and Forecasting*, **5**, 570–575.

Schönhuber, M., H. E. Urban, J. P. V. P. Baptista, W. L. Randeu, and W. Riedler, 1997. Weather radar versus 2D-video disdrometer data. In *Weather radar technology for water resources management*, B. P. F. Braga, Jr., and O. Massambani, Eds., UNESCO Press, pp. 159–171.

Schuur, T., A. Ryzhkov, P. Heinselman, D. Zrnic, D. Burgess, and K. Scharfenberg, 2003. *Observations and classification of echoes with the polarimetric WSR-88D radar.* Report of the National Severe Storms Laboratory, Norman, OK, 73069, 46 pp.

Seliga, T., and V. Bringi, 1978. Differential reflectivity and differential phase shift: Applications in radar meteorology. *Radio Science*, **13**, 271–275.

Seliga, T., and E. A. Muller, 1982. Implementation of a fast-switching differential reflectivity dual-polarization capability on the CHILL radar: First observations. In *Preprints, URSI*, August, 23–27, Bournemouth, United Kingdom.

Seliga, T., R. Humphries, and J. Metcalf, 1990. Polarization diversity in radar meteorology: Early developments. In *Radar in Meteorology*, D. Atlas, Ed., American Meteorological Society, 109–114.

Seliga, T. A., K. Aydin, and H. Direskeneli, 1986. Disdrometer measurements during an intense rainfall event in Central Illinois: Implications for differential reflectivity radar observations. *Journal of Climate and Applied Meteorology*, **25**, 835–846.

Seliga, T. A., and V. N. Bringi, 1976. Potential use of radar differential reflectivity measurements at orthogonal polarizations for measuring precipitation. *Journal of Applied Meteorology*, **15**, 69–76.

Seliga, T. A., V. N. Bringi, and H. H. Al-Khatib, 1979. Differential reflectivity measurements in rain: First experiments. *IEEE Transactions on Geoscience Electronics*, **17**, 240–244.

Shen, L. C., and J. A. Kong, 1983. *Applied electromagnetism.* Brooks/Cole Engineering Division, 507 pp.

Siggia, A., and R. Passarelli, Jr, 2004. Gaussian model adaptive processing (GMAP) for improved ground clutter cancellation and moment calculation. *Proceedings of the ERAD*, 421–424.

Smith, P., Jr, C. Myers, and H. Orville, 1975. Radar reflectivity factor calculations in numerical cloud models using bulk parameterization of precipitation. *Journal of Applied Meteorology*, **14**, 1156–1165.

Snyder, J. C., H. B. Bluestein, G. Zhang, and S. J. Frasier, 2010. Attenuation correction and hydrometeor classification of high-resolution, X-band, dual-polarized mobile radar measurements in severe convective storms. *Journal of Atmospheric and Oceanic Technology*, **27**, 1979–2001.

Srivastava, R., A. Jameson, and P. Hildebrand, 1979. Time-domain computation of mean and variance of Doppler spectra. *Journal of Applied Meteorology*, **18**, 189–194.

Steiner, M., and J. A. Smith, 2002. Use of three-dimensional reflectivity structure for automated detection and removal of nonprecipitating echoes in radar data. *Journal of Atmospheric and Oceanic Technology*, **19**, 673–686.

Stewart, R. E., J. D. Marwitz, J. C. Pace, and R. E. Carbone, 1984. Characteristics through the melting layer of stratiform clouds. *Journal of the Atmospheric Sciences*, **41**, 3227–3237.

Stokes, G. G., 1852a. On the composition and resolution of streams of polarized light from different sources. *Transactions of the Cambridge Philosophical Society*, **9**, 399–461.

Stokes, G. G., 1852b. Ueber die Veränderung der Brechbarkeit des Lichts. *Annalen der Physik*, **163**, 480–490.

Straka, J. M., D. S. Zrnić, and A. V. Ryzhkov, 2000. Bulk hydrometeor classification and quantification using polarimetric radar data: Synthesis of relations. *Journal of Applied Meteorology*, **39**, 1341–1372.

Stratton, J. A., 1941. *Electromagnetic theory*. 1st ed. Mcgraw-Hill College, 615 pp.

Sun, J., and N. A. Crook, 1997. Dynamical and microphysical retrieval from Doppler radar observations using a cloud model and its adjoint. Part I: Model development and simulated data experiments. *Journal of the Atmospheric Sciences*, **54**, 1642–1661.

Tatarskii, V. I., 1971. *The effects of the turbulent atmosphere on wave propagation*. Israel Program for Scientific Translations, Jerusalem.

Testud, J., E. Le Bouar, E. Obligis, and M. Ali-Mehenni, 2000. The rain profiling algorithm applied to polarimetric weather radar. *Journal of Atmospheric and Oceanic Technology*, **17**, 332–356.

Thompson, G., P. R. Field, R. M. Rasmussen, and W. D. Hall, 2008. Explicit forecasts of winter precipitation using an improved bulk microphysics scheme. Part II: Implementation of a new snow parameterization. *Monthly Weather Review*, **136**, 5095–5115.

Thurai, M., and V. N. Bringi, 2005. Drop axis ratios from a 2D video disdrometer. *Journal of Atmospheric and Oceanic Technology*, **22**, 966–978.

Toll, J. S., 1956. Causality and the dispersion relation: Logical foundations. *Physical Review*, **104**, 1760.

Tomasic, B., J. Turtle, and S. Liu, 2002. A geodesic sphere phased array antenna for satellite control and communication. *XXVIIth General Assembly of the International Union of Radio Science*, Maastricht.

Torres, S. M., and D. A. Warde, 2014. Ground clutter mitigation for weather radars using the autocorrelation spectral density. *Journal of Atmospheric and Oceanic Technology*, **31**, 2049–2066.

Tsang, L., K.-H. Ding, G. Zhang, C. Hsu, and J. A. Kong, 1995. Backscattering enhancement and clustering effects of randomly distributed dielectric cylinders overlying a dielectric half space based on Monte-Carlo simulations. *IEEE Transactions on Antennas and Propagation*, **43**, 488–499.

Tsang, L., J. A. Kong, and K.-H. Ding, 2000. *Scattering of electromagnetic waves, theories and applications*. Vol. 27, Wiley.

Tsang, L., J. A. Kong, and R. T. Shin, 1985. *Theory of microwave remote sensing*. Wiley, New York.

Tuttle, J. D., and R. E. Rinehart, 1983. Attenuation correction in dual-wavelength analyses. *Journal of Climate and Applied Meteorology*, **22**, 1914–1921.

Twersky, V., 1964. On propagation in random media of discrete scatterers. *Proceedings of Symposia in Applied Mathematics*, 84–116.

Ulaby, F. T., and C. Elach, 1990. *Radar polarimetry for geoscience applications*. Artech House, 364 pp.

Ulbrich, C. W., 1983. Natural variations in the analytical form of the raindrop size distribution. *Journal of Climate and Applied Meteorology*, **22**, 1764–1775.

Van De Hulst, H., 1957. *Light scattering by small particles*. Wiley, 470 pp.

Vivekanandan, J., W. Adams, and V. Bringi, 1991. Rigorous approach to polarimetric radar modeling of hydrometeor orientation distributions. *Journal of Applied Meteorology*, **30**, 1053–1063.

Vivekanandan, J., S. M. Ellis, R. Oye, D. S. Zrnic, A. V. Ryzhkov, and J. Straka, 1999. Cloud microphysics retrieval using S-band dual-polarization radar measurements. *Bulletin of the American Meteorological Society*, **80**, 381–388.

Vivekanandan, J., G. Zhang, and E. Brandes, 2004. Polarimetric radar estimators based on a constrained gamma drop size distribution model. *Journal of Applied Meteorology*, **43**, 217–230.

Wang, Y., and V. Chandrasekar, 2006. Polarization isolation requirements for linear dual-polarization weather radar in simultaneous transmission mode of operation. *IEEE Transactions on Geoscience and Remote Sensing*, **44**, 2019–2028.

Waterman, P., 1965. Matrix formulation of electromagnetic scattering. *Proceedings of the IEEE*, **53**, 805–812.

Waterman, P., 1969. New formulation of acoustic scattering. *The Journal of the Acoustical Society of America*, **45**, 1417–1429.

Weber, M. E., J. Y. N. Cho, J. S. Herd, J. M. Flavin, W. E. Benner, and G. S. Torok, 2007. The next-generation multimission U.S. Surveillance Radar Network. *Bulletin of the American Meteorological Society*, **88**, 1739–1751.

Williams, C. R., V. Bringi, L. Carey, V. Chandrasekar, P. Gatlin, Z. Haddad, R. Meneghini, et al., 2014. Describing the shape of raindrop size distributions using uncorrelated raindrop mass spectrum parameters. *Journal of Applied Meteorology and Climatology*, **53**, 1282–1296.

Wurman, J., 2003. Preliminary results from the rapid-DOW, a multi-beam inexpensive alternative to phased arrays. *Preprints, 31st Conference on Radar Meteorology*, American Meteor Society, Seattle, WA.

Wurman, J., S. Gill, and M. Randall, 2001. An inexpensive, mobile, rapid-scan radar. *Preprints, 30th International Conference on Radar Meteorology*, American Meteor Society, Munich, Germany, CD-ROM P.

Wurman, J., S. Heckman, and D. Boccippio, 1993. A bistatic multiple-Doppler network. *The Journal of Applied Meteorology*, **32**, 1802–1814.

Xue, M., K. K. Droegemeier, and V. Wong, 2000. The Advanced Regional Prediction System (ARPS)—A multi-scale nonhydrostatic atmospheric simulation and prediction model. Part I: Model dynamics and verification. *Meteorology and Atmospheric Physics*, **75**, 161–193.

Xue, M., K. K. Droegemeier, V. Wong, A. Shapiro, K. Brewster, F. Carr, D. Weber, Y. Liu, and D. Wang, 2001. The Advanced Regional Prediction System (ARPS)—A multi-scale nonhydrostatic atmospheric simulation and prediction tool. Part II: Model physics and applications. *Meteorology and Atmospheric Physics*, **76**, 143–165.

Xue, M., Y. Jung, and G. Zhang, 2010. State estimation of convective storms with a two-moment microphysics scheme and an ensemble Kalman filter: Experiments with simulated radar data. *Quarterly Journal of the Royal Meteorological Society*, **136**, 685–700.

Xue, M., M. Tong, and G. Zhang, 2009. Simultaneous state estimation and attenuation correction for thunderstorms with radar data using an ensemble Kalman filter: Tests with simulated data. *Quarterly Journal of the Royal Meteorological Society*, **135**, 1409–1423.

Zahrai, A., and D. a. S. Zrnić, 1993. The 10-cm-wavelength polarimetric weather radar at NOAA's National Severe Storms Laboratory. *Journal of Atmospheric and Oceanic Technology*, **10**, 649–662.

Zhang, G., 1998. Detection and imaging of targets in the presence of clutter based on angular correlation function. PhD dissertation, Department of Electrical Engineering, University of Washington.

Zhang, G., 2015. Comments on "Describing the shape of raindrop size distributions using uncorrelated raindrop mass spectrum parameters". *Journal of Applied Meteorology and Climatology*, **54**, 1970–1976.

Zhang, G., and R. J. Doviak, 2007. Spaced-antenna interferometry to measure crossbeam wind, shear, and turbulence: Theory and formulation. *Journal of Atmospheric and Oceanic Technology*, **24**, 791–805.

Zhang, G., and R. J. Doviak, 2008. Spaced-antenna interferometry to detect and locate subvolume inhomogeneities of reflectivity: An analogy with monopulse radar. *Journal of Atmospheric and Oceanic Technology*, **25**, 1921–1938.

Zhang, G., R. J. Doviak, J. Vivekanandan, W. O. Brown, and S. A. Cohn, 2004b. Performance of correlation estimators for spaced-antenna wind measurement in the presence of noise. *Radio Science*, **39**, RS3017, doi:10.1029/2003RS003022.

Zhang, G., R. J. Doviak, D. S. Zrnic, J. Crain, D. Staiman, and Y. Al-Rashid, 2009. Phased array radar polarimetry for weather sensing: A theoretical formulation for bias corrections. *IEEE Transactions on Geoscience and Remote Sensing*, **47**, 3679–3689.

Zhang, G., R. J. Doviak, D. S. Zrnic, R. Palmer, L. Lei, and Y. Al-Rashid, 2011c. Polarimetric phased-array radar for weather measurement: A planar or cylindrical configuration? *Journal of Atmospheric and Oceanic Technology*, **28**, 63–73.

Zhang, G., J. Hou, S. Ito, and T. Oguchi, 1990. Optical wave propagation in random media composed of both turbulence and particles: Radiative transfer equation approach. *Journal of the Communications Research Laboratory*, **37**, 43–62.

Zhang, G., S. Luchs, A. Ryzhkov, M. Xue, L. Ryzhkova, and Q. Cao, 2011b. Winter precipitation microphysics characterized by polarimetric radar and video disdrometer observations in Central Oklahoma. *Journal of Applied Meteorology and Climatology*, **50**, 1558–1570.

Zhang, G., S. Luchs, and M. Xue, 2011a. Melting models for winter precipitation microphysics. *35th Conference on Radar Meteorology*, AMS, Omni William Penn Hotel, Pittsburgh, PA, 26–30 September 2011.

Zhang, G., S. Karimkashi, L. Lei, R. Kelley, J. P. Meier, R. Palmer, C. Futon, et al. 2013. A cylindrical polarimetric phased array radar concept—A path to multi-mission capability. *2013 IEEE International Symposium on Phased Array Systems & Technology*, IEEE, pp. 481–484.

Zhang, G., J. Sun, and E. A. Brandes, 2006. Improving parameterization of rain microphysics with disdrometer and radar observations. *Journal of the Atmospheric Sciences*, **63**, 1273–1290.

Zhang, G., L. Tsang, and Z. Chen, 1996. Collective scattering effects of trees generated by stochastic Lindenmayer systems. *Microwave and Optical Technology Letters*, **11**, 107–111.

Zhang, G., J. Vivekanandan, and E. Brandes, 2001. A method for estimating rain rate and drop size distribution from polarimetric radar measurements. *IEEE Transactions on Geoscience and Remote Sensing*, **39**, 830–841.

Zhang, G., J. Vivekanandan, E. A. Brandes, R. Meneghini, and T. Kozu, 2003. The shape–slope relation in observed gamma raindrop size distributions: Statistical error or useful information? *Journal of Atmospheric and Oceanic Technology*, **20**, 1106–1119.

Zhang, G., J. Vivekanandan, and M. K. Politovich, 2004a. Radar/radiometer combination to retrieve cloud characteristics for icing detection. *11th Conference on Aviation, Range, and Aerospace*.

Zhang, G., M. Xue, Q. Cao, and D. Dawson, 2008. Diagnosing the intercept parameter for exponential raindrop size distribution based on video disdrometer observations: Model development. *Journal of Applied Meteorology and Climatology*, **47**, 2983–2992.

Zhang, G., D. S. Zrnić, L. Borowska, and Y. Al-Rashid, 2015. Hybrid scan and joint signal processing for a high efficient MPAR. *AMS Annual Meeting 31st Conference on Environmental Information Processing Technologies*, Phoenix, AZ.

Ziegler, C. L., P. S. Ray, and N. C. Knight, 1983. Hail growth in an Oklahoma multicell storm. *Journal of the Atmospheric Sciences*, **40**, 1768–1791.

Zrnić, D. a. S., 1975. Simulation of weatherlike Doppler spectra and signals. *Journal of Applied Meteorology*, **14**, 619–620.

Zrnić, D. S., 1977. Spectral moment estimates from correlated pulse pairs. *IEEE Transactions on Aerospace and Electronic Systems*, **AES-13**, 344–354.

Zrnić, D. S., 1991. Complete polarimetric and Doppler measurements with a single receiver radar. *Journal of Atmospheric and Oceanic Technology*, **8**, 159–165.

Zrnić, D. S., R. Doviak, G. Zhang, and A. Ryzhkov, 2010a. Bias in differential reflectivity due to cross coupling through the radiation patterns of polarimetric weather radars. *Journal of Atmospheric and Oceanic Technology*, **27**, 1624–1637.

Zrnić, D. S., Z. Guifu, and R. J. Doviak, 2011. Bias correction and Doppler measurement for polarimetric phased-array radar. *IEEE Transactions on Geoscience and Remote Sensing*, **49**, 843–853.

Zrnić, D. S., J. F. Kimpel, and D. E. Forsyth, 2007. Agile-beam phased array radar for weather observations. *Bulletin of the American Meteorological Society*, **88**, 1753–1766.

Zrnić, D. S., V. M. Melnikov, and A. V. Ryzhkov, 2006. Correlation coefficients between horizontally and vertically polarized returns from ground clutter. *Journal of Atmospheric and Oceanic Technology*, **23**, 381–394.

Zrnić, D. S., A. Ryzhkov, J. Straka, Y. Liu, and J. Vivekanandan, 2001. Testing a procedure for automatic classification of hydrometeor types. *Journal of Atmospheric and Oceanic Technology*, **18**, 892–913.

Zrnić, D. S., G. Zhang, V. Melnikov, and J. Andric, 2010b. Three-body scattering and hail size. *Journal of Applied Meteorology and Climatology*, **49**, 687–700.

Index

Note: Page numbers in *italics* indicate figures and tables.

Printed and bound by CPI Group (UK) Ltd, Croydon, CR0 4YY

01/11/2024

01782624-0007